AF548694

Plant Transposable Elements

BASIC LIFE SCIENCES

Alexander Hollaender, Founding Editor

Recent volumes in the series:

Volume 34 BASIC AND APPLIED MUTAGENESIS: With Special Reference to Agricultural Chemicals in Developing Countries
Edited by Amir Muhammed and R. C. von Borstel

Volume 35 MOLECULAR BIOLOGY OF AGING
Edited by Avril D. Woodhead, Anthony D. Blackett, and Alexander Hollaender

Volume 36 ANEUPLOIDY: Etiology and Mechanisms
Edited by Vicki L. Dellarco, Peter E. Voytek, and Alexander Hollaender

Volume 37 GENETIC ENGINEERING OF ANIMALS: An Agricultural Perspective
Edited by J. Warren Evans and Alexander Hollaender

Volume 38 MECHANISMS OF DNA DAMAGE AND REPAIR: Implications for Carcinogenesis and Risk Assessment
Edited by Michael G. Simic, Lawrence Grossman, and Arthur C. Upton

Volume 39 ANTIMUTAGENESIS AND ANTICARCINOGENESIS MECHANISMS
Edited by Delbert M. Shankel, Philip E. Hartman, Tsuneo Kada, and Alexander Hollaender

Volume 40 EXTRACHROMOSOMAL ELEMENTS IN LOWER EUKARYOTES
Edited by Reed B. Wickner, Alan Hinnebusch, Alan M. Lambowitz, I. C. Gunsalus, and Alexander Hollaender

Volume 41 TAILORING GENES FOR CROP IMPROVEMENT: An Agricultural Perspective
Edited by George Bruening, John Harada, Tsune Kosuge, and Alexander Hollaender

Volume 42 EVOLUTION OF LONGEVITY IN ANIMALS: A Comparative Approach
Edited by Avril D. Woodhead and Keith H. Thompson

Volume 43 PHENOTYPIC VARIATION IN POPULATIONS: Relevance to Risk Assessment
Edited by Avril D. Woodhead, Michael A Bender, and Robin C. Leonard

Volume 44 GENETIC MANIPULATION OF WOODY PLANTS
Edited by James W. Hanover and Daniel E. Keathley

Volume 45 ENVIRONMENTAL BIOTECHNOLOGY: Reducing Risks from Environmental Chemicals through Biotechnology
Edited by Gilbert S. Omenn

Volume 46 BIOTECHNOLOGY AND THE HUMAN GENOME: Innovations and Impact
Edited by Avril D. Woodhead and Benjamin J. Barnhart

Volume 47 PLANT TRANSPOSABLE ELEMENTS
Edited by Oliver Nelson

A Continuation Order Plan is available for this series. A continuation order will bring delivery of each new volume immediately upon publication. Volumes are billed only upon actual shipment. For further information please contact the publisher.

Plant Transposable Elements

Edited by

Oliver Nelson
University of Wisconsin-Madison
Madison, Wisconsin

Technical Editors

Claire M. Wilson and Cosette G. Saslaw
The Council for Research Planning in Biological Sciences, Inc.
Washington, D.C.

Plenum Press • New York and London

Library of Congress Cataloging in Publication Data

International Symposium on Plant Transposable Elements (1987: University of Wisconsin-Madison)
Plant transposable elements / edited by Oliver Nelson.
/ p. cm.—(Basic life sciences: v. 47)
"Proceedings of an International Symposium on Plant Transposable Elements, held August 22–26, 1987, at the University of Wisconsin-Madison"—T.p. verso.
Includes bibliographical references and index.
ISBN 0-306-43001-0
1. Plant molecular genetics—Congresses. 2. Insertion elements, DNA—Congresses. 3. Mutagens—Congresses. 4. Genetic regulation—Congresses. I. Nelson, Oliver. II. Title. III. Series.
QK981.4.I57 1987 88-21917
581.1'5—dc19 CIP

This book was copyedited and entirely retyped by the staff of the Council for Research Planning in Biological Sciences, Inc., located on the premises of Associated Universities, Inc., of which the Council is a guest.

Proceedings of an International Symposium on Plant Transposable Elements, held August 22–26, 1987, at the University of Wisconsin-Madison, Madison, Wisconsin.

A Division of Plenum Publishing Corporation
233 Spring Street, New York, N.Y. 10013

Printed in the United States of America

DEDICATION

There are few instances in science in which the genesis of a field of research is identified as closely with one person as is that of transposable elements and Barbara McClintock. Our meeting was still another tribute to her brilliant analyses of mosaic kernel phenotypes that arose initially in the progeny of plants subjected to systematic genome disruption.

The last few years have witnessed continuous advances in our understanding of the structure of these elements and their effects on gene function as the techniques of molecular biology have been brought to bear, and it is these recent advances that were highlighted here against a background of the taxonomy and formal genetics of the transposable elements. The Scientific Organizing Committee hoped in this manner to present a coherent and integrated view of transposable elements that would open the area to those not already intimately familiar with the subject, while simultaneously permitting substantive interchanges among researchers in the field.

The suggestion that we organize this symposium to explore advances in our knowledge of plant transposable elements came from Alexander Hollaender of the Council for Research Planning in Biological Sciences. He participated in our initial planning as a member of the Scientific Organizing Committee, but he was not to see those plans brought to fruition. Alexander Hollaender died in December, 1986 after an illustrious career in research and administration. Claire Wilson, his associate at the Council for Research Planning, assisted in plans for this symposium and prepared, with the help of Cosette Saslaw, the manuscripts presented here for publication.

We have dedicated this symposium to Barbara McClintock, whose research opened new horizons in our knowledge of genome organization, and to the memory of Alexander Hollaender.

ACKNOWLEDGEMENTS

It is a pleasure to acknowledge the generous support of the sponsors whose grants made possible this symposium with participation of speakers from the United States and abroad. The sponsoring organizations were Agracetus, Agrigenetics, Argonne Universities Association Trust Fund, Bethesda Research Laboratories, the U.S. Department of Energy, DuPont, the International Society of Plant Molecular Biology, the Midwest Plant Biotechnology Consortium, Monsanto, the National Institutes of Health, the National Science Foundation, Pfizer-DeKalb, Pioneer Hi-Bred, the U.S. Department of Agriculture, and the University of Wisconsin Graduate School.

The Scientific Organizing Committee, which was responsible for the program of the symposium, consisted of Benjamin Burr, Nina Fedoroff, Michael Freeling, Oliver Nelson, Virginia Walbot, and Susan Wessler, together with the initial participation of Alexander Hollaender. We appreciate the participation of the speakers, their thoughtful presentations, and the timely arrival of their manuscripts.

I should also like to acknowledge the major editorial role played by E.T. Bingham. Although he handled nearly one-half of the manuscripts in an editorial capacity, he declined to take the editorial bow to which he was clearly entitled.

Oliver Nelson

CONTENTS

POSITIVE AND NEGATIVE REGULATION OF THE SUPPRESSOR-MUTATOR ELEMENT

N. Fedoroff, P. Masson, J. Banks,
and J. Kingsbury

Carnegie Institution of Washington
Department of Embryology
Baltimore, Maryland 21210

INTRODUCTION

The Suppressor-mutator (Spm) element is one of the most extensively studied maize transposable elements. It is the fully functional member of a transposable element family containing many functionally altered and mutant members. McClintock's genetic studies on the Spm family defined the functional interrelationships both among elements of the family and between the elements and genes with Spm insertion mutations (12-27). Rhoades and Dempsey (35) independently identified mutations caused by the same element family; these were subsequently studied by Peterson (30-34), who named the element Enhancer (En).

Many elements of the Spm family have recently been cloned and analyzed, as have several genes with Spm insertion mutations (1,8,10,28,29, 38,39,41). The Spm and En elements are very similar 8.3-kb elements: the two that have been sequenced differ in length by four nucleotides, and in sequence by an additional six nucleotides (11,28). None of the differences appears to affect element function. Many mutant elements have also been analyzed. The predominant class of mutant element comprises transposition-defective, internally deleted Spm elements (dSpm). There exist several additional types of Spm element designated Spm-weak (Spm-w), inactive Spm (Spm-i), and Modifier (Mod).

There are two types of mutant alleles with dSpm insertions in which expression of the gene has come under the control of the inserted element. These have been designated Spm-suppressible and Spm-dependent (11). In an Spm-suppressible allele, the gene with the dSpm insertion continues to be expressed in the absence, but not in the presence, of a trans-acting Spm. By contrast, an Spm-dependent allele is expressed only in the presence of a functional Spm element elsewhere in the genome. We have interpreted such interactions between genes and Spm elements as a reflection of the element's regulatory system (11).

In this chapter we review the properties of the Spm-dependent a-m2 alleles of the maize a locus that have led us to postulate the existence of both a positive, element-encoded, autoregulatory function, as well as a negative, regulatory mechanism that is probably not element-encoded. We also summarize evidence that the Spm element can exist in at least two genetically distinguishable inactive states and show that these differ at the molecular level by the extent of element sequence methylation. We further suggest that the Spm element's ability to be mobilized in trans is controlled to a large extent by a sequence near, but not at, the 5' end of its transcription unit, whose deletion or methylation substantially reduces the element's mobility.

Spm HAS A POSITIVE AUTOREGULATORY MECHANISM

McClintock reported many years ago that an Spm element could become reversibly inactivated (16-18,27). She also showed that an active element present in the same genome with an inactive one could transiently activate it. This implies the existence of an element-encoded molecule that can trans-activate expression of an inactive element. Our studies on the Spm-dependent a-m2 alleles of the a locus indicate that in these alleles, the locus has come under the control of the element-encoded regulatory system and have provided further insight into the nature of the regulatory mechanism.

The a-m2 alleles constitute a series of derivatives of a mutant allele in which an Spm element inserted just upstream (0.1 kb) of the transcription start site at the a locus (40). McClintock originally selected the derivatives on the basis of phenotypic changes affecting either the element's excision or expression of the a gene. The relationship among the a-m2 alleles is shown in Fig. 1 (11). The figure also shows the type of Spm insertion in each allele, as defined by genetic criteria, as well as the physical size of the insertion. The insertions are of three types: the

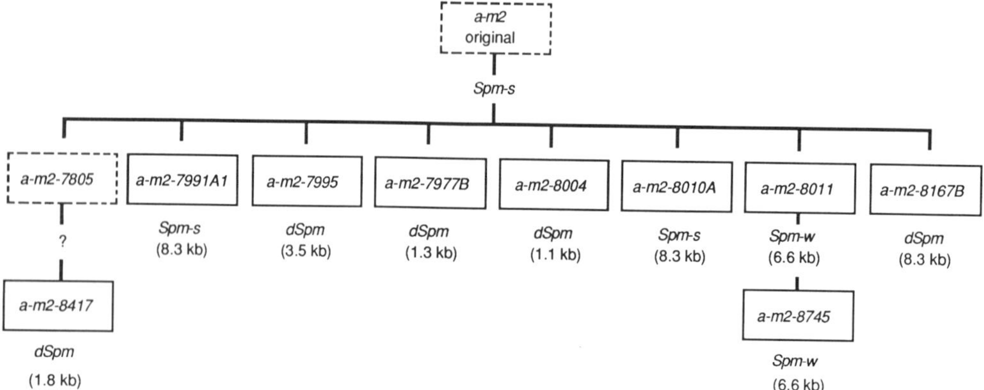

Fig. 1. Derivation of the a-m2 alleles. The alleles analyzed are designated by numbers enclosed in boxes. Those that are in the lineage, but not analyzed, are indicated by discontinuous boxes. The type and size of the Spm insertion is indicated for each allele (from Ref. 11).

standard, fully functional Spm, Spm-s; weakly active Spm elements, Spm-w, that excise less frequently and later in development than the Spm-s; and transposition-defective Spm elements, dSpm.

The a gene of the a-m2 alleles is co-expressed with the resident Spm element. The original a-m2 allele (now lost) contained a complete Spm element at the locus and gave a palely pigmented phenotype with many small revertant somatic sectors. That is, gene expression was not completely disrupted by the insertion, but was reduced. That expression of the a gene is under the control of the Spm element in the a-m2 alleles is most clearly demonstrated by the properties of the several dSpm derivatives of the original insertion mutation. All of the dSpm derivatives studied give a completely colorless phenotype in the absence of a trans-acting Spm element (Fig. 2a), indicating that the gene is not expressed. All but one of the dSpm derivatives studied show a gene expression throughout the kernel aleurone layer in the presence of an Spm element elsewhere in the genome (Fig. 2). Spm-dependent a gene expression is observed only with alleles that have inserted Spm elements; the capacity for Spm-dependent expression of the gene is lost when the insertion is excised from the locus. Thus, expression of the a gene in the a-m2 alleles with a dSpm element is mediated by an interaction between an element-encoded gene product and the element sequence at the locus.

We have learned from the molecular analysis of the dSpm elements in the a-m2 alleles that the element sequences that mediate a gene expression are at the element ends (11). Almost all of the a-m2 alleles have internal deletions. The results of our structural analyses of several of the alleles are represented diagrammatically in Fig. 3, which also shows the structural relationship between the a gene (40) and the element and the structure of the major element-encoded transcript (28). The ability of the dSpm element to mediate a gene expression is unaffected by deletion of most of the internal element sequence. The level of gene expression is approximately the same in an allele carrying a complete 8.3-kb Spm-s, a 6.7-kb Spm-w (not shown), and either the 3.4-kb dSpm element of the a-m2-7995 or the

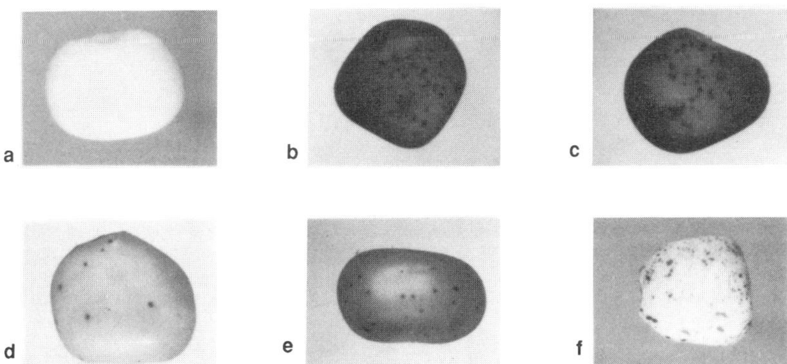

Fig. 2. Phenotypes of the a-m2 alleles with dSpm insertions. a. Colorless phenotype of all of the alleles in the absence of an Spm element; b. a-m2-7995; c. a-m2-7977B; d. a-m2-8004; e. a-m2-8167B; f. a-m2-8417. The kernels shown in b-d all contain an Spm-s element (from Ref. 11).

1.3-kb dSpm element of the a-m2-7977B alleles (Fig. 2b and c). Thus, sequences mediating a gene expression are near element ends.

The element sequences involved in a gene expression are probably near its left end. This follows from the observation of similar levels of gene expression with a complete Spm-s element, in which the right element end is at a distance of more than 8 kb from the gene, and in alleles with much shorter elements, but with deletions extending near the element's right end. This inference is further strengthened by the observation that the a-m2-8004 allele, which shows a markedly lower level of Spm-dependent a gene expression than do the other alleles (Fig. 2d), contains an element that is missing more of the sequence at the element's left end, and retains more of the right end, than the element in the a-m2-7977B allele (Fig. 3).

Because the a gene is expressed only in the presence of a fully functional Spm element (either Spm-s or Spm-w), it follows that gene expression depends on an interaction between a diffusible, element-encoded gene product and sequences near the element's own transcription initiation site, identified by Pereira et al. (28) at nucleotide 209 from the element's left end. Thus, sequences near the element's transcription start site function either as an enhancer or a bidirectional promoter for the a gene. We believe that these interactions between the trans-acting Spm element and the dSpm elements of the a-m2 alleles reveal the mechanism by which the element normally regulates its own expression. According to this interpretation, an element-encoded gene product interacts with sequences near the element's transcription initiation site to promote expression of the element. Further evidence that the element encodes a positive autoregulatory function is provided by the analysis of the genetic interactions between active and inactive elements, detailed below.

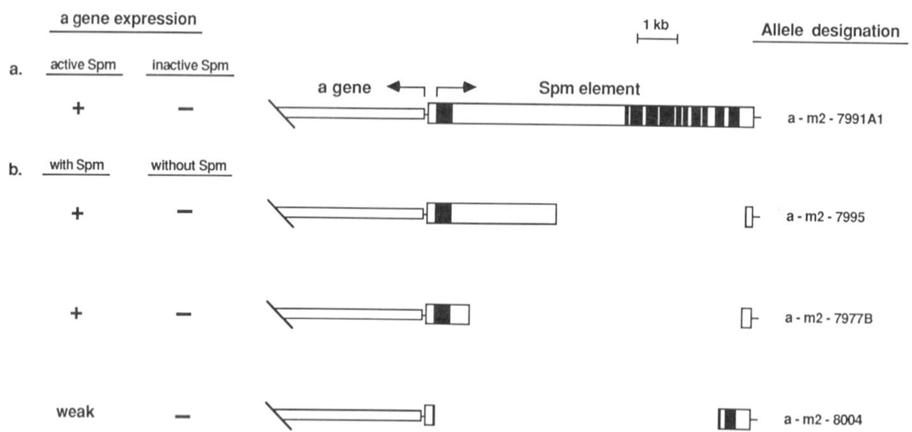

Fig. 3. The complete 8.3-kb Spm element is inserted 0.1 kb upstream of the a gene. As indicated by the arrows, the gene and the element are transcribed divergently. The Spm element's intron/exon structure is assumed to be identical to that of the similar En element (28) and is represented by the unshaded (intron) and shaded blocks (exons) within the element. The structure of the element in several of the a-m2 alleles is shown and the corresponding level of a gene expression is noted (from Ref. 8).

GENETIC INACTIVATION OF THE Spm ELEMENT

Spm elements can undergo a reversible genetic inactivation (16-18, 21,27). McClintock (21) reported that when the Spm element of the original a-m2 allele became inactive, the a gene was no longer expressed. We have selected a similar inactive derivative of the a-m2-7991A1 allele (designated a-m2-7991A1-i), which has an Spm-s at the locus, but gives a less dense variegation pattern than was characteristic of the original allele (11). The phenotypes of kernels with active and inactive Spm elements are shown in Fig. 4a and c, respectively. The colorless phenotype of the kernel in Fig. 4c shows that both gene and element are inactivated together (Fig. 3). Figure 4b shows the phenotype of a kernel in which the resident Spm element underwent a transition from the inactive to the active form during development, and it can be seen that both the a gene, as judged by the pale pigmentation of the kernel, and the element, as judged by the presence of small revertant sectors, are reactivated simultaneously.

We have begun a genetic analysis of the inactive derivative of the a-m2-7991A1 allele, and the first results of that analysis are summarized in Tab. 1 (8). Three tentative conclusions can be drawn from the results of the crosses shown in the table. First, the stability of the inactive state differs among the sample of plants tested. Second, the element is much more frequently reactivated when propagated through the male than when propagated through the female. Third, the inactive element is always active in the presence of an Spm-w element located elsewhere in the genome, either on the homolog or on another chromosome.

We have designated elements in this relatively unstable quiescent state as Spm-i elements. This genetic state resembles that identified by McClintock using alleles of the a2 locus to monitor the activity of the element (16-18,27). Not only is the Spm-i element of the present derivatives activated by a trans-acting Spm, but expression of the a gene is similarly reactivated. Thus, as outlined above, the trans-acting element must provide a diffusible gene product necessary for expression of both element and gene. It should be noted that the genetic properties of an Spm-i element resemble those of what McClintock designated Mod (15-17, 25). Mod was identified as a member of the Spm family by its mobility in the presence of an Spm element. The Mod element enhances the somatic excision frequency promoted by either an Spm-s or an Spm-w element, but has no genetic effect in the absence of another active element.

We have identified and undertaken the investigation of a second, more stable inactive state of the Spm element, which we designate cryptic

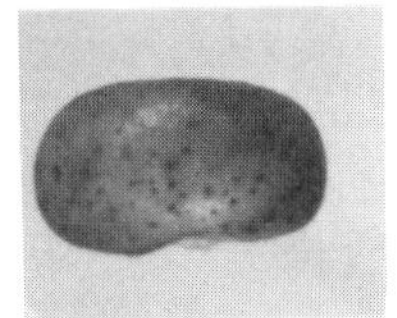

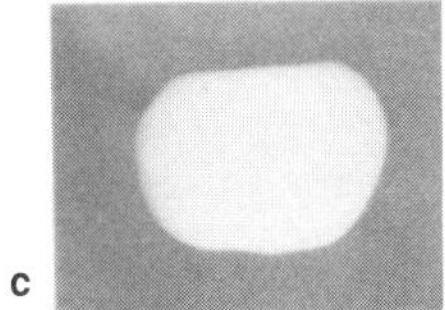

Fig. 4. The inactivation and reactivation of the Spm element. Phenotypes of the a-m2-7991A1-i kernels in which the Spm-s element is a, active; b, inactive (colorless sector) and active (pigmented sector containing deeply pigmented spots) in different areas of the kernel; and c, inactive (from Ref. 11).

Tab. 1. Spontaneous and Spm-promoted reactivation of an inactive Spm element.

Plant number	Parent carrying inactive Spm	Percent of active Spm elements[a]	Percent of active Spm in presence of weak Spm element[b]
1	female	0	---[c]
2	male	51.7	100
3	female	0	---
4	female	0	---
	male	0	---
5	female	0.4	---
	male	73.5	---
6	female	0	---
	male	30.1	100
7	female	0.6	---
8	female	8.2	---
	male	--	100
9	female	19.4	---
	male	42.9	100

[a]Expressed as percent of kernels receiving the inactive Spm element that exhibit activity of the element, as judged by expression of the a gene of the a-m2-7991A1 allele and somatic excision of the Spm element.

[b]Expressed as percent of kernels receiving the inactive Spm element that exhibit full Spm activity.

[c]Not determined.

(Spm-cr). Unlike an Spm-i element, an Spm-cr element is not active in the presence of another Spm element. In initial studies on the Spm-cr element, it was noted that new, independently segregating Spm-s and Mod elements can be recovered from strains carrying an Spm-w element (6). The appearance of new elements has been observed in the presence of several different Spm-w elements, some of which contain internal deletions. The frequency with which new elements appear in such strains is higher, by an order of magnitude or more, than the frequency of germline transposition of the resident Spm-w element (6), indicating that the new elements are not the result of transposition events that permit a change in the phenotypic expression of the element. Moreover, the frequency with which new elements appear is at least two orders of magnitude higher in the presence of an Spm-w element than in the absence of an active Spm (6). Thus, an active element promotes the activation of a cryptic element.

Newly activated elements differ in two genetic properties. They either resemble the Mod element or an Spm-s element. Moreover, the Spm-s elements differ in the heritability of the active state. Both of these points are illustrated by the data presented in Tab. 2. The progeny of plant 1466A-2 are especially interesting in this regard. The newly activated element in this plant was inherited almost exclusively through the female and resembles the Mod element, since it is active in the presence, but not in the absence, of an Spm-w element. However, of the small number of kernels showing the presence of a second element obtained when plant 1466A-2 was used as a male, one carried an Spm-s element, suggesting that the same cryptic element can be converted to either a Mod or an Spm-s element.

The newly activated state of the initially cryptic element is, in some cases, less than fully heritable. This is revealed by the underrepresentation of kernels exhibiting the element in some crosses carrying both the

Tab. 2. Genetic analysis of maize plants with newly activated Spm elements.

Plant	Used as	Percent of kernels with newly activated element	Element[a]
1466A-1	female (ear 1)	50.8	Spm-s
	male	43.1	Spm-s
	male	49.5	Spm-s
	male	48.2	Spm-s
	male	52.9	Spm-s
1466A-2	female (ear 1)	45.9	Mod[b]
	female (ear 2)	16.2	Mod
	male	0.4	Mod, Spm-s[c]
1857A-1	female	35.5	Spm-s
	male	28.9	Spm-s
1888-2	female	48.2	Spm-s
	male	53.1	Spm-s

[a]The activity of the element was determined by its ability to trans-activate excision of a dSpm element both in the presence and absence of an independently segregating Spm-w element.

[b]Mod designates an element that is active only in the presence of an Spm.

[c]Of the two kernels analyzed, one proved to have an Spm-s and the other had a Mod element.

Spm-w and the newly activated element (Tab. 2). It was also noted that among the initial kernels selected and analyzed, only about half gave evidence of a heritably activated element (6). It appears, therefore, that newly activated elements have a tendency to revert to the inactive condition.

We have recently undertaken the study of the a-m2-8167B allele, which may contain an Spm-cr. We initially reported that the Spm element in this allele, designated dSpm-8167B, exhibits the properties of a dSpm element, although the structure of the cloned element has so far proved indistinguishable from that of a cloned Spm-s element (11). The phenotype of this allele is shown in Fig. 2e and resembles that of the a-m2-8004 allele. Both exhibit a low level of Spm-dependent a gene expression, as well as a low transposition frequency, in the presence of a trans-acting Spm-s. Moreover, the element exhibits no activity of its own (11).

The results of recent molecular analyses of the dSpm-8167B element in genomic DNA have revealed that it is more resistant to cleavage with methylation-sensitive restriction endonucleases than is the Spm-s element of the a-m2-7991A1 allele. This is illustrated in Fig. 5a, using genomic DNA digested with SalI and BamHI. The results of analyses with several enzymes are represented diagrammatically in Fig. 6 and indicate that the dSpm-8167B element is resistant to digestion with all of the tested enzymes, while the Spm-s-7991A1 element is sensitive to cleavage at certain sites within a GC-rich region near the element's left end. It should be noted, moreover, that the lengths of the restriction fragments detected in genomic DNA are those anticipated if the target sites for a given enzyme within the element are resistant to cleavage, while those outside of the element within the a gene are sensitive to cleavage.

The foregoing observations have several implications. First, because the phenotype of the a-m2-8167B allele resembles that of the a-m2-8004 allele, rather than that of alleles with longer Spm elements, there must be something about the element that alters its ability both to mediate a gene expression and to excise in the same way as does the extensive deletion in the dSpm-8004 element (Fig. 2 and 3). The ends of the cloned dSpm-8167B element have been sequenced and found to be the same as those of longer elements that excise at a higher frequency and support a higher level of a gene expression (Fig. 2 and 3; Ref. 11).

The sequence whose deletion distinguishes the behavior of the dSpm-8004 element from that of the other elements is located between 0.28 and 0.86 kb from the element's left end (Fig. 3; Ref. 11). More than half of the implicated sequence corresponds to the GC-rich first intron of the element (28). The distribution of methylatable C residues in CG dinucleotides and CNG trinucleotides is represented schematically for this portion of the element in Fig. 7, and their clustering in this region is evident.

Also indicated in Fig. 7 is the location of the several methylation-sensitive restriction endonuclease cleavage sites that have been tested and found to be fully methylated in the dSpm-8167B element. We tender the tentative interpretation that the relative immobility of the dSpm-8167B element, as well as its reduced capacity to mediate the Spm-dependent activation of a gene expression, are attributable to methylation of sites near, but not at, the element's left end, within the GC-rich region whose importance in both functions has been defined by the dSpm-8004 deletion. Thus, we suggest that the phenotypic similarity between the a-m2-8004 and a-m2-8167B alleles is attributable to the methylation of sites in the dSpm-8167B element that are deleted in the dSpm-8004 element.

Because of the parallel between Spm control of a gene expression and its control of its own expression, the possibility arises that the dSpm-8167B element is actually a cryptic element and that some or all of the apparent methylation is responsible for its inactivity. If this is the case,

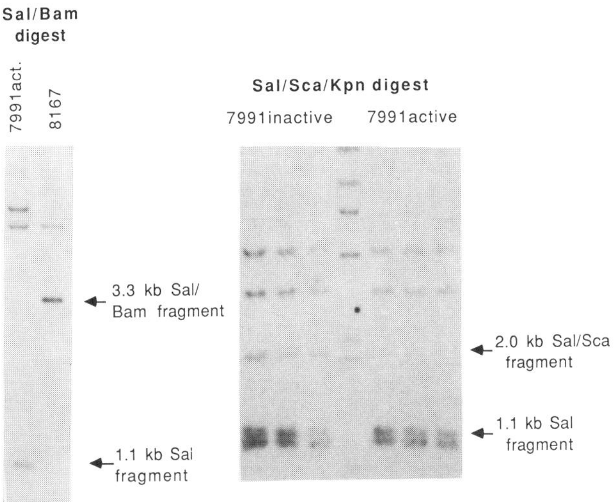

Fig. 5. Blot hybridization analysis of the a gene in maize genomic DNA cleaved with methylation-sensitive restriction enzymes. a. DNA isolated from plants homozygous for the a-m2-7991A1 and a-m2-8167B alleles was digested with SalI and BamHI, fractionated on an agarose gel, transferred to a filter membrane, and probed with a fragment of the a gene immediately to the left of the Spm insertion site in the a-m2 alleles (11). The length of the fragment containing part of the a gene and extending into the element to the first internal SalI site is 1.1 kb; if neither SalI site within the element is cleaved, the anticipated length of the fragment extending from the SalI site in the a gene to the BamHI site within the element is 3.3 kb (11). Additional fragments detected with the a gene probe represented homology to one or more independently segregating sequences present in some genetic backgrounds. b. DNA was isolated from plants that were heterozygous for the a-m2-7991A1 allele and a recessive mutant a allele. The three lanes designated 7991 inactive contain DNA from plants grown from kernels in which the Spm element was genetically inactive, while those designated 7991 active contain DNA from plants grown from kernels in which the resident Spm element had returned to an active state. All of the DNA samples were digested with SalI and ScaI. The a-m2-7991A1 allele yields a 2.2 kb SalI fragment, while the a locus of the homolog yields the slightly shorter SalI fragment. If the first SalI site within the element is not cleaved, then the length of the anticipated fragment extending to the ScaI site is 2.0 kb. Additional fragments of larger size represent a homologous sequence that is not linked to the a gene.

then an active element may promote its reactivation, a possibility we are presently investigating. Moreover, if genetic inactivation is caused by methylation, then the Spm-i derivatives analyzed in Tab. 1 should be methylated. The results of such an experiment are shown in Fig. 5b, and the results of several experiments are represented diagrammatically in Fig. 6. For each enzyme tested, we observe that the Spm-i element exhibits the DNA fragments expected if a fraction of each target site within the GC-rich region is methylated in a given plant. For each of the sites examined so far, the relative abundance of the fragments suggests that 5-20% of a given target site is methylated. Thus, the unstably inactive Spm-i element gives evidence of partial methylation.

We do not yet know whether the Spm element population in the Spm-i plants represents a mixture of fully modified and completely unmodified elements or, alternatively, whether a given element is modified at a subset of the target sites within its sequence. Nonetheless, there is a clear correlation between the acquisition of resistance to cleavage by methylation-sensitive restriction endonucleases and the genetic change from an active to an inactive condition. Moreover, genetic reactivation of the Spm-i element is accompanied by a reduction in the intensity or disappearance of the characteristic restriction endonuclease fragment containing uncleaved sites (Fig. 5b).

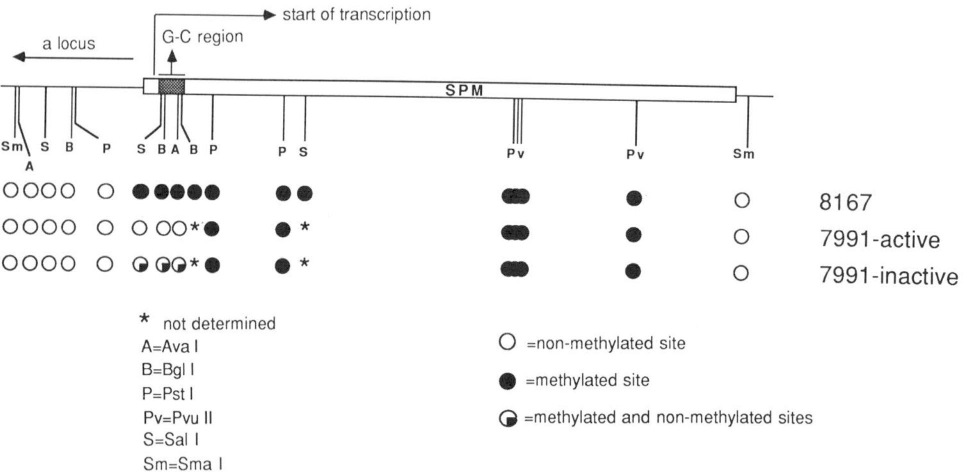

Fig. 6. Diagrammatic representation of the sensitivity of the various Spm elements analyzed to cleavage with methylation-sensitive restriction endonucleases. The diagram summarizes the results of blot hybridization analysis of genomic DNA from plants containing the indicated a-m2 alleles with an Spm element that is either active or inactive. The Spm element of the a-m2-8167B allele behaves genetically as if it were a defective element, but is structurally indistinguishable from the active Spm-s of the a-m2-7991A1 allele. Empty circles represent cleavable (unmethylated) sites, filled circles represent sites that are fully resistant to cleavage (methylated), and partly filled circles represent sites that are partly cleaved by a given restriction enzyme in plant genomic DNA.

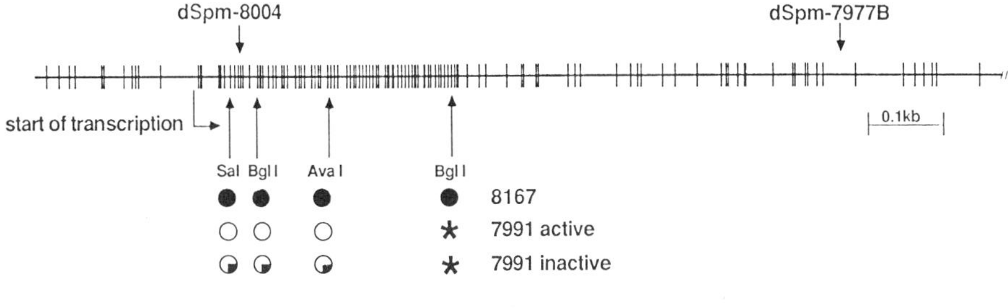

Fig. 7. Distribution of CG dinucleotides and CNG trinucleotides at the left end of the Spm element. The horizontal line represents the left end of the Spm element and the vertical lines represent the positions of CGs and CNGs (11). The transcription start site identified by Pereira et al. (28) is indicated by the arrow below the diagram. The positions of the left deletion end points in the a-m2-8004 and a-m2-7977B alleles are indicated by the arrows above the diagram (11). The use of empty, filled, and partly filled circles is the same as in Fig. 6.

DISCUSSION

Genetic and molecular studies on the unique a-m2 alleles have allowed us to extend our understanding of the mechanisms that regulate expression of the Spm element. We infer from the results of these studies that the Spm element itself encodes a positive autoregulatory gene product. Evidence for its existence is of two kinds. First, an active Spm element can trans-activate a genetically inactive Spm element (Tab. 1; Ref. 16-18,27). Second, Spm can trans-activate expression of the a gene in the a-m2 alleles with a dSpm insertion just upstream of the gene's transcription initiation site (11). We have interpreted such element-mediated gene expression to be a reflection of the element's own regulatory mechanism (11).

Analysis of the phenotypic consequences of deletions within the inserted element leads to the conclusion that the element sequence that mediates element-dependent a gene expression is near its left end, in the vicinity of its site of transcription initiation. Maximal a gene expression and element excision both require element termini, as well as a sequence within the element's transcription unit near its 5' end. Moreover, similar phenotypes are observed with an element from which the sequence has been deleted and an intact element in which the sequence is present, but extensively modified by methylation in genomic DNA, rendering it resistant to cleavage by methylation-sensitive restriction endonucleases. These observations suggest that an Spm-encoded protein must interact with the internal sequence in both the excision process and in activating expression of the adjacent gene, which we believe to be a reflection of its role in activating element gene expression.

It is not yet clear whether the element-encoded trans-acting transposition function and its positive regulatory function reside in the same or different gene products. A single major transcript has been identified for the element, as have several minor transcripts (11,28). The analysis of the structure of mutant elements suggests that the various element-encoded functions do not reside in separate, complementing transcription units (11), but does not eliminate the possibility that they reside in different proteins encoded by alternatively spliced mRNAs encoding by a single transcription unit.

Genetic studies on reversibly inactivated elements have permitted the identification of at least two clearly distinguishable inactive states, which we have termed inactive and cryptic. We have isolated a derivative of the a-m2-7991A1 (Spm-s) allele in which the resident element has become inactive, but undergoes reactivation rather frequently in progeny. Such an element is fully active in the presence of a trans-acting Spm-w element and resembles the behavior of the Mod element identified and studied by McClintock. We have identified a different, silent form of the Spm element, designated cryptic, which is not always active in the presence of a trans-acting Spm element, but whose conversion to either a Mod element or an Spm-s is stimulated by the Spm element. Our present view of the interrelationships among the different forms of Spm is depicted diagrammatically in Fig. 8.

We have begun to analyze the molecular basis of element inactivation using the a-m2 alleles. We have studied one dSpm element, the dSpm-8167B element, in some detail and found it to be structurally intact, but highly modified in genomic DNA, rendering it resistant to cleavage by a variety of methylation-sensitive restriction endonucleases. To test the hypothesis that the dSpm-8167B element is an Spm-cr, we are presently seeking to reactivate the element genetically. We have also begun an analysis of the structure of the newly arisen Spm-i element at the same insertion site and have found that there is a clear correlation between genetic inactivity of the element and the presence of genomic DNA fragments that cannot be cleaved by methylation-sensitive restriction endonucleases. The genomic DNA from a plant with an unstably inactive Spm-i element comprises a mixture of modified and unmodified sequences, while the genomic DNA from a plant in which the element has returned to an active state contains only unmodified sequences in the GC-rich region near the element's left end. We do not yet know whether the Spm-i elements in a given plant comprise a mixture of fully modified and unmodified elements, or whether each Spm-i element is modified at a subset of sites within the GC-rich region.

Although our studies on cryptic and inactive elements are not yet far advanced, they allow us to draw several tentative conclusions. Element inactivity is associated with increased methylation of certain sites in the GC-rich region near the left end of the element, but not with methylation of adjacent a gene sequences. Stable inactivation may be attributable to extensive methylation of elements at most or all methylation-sensitive sites. An Spm-encoded gene product can not only activate the expression of an Spm-i, but can stimulate the conversion of a cryptic element to a heritably active form.

Methylation of element sequences has been implicated in the genetic inactivation of other maize transposable elements (2-5,7,37). Studies on the Spm element indicate that inactivation is a more complex phenomenon than anticipated. First, methylation affects the mobility of an element, as judged by its ability to be excised by an active element located elsewhere in the genome. Observations on dSpm elements similar to ours have been made by V. Raboy and O. Nelson (pers. comm.) using dSpm alleles of the

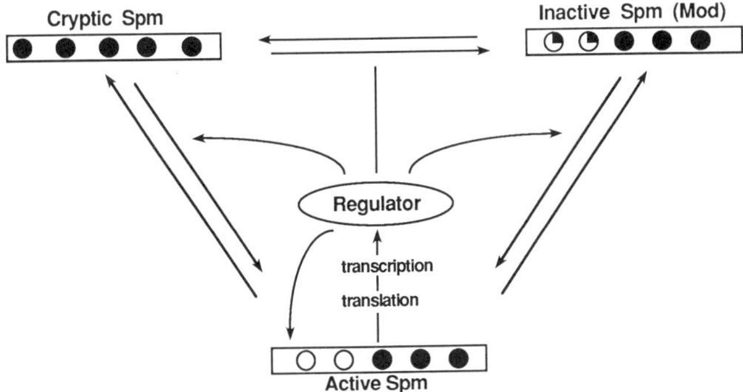

Fig. 8. Diagrammatic representation of the interrelationships among active, inactive, and cryptic Spm elements. Empty and filled circles represent unmethylated and methylated sites, respectively. Partly filled circles represent sites that may or may not be methylated in a given element.

bronze locus. Both studies implicate a GC-rich sequence within the transcription unit as the important target sequence whose methylation reduces element mobility. Second, element sequence methylation is correlated with the genetic inactivation of the Spm element, as judged by its ability to trans-activate excision of elements located elsewhere in the genome. The ability of methylation at specific sites to reduce mobility and element expression finds precedent in the bacterial transposon Tn10 (36). Third, an Spm-encoded gene product can both promote expression of an inactive element and promote the conversion of a cryptic element to a heritably active form. We have proposed that these interactions between active and silent elements reflect the normal functions of the Spm's regulatory gene product in maintaining element gene expression (11).

Note added in proof: Because of adverse weather conditions, pollen was collected only from the tillers of the plants for which data are given in Tab. 1. More extensive analysis of sibling plants was carried out in the following year and revealed that the differences between female and male transmission are attributable to the extensive reactivation of inactive elements in tillers and their maintenance in an inactive state when transmitted through gametes produced on the plant's main stalk.

REFERENCES

1. Banks, J., J. Kingsbury, V. Raboy, J.W. Schiefelbein, O. Nelson, Jr., and N. Fedoroff (1985) The Ac and Spm controlling element families in maize. Cold Spring Harbor Symp. Quant. Biol. 50:307-311.
2. Bennetzen, J.L. (1985) The regulation of Mutator function and Mu1 transposition. UCLA Symp. Molec. Cell. Biol. 35:343-354.
3. Chandler, V.L., and V. Walbot (1986) DNA modification of a maize transposable element correlates with loss of activity. Proc. Natl. Acad. Sci., USA 83:1767-1771.

4. Chomet, P.S., S. Wessler, and S.L. Dellaporta (1987) Inactivation of the maize transposable element Activator (Ac) is associated with DNA modification. EMBO J. 6:295-302.
5. Dellaporta, S.L., and P.S. Chomet (1985) The activation of maize controlling elements. In Genetic Flux in Plants, B. Hohn and E.S. Dennis, eds. Springer-Verlag, New York, pp. 169-216.
6. Fedoroff, N.V. (1986) Activation of Spm and modifier elements. Maize Gen. Coop. Newslet. 60:18-20.
7. Fedoroff, N., S. Wessler, and M. Shure (1983) Isolation of the transposable maize controlling elements Ac and Ds. Cell 35:235-242.
8. Fedoroff, N., P. Masson, and J. Banks (1987) Regulation of the maize Suppressor-mutator element. In Eukaryotic Transposable Elements as Mutagenic Agents. Cold Spring Harbor (in press).
9. Fedoroff, N., M. Shure, S. Kelly, M. Johns, D. Furtek, J. Schiefelbein, and O. Nelson, Jr. (1984) Isolation of Spm controlling elements from maize. Cold Spring Harbor Symp. Quant. Biol. 49:339-345.
10. Gierl, A., Zs. Schwarz-Sommer, and H. Saedler (1985) Molecular interactions between the components of the En-I transposable element system of Zea mays. EMBO J. 4:579-583.
11. Masson, P., R. Surosky, J.A. Kingsbury, and N.V. Fedoroff (1987) Genetic and molecular analysis of the Spm-dependent a-m2 alleles of the maize a locus. Genetics 177:117-137.
12. McClintock, B. (1951) Mutable loci in maize. Carnegie Institution of Washington Yearbook 50:174-181.
13. McClintock, B. (1954) Mutations in maize and chromosomal aberrations in Neurospora. Carnegie Institution of Washington Yearbook 53:254-260.
14. McClintock, B. (1955) Controlled mutation in maize. Carnegie Institution of Washington Yearbook 54:245-255.
15. McClintock, B. (1956) Mutatuions in maize. Carnegie Institution of Washington Yearbook 55:323-332.
16. McClintock, B. (1957) Genetic and cytological studies of maize. Carnegie Institution of Washington Yearbook 56:393-401.
17. McClintock, B. (1958) The Suppressor-mutator system of control of gene action in maize. Carnegie Institution of Washington Yearbook 57:415-429.
18. McClintock, B. (1959) Genetic and cytological studies of maize. Carnegie Institution of Washington Yearbook 58:452-456.
19. McClintock, B. (1961) Further studies of the Suppressor-mutator system of control of gene action in maize. Carnegie Institution of Washington Yearbook 60:469-476.
20. McClintock, B. (1961) Some parallels between gene control systems in maize and bacteria. Am. Nat. 95:265-277.
21. McClintock, B. (1962) Topographical relations between elements of control systems in maize. Carnegie Institution of Washington Yearbook 61:448-461.
22. McClintock, B. (1963) Further studies of gene-control systems in maize. Carnegie Institution of Washington Yearbook 62:486-493.
23. McClintock, B. (1964) Aspects of gene regulation in maize. Carnegie Institution of Washington Yearbook 63:592-602.
24. McClintock, B. (1965) Components of action of the regulators Spm and Ac. Carnegie Institution of Washington Yearbook 64:527-536.
25. McClintock, B. (1965) The control of gene action in maize. Brookhaven Symp. Quant. Biol. 18:162-184.
26. McClintock, B. (1986) The states of a gene locus in maize. Carnegie Institution of Washington Yearbook 66:20-28.

27. McClintock, B. (1971) The contribution of one component of a control system to versatility of gene expression. Carnegie Institution of Washington Yearbook 70:5-17.
28. Pereira, A., H. Cuypers, A. Gierl, Zs. Schwarz-Sommer, and H. Saedler (1986) Molecular analysis of the En/Spm transposable element system of Zea mays. EMBO J. 5:835-841.
29. Pereira, A., Zs. Schwarz-Sommer, A. Gierl, I. Bertram, P.A. Peterson, and H. Saedler (1985) Genetic and molecular analysis of the Enhancer (En) transposable element system of Zea mays. EMBO J. 4:17-23.
30. Peterson, P.A. (1953) A mutable pale green locus in maize. Genetics 38:682.
31. Peterson, P.A. (1960) The pale green mutable system in maize. Genetics 45:115-133.
32. Peterson, P.A. (1961) Mutable a1 of the En system in maize. Genetics 46:759-771.
33. Peterson, P.A. (1965) A relationship between the Spm and En control systems in maize. Am. Nat. 99:391-398.
34. Peterson, P.A. (1966) Phase variation of regulatory elements in maize. Genetics 54:249-266.
35. Rhoades, M.M., and E. Dempsey (1950) New mutable loci. Maize Gen. Coop. Newslet. 24:50.
36. Roberts, D., B.C. Hoopes, W.R. McClure, and N. Kleckner (1985) IS10 transposition is regulated by DNA adenine methylation. Cell 43:117-130.
37. Schwartz, D., and E. Dennis (1986) Transposase activity of the Ac controlling element in maize is regulated by its degree of methylation. Molec. Gen. Genet. 205:476-482.
38. Schwarz-Sommer, Zs., A. Gierl, R. Berndtgen, and H. Saedler (1985) Sequence comparison of "states" of a1-m1 suggests a model of Spm (En) action. EMBO J. 4:2439-2443.
39. Schwarz-Sommer, Zs., A. Gierl, R.B. Klösgen, U. Wienand, P.A. Peterson, and H. Saedler (1984) The Spm (En) transposable element controls the excision of a 2-kb DNA insert at the wx-m8 allele of Zea mays. EMBO J. 3:1021-1028.
40. Schwarz-Sommer, Zs., N. Shepherd, E. Tacke, A. Gierl, W. Rohde, L. Leclercq, M. Mattes, R. Berndtgen, P.A. Peterson, and H. Saedler (1987) Influence of transposable elements on the structure and function of the A1 gene of Zea mays. EMBO J. 6:287-294.
41. Tacke, E., Zs. Schwarz-Sommer, P.A. Peterson, and H. Saedler (1986) Molecular analysis of states of the A locus of Zea mays. Maydica 31:83-91.

GERM LINE AND SOMATIC MUTATOR ACTIVITY: ARE THEY FUNCTIONALLY RELATED?*

Donald S. Robertson,[1] David W. Morris,[2]
Philip S. Stinard,[1] and Bradley A. Roth[3]

[1]Department of Genetics
Iowa State University
Ames, Iowa 50011

[2]Department of Biological Sciences
George Washington University
Washington, D.C. 20052

[3]Plant Gene Expression Center
U.S. Department of Agriculture
Agricultural Research Service
Albany, California 94710

ABSTRACT

Results of experiments to test for a relationship, if any, between somatic Mutator (Mu) activity (somatic mutability) and germline Mu activity are described. Somatic Mu activity was measured by the intensity of the mutable pattern in unstable a1-Mum aleurone mutants. Germinal activity was determined by the frequency with which a Mu plant transmitted new mutants to its outcross progeny plants. None of the experiments demonstrated tight linkage between these two Mu-regulated phenomena. Not all plants with intense somatic mutability patterns exhibited germline activity. Nor did all plants that had little or no somatic mutability lack germinal activity. In one inbred lineage, which had retained an intense somatic mutability pattern for four generations, there was a sudden shift to low somatic mutability in the progeny of the fifth generation. Some representatives of this lineage, taken from different generations but all with intense mutability, had germinal activity, but others did not. The same was true for the low-intensity lines observed in the progeny of the fifth generation. The lines with a low-intensity somatic mutability pattern had Mu elements that were modified at their HpaII sites. This modification was

*Journal paper No. J-12799 of the Iowa Agriculture and Home Economics Experiment Station, Ames, Iowa, Projects No. 2623 and 2707.

found in "low" somatic mutability lines with and "low" somatic mutability lines without germinal Mu activity.

At present, there is insufficient evidence to establish how germline activity and somatic mutability in Mutator stocks are regulated and if they are regulated via a common mechanism. There is evidence that DNA modification is involved in both phenomena, but more information on the state of modification of the element inserted at the mutant aleurone locus in different situations with respect to levels of somatic mutability and germinal activity is needed before the role of modification in these two aspects of the Mutator system can be fully understood. What role, if any, putative regulators of somatic activity have in germinal activity is yet to be determined.

INTRODUCTION

The maize Mutator system was first described by Robertson (14) as a system that induced a high frequency of germinally transmitted mutants (50- to 150-fold above that of controls). Research carried out in Michael Freeling's laboratory established that a Mutator-induced Adh1 allele had a 1.4-kb segment of DNA (Mu1) inserted in the first intron (4,22). Mu1 and a 300-bp larger version (Mu1.7) were found in copy numbers of from 10 to 70 in Mutator lines (3). The Mu1 element was sequenced by Barker et al. (2), and the 1.7-kb element was sequenced by Taylor and Walbot (23,24). Most corn lines had sequences with some homology to the terminal repeated ends and internal sequences of the Mu1 element, and some had been demonstrated to have intact elements (9,20).

These elements, found in non-Mutator lines, did not undergo transposition. However, the numerous Mu elements in Mutator lines underwent frequent transpositions each generation. Outcross offspring had, on average, the same number of copies of these elements as their parents. Nevertheless, considerable variation in copy number could be found in individual plants of an outcross progeny, with both greater and lesser numbers observed (1,7,26). Both Bennetzen (4,5) and researchers in Virginia Walbot's laboratory (8,25,26) have demonstrated a high degree of modification of the Mu elements in some stable derivatives of Mu-induced mutable aleurone mutants. Potential C-methylation sites in the Mu elements, such as EcoRII, HinfI, HpaII, and AvaII, have been shown to be subject to such modification, and to have become impervious to digestion by these restriction endonucleases.

With the discovery of mutable Mu-induced aleurone mutants (e.g., a1, 17; bz1, 5; bz2, 25), Mutator activity could be studied at the somatic level. Early analysis of these mutants did not reveal a simple dual element (i.e., receptor-regulator) control of somatic mutability. Although stable mutant derivatives of mutable alleles were found, they did not seem to occur in the Mendelian ratios observed for the mutable controlling-element systems previously described [see Fedoroff (10) for a review]. Robertson et al. (19), Bennetzen (5), and Walbot (25) have demonstrated that such stable mutant derivatives can be reactivated by crossing with plants that have an active Mutator system. These crosses, in which reactivation occurred, did not give evidence of simple Mendelian ratios expected for two-element systems. Robertson (18) reported the first systematic occurrence of reactivation of stable derivatives in Mu-induced aleurone mutants in which 1:1 ratios consistently were observed. Such stocks are presently under intensive investigation.

As a result of genetic and molecular studies, various criteria have been used to define an active Mutator system. The original criterion, used by Robertson (14), was the demonstration that a Mu plant transmitted new mutant genes to its outcross progeny. Such Mu activity is termed germinal activity. Another criterion, which has been suggested by Robertson et al. (19) to define an active Mutator system, is the copy number of Mu1-like elements. Active Mu plants had relatively more Mu1-like elements than inactive plants. Bennetzen et al. (7) and Bennetzen (4), however, found no correlation between Mu-element copy number and germinal activity and/or the intensity of mutability in unstable Mu-induced aleurone mutants. Somatic mutability is a third criterion which has been used by some as a measure of an active Mutator system. For Walbot (25), and Walbot and Warren (26), the presence of somatic mutability was considered an indicator of the presence of an active Mutator system. Stable mutant derivatives were considered to have lost all Mu activity. In support of this idea, Robertson et al. (19) showed that, for 11 stable derivative lines, Mu germinal activity had been lost. Walbot and co-workers, however, have not presented extensive data supporting their conclusion that somatic mutability is a good indicator of germinal activity. A fourth criterion for determining the activity of a Mutator system is the degree of DNA modification present in the Mu elements. Walbot (25) and Chandler and Walbot (8) studied the correlation of somatic mutability with the modification of Mu elements. They concluded that there was a strong correlation between the loss of somatic mutability and modification of the Mu elements.

In these papers, the authors do not suggest that somatic mutability and/or modification are accurate indicators of germinal Mu activity. Walbot and Warren (26), however, do suggest that the maintenance of Mu-element copy number, somatic mutability, and the extent of modification of HinfI sites are equally good indicators of an active Mutator system. Bennetzen (5) suggests that Mutator germinal activity is "associated with transposition and high copy number of Mu1-like elements...and the somatic mutability of Mutator-induced lesions."

In this paper, we present extensive results bearing on the correlation of Mutator somatic mutability and Mutator germinal activity. In addition, we present more limited molecular characterizations dealing with Mu-element copy number, somatic mutability, germinal activity, and DNA modification.

MATERIALS AND METHODS

Definitions of terms and symbols used in this report are found in Tab. 1.

Genetic Studies

Three types of genetic studies were undertaken. The most straightforward study for the correlation of somatic mutability with germinal activity involved outcrossing stocks of three Mu-induced a1 mutant alleles (a1-Mum1, a1-Mum2, and a1-Mum3) for several generations, while scoring for mutability and germinal activity. A five-class scale was used as a

Tab. 1. Definition of symbols and terms.

Symbol/term	Definition
Al	One of many complementary genes necessary for the production of color in the aleurone.
al	The recessive allele of Al, which when homozygous will result in colorless aleurone.
sh2	A recessive allele responsible for shrunken and shriveled seeds.
al sh2	An al tester stock homozygous for recessive al, but with all other genes necessary for aleurone color. This stock is also homozygous for the closely linked sh2 allele.
al-Mum	A generic symbol for any of three Mutator-induced mutable al mutants.
al-Mum1 al-Mum2 al-Mum3	Three independent Mutator-induced mutable al mutants.
al-Mum stable	A generic symbol for the stable derivative of an al-Mum allele.
al-Mum1 stable al-Mum2 stable al-Mum3 stable	The three possible stable derivatives.
Mu element	Any one of a number of different transposable elements involved in the Mutator system.
Mu1 element	The specific Mu element first described by Bennetzen et al. (6) and Barker et al. (2).
Mu1.7 element	The larger Mu element characterized by Taylor and Walbot (23).
Pl. Mu	A purple aleurone stock with germinal Mutator activity.
Germinal activity	Mutator activity in the germline of a plant that results in new mutant genes that are transmitted to the outcross progeny. There is evidence that some mutants in these outcross progeny may be induced in other than germline tissue (e.g., in the gametophytes). The term "germinal activity," as used in this paper, includes all new mutants found in the outcross progeny, regardless of their origin.
Somatic mutability	The mutable (spotted) aleurone pattern found in Mutator-induced al mutants.
Somatic mutability scale	Class 1, stable mutant; class 2, low mutability; class 3, medium mutability; class 4, high mutability; class 5, purple.
1.1 Ratio	Ears with a 1:1 ratio of (3 plus 4) to (2 plus 1) classes of mutable seeds. (See Tab. 5, footnote a, for more details.)
Modification	The DNA of Mu elements is said to be modified if it cannot be cleaved by restriction endonucleases such as MspI, HpaII, EcoRII, BglI, and AvaII, which do not cut DNA containing 5-methylcytosine at their target sites.
Intracross	Cross among descendants of a common progenitor plant.

measure of mutability (i.e., class 1, stable mutant; class 2, light mutable pattern; class 3, medium mutable pattern; class 4, heavy mutable pattern; and class 5, stable purple). In each generation, ears were selected that had predominantly seeds of one of the classes 2 through 4. The mutant seeds on these ears were scored for mutability, and an average mutability score was determined for each ear. From these ears, seeds of the predominant class were planted for the next generation's outcross. A sample of 50 seeds from each outcross ear also was selected and sown, and the resulting plants self-pollinated. The seedlings of the selfed ears were scored for the presence of new mutants (germinal Mutator activity). (See Fig. 1 for a schematic diagram of this procedure.)

The second series of experiments involved the testing for germinal activity of stable derivative lines in which somatic reactivation (restoration of somatic mutability) had occurred as a result of crossing to germinally active purple aleurone Mu lines (Pl. Mu). These tests are an extension of experiments first described by Robertson et al. (19). The F_1's between stable a1-Mum derivatives and Pl. Mu lines, which upon selfing segregated for mutable seeds, were crossed to non-Mu purple aleurone stocks, and 50 of the outcross plants were selfpollinated. The seeds of the selfed ears were scored for somatic mutability, and the seedlings from these ears were tested for evidence of germinal Mu activity. An estimated average mutability score for each outcross family was obtained. To determine this score, an average mutability value was estimated for each ear, and from these ear values an average mutability value was determined for the whole family. Although these are only estimated values, they agree reasonably well with values determined by actually classifying each seed on each ear of an outcross family. Germinal activity was determined by seedling testing of each selfed ear. F_2 plants from mutable seeds of the a1-Mum stable derivative by Pl. Mu cross were also tested in the same manner. (See Fig. 2 for a schematic diagram of this procedure.)

The third set of experiments involved inbreeding a stock of a1-Mum2 with high somatic mutability (i.e., predominantly class 4 seeds). This stock had been selfed or intracrossed for eight generations through the 1986-87 winter nursery (Roth, Ref. 20). It has been possible to retain high somatic mutability in all generations. Beginning with the sixth generation, however, some crosses gave rise to ears with predominantly low mutability or stable seeds. Some individuals in this pedigree have also been tested for germinal activity. In later generations, in which the loss of somatic mutability has been observed, both highly mutable lines and low-mutability or stable lines have been tested for germinal activity.

Some of the individuals in this pedigree (see Fig. 3) also have been molecularly tested to determine the copy number of Mu1-like elements and their degree of modification. The procedures that follow were used for isolating and characterizing the DNA in this lineage.

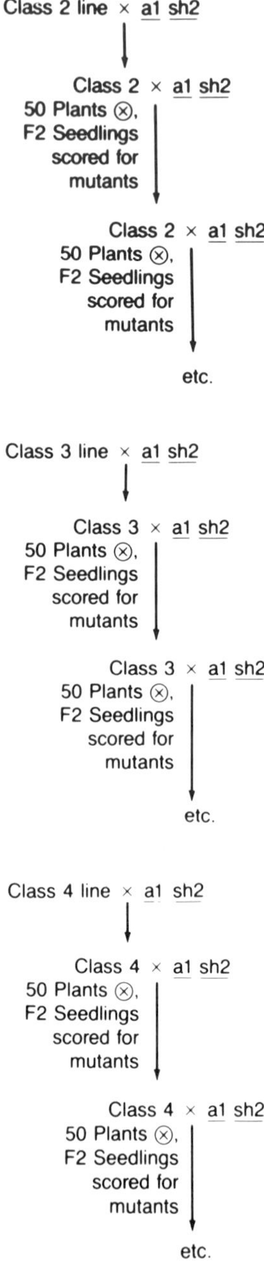

Fig. 1. Schematic diagram of serial crosses made to test the correlation of Mu somatic mutability and Mu germinal activity. (Example is for a1-Mum2.)

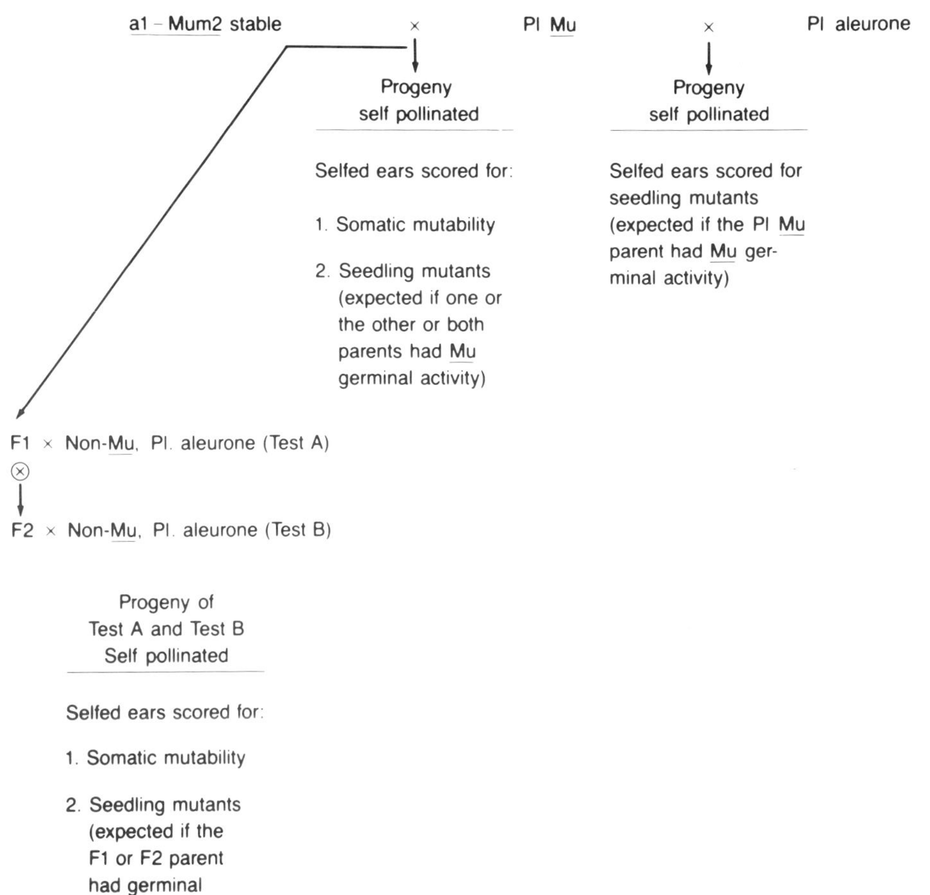

Fig. 2. A schematic diagram of: 1) The crosses by which somatic mutability was restored to a1-Mum stable alleles, and 2) the crosses by which these a1-Mum alleles, which had recovered somatic mutability, were tested for Mu germinal activity and inheritance of the recovered somatic mutability. The recovery of somatic mutability was determined by the a1-Mum2 stable X Pl. Mu cross and the selfed progeny of this cross. The germinal activity of the Pl. Mu parents was determined by the Pl. Mu X Pl. aleurone cross, and its selfed progeny. A cross of a1-Mum2 stable X Pl. aleurone, and its selfed progeny, were used as controls to ensure that the a1-Mum2 stable parent contributed no germinal activity or somatic mutability (these tests were negative). The outcrosses of the F_1 (Test A) and F_2 (Test B) plants to non-Mu Pl. aleurone plants, and the selfed progeny of these crosses, were used to determine the Mu germinal activity of the F_1 and F_2 parents and the transmission of somatic mutability.

Molecular Studies

Maize DNA was isolated according to Rivin et al. (13), digested with restriction endonucleases, fractionated on agarose gels, and transferred to Genatran (Plasco) using Southern's procedure (21). Filters were hybridized to a gel-purified (12) AvaI-TaqI internal Mu1 fragment (6) labeled to high specific activity by random hexamer priming (11), stringently washed, and exposed to Kodak XAR-5 film with one Lightning-Plus (Dupont) intensifying screen. Filters were stripped by two 20 min washes with 2,000 ml of wash buffer (0.1% SDS, 0.1 x SSC) heated to 95-100°C. Rehybridization with a 9-kb maize ribosomal tandem repeat probe (obtained from David Grant, Pioneer Hi-Bred International, Johnston, Iowa) was used to test for complete digestion and for the standardization of DNA samples in each lane. Mu element copy number was determined by counting individual Mu1 bands in an EcoRI digest, and by comparing a densitometric scan of Mu elements in a Tth111I digest to copy number reconstructions.

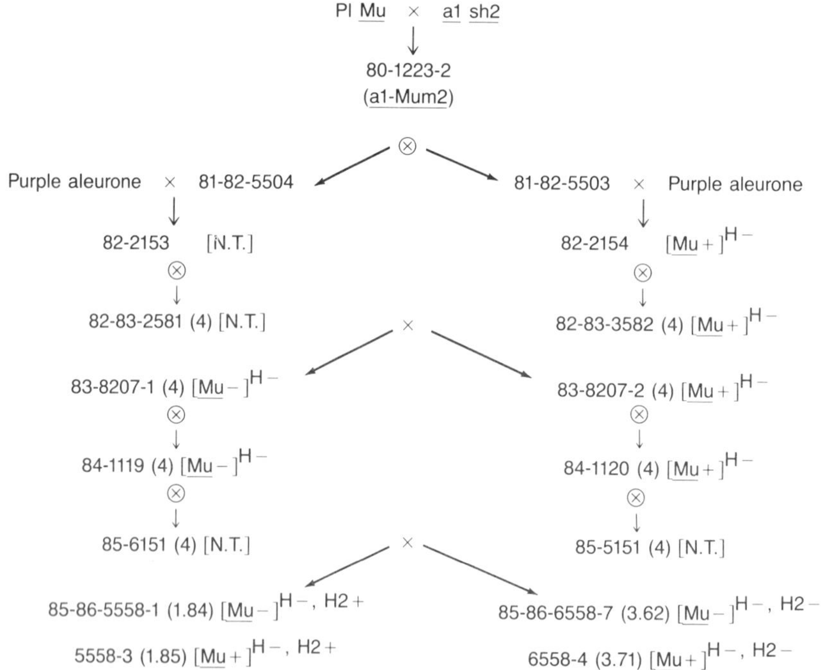

Fig. 3. A partial pedigree of the inbreeding regime of an a1-Mum2 stock, which retained high somatic mutability. The information in () is the somatic mutability score, while that in [] indicates germinal activity for a particular plant or generation (Mu+, germinal activity; Mu−, no germinal activity). H−, HinfI sites not modified. H+, HinfI sites modified. H2−, HpaII sites not modified. H2+, HpaII sites modified. Data modified from Roth (20).

RESULTS

In the first-generation tests for the correlation of somatic mutability and germinal activity (Tab. 2), lines with low somatic activity (i.e., mutability scores in the range 1.00-2.24) did not have germinal activity. Among the lines with medium mutability scores (in the range 2.25-3.24), some had germinal activity and others did not. Results similar to those observed for the medium lines were found for plants with high mutability (i.e., mutability scores in the range 3.25-4.00). In the second generation (Tab. 3) there was a rather marked loss of germinal activity. There were six instances in which plants from first-generation families with germinal activity did not show germinal activity in the second generation. In only two outcrosses of first-generation germinally active plants was germinal activity retained in the second generation, in spite of somatic mutability levels remaining relatively constant. In these two instances, the plants that retained germinal activity were crosses in which a medium- and a high- somatic-mutability line shifted to lines with low somatic mutability.

Thus, it was possible to have a stock with low somatic mutability that still showed germinal activity, contrary to the first-generation observations and previously published observations (19). One second-generation high-mutability line (86-1206), which descended from a first-generation line with high mutability but no germinal activity (85-3164), gained germinal activity. Whether this was a gain of Mutator activity or a misclassification of the first-generation family (85-3164) due to sampling error (i.e., a chance exclusion of a plant with a Mu-induced mutant in the 47-plant sample tested the first generation) could not be determined from the data. There were, however, two other second-generation outcross families descended from 85-3164, as well as one third-generation family, none of which showed germinal activity.

In the third generation (Tab. 4), the trend toward the loss of germinal activity continues. The two lines that now show germinal activity did not show germinal activity in the previous generation, although in one instance (family 1485) the number of plants tested in the second generation was small (25 plants); thus, it may have been just a chance event that no mutations were observed. In the other outcross family where germinal activity occurred (1501), there had been no observed germinal activity for the previous two generations. Accompanying this initiation of germinal activity was a concomitant increase in somatic mutability.

In some outcrosses to a1 sh2, there were ears that segregated in a 1:1 ratio for classes 3 plus 4 and classes 1 plus 2. Robertson (18) has presented evidence that some of these 1:1 ratios might be due to the segregation of a regulatory factor controlling the expression of somatic mutability. Thus, putative dual-element systems controlling the expression of aleurone mutability, similar to Ac and Ds, seemed to be present in these crosses. The tests of such stocks for germinal activity (Tab. 5) indicate that germinal activity was not consistently found. In families 85-86-3514 and 85-86-3566, stable seeds and highly mutable seeds from the same ear were tested separately (in all other families in Tab. 5, except 85-1503, equal numbers of seeds with high and low mutability were intermingled when planted, and mutants occurred only within the highly mutable class. This result would be expected if germinal activity is somehow associated with highly mutable seeds. In family 85-1503, however, only high-mutability seeds (classes 3 or 4) were planted, and no germinal activity was observed; in most other crosses, at least half the progeny consisted of plants from highly mutable seeds.

Tests of a1-Mum stable lines, which were reactivated by crossing to Pl. Mu plants (Tab. 6), indicate that, although all regained considerable somatic mutability, some lost germinal activity that was contributed by the Pl. Mu parent.

High somatic mutability was maintained in a line with the a1-Mum2 allele, which had been maintained by selfing or intracrossing within the line (Fig. 3). The fifth-generation intracrosses gave rise to lines that retained high somatic mutability and those that suddenly (within one generation) shifted to low mutability. These latter lines, when crossed an additional generation, either among themselves or outcrossed to a1 sh2, lost all, or almost all, somatic mutability. (Data not shown. See Roth, Ref. 20, for more details.) Germinal activity was lost early from half of this pedigree while it was retained in the other half. Before the fifth generation, however, both sides of the pedigree retained high somatic mutability. The 1985-86 low-mutability and high-mutability sibling lines, when tested for germinal activity, revealed that both germinally active and germinally inactive plants were found among the low as well as the high somatically mutable lines.

Roth (Ref. 20 and ms. in prep.) determined the copy numbers of Mu elements for single plants from 1982, 1984, and 1985 generations by counting bands in an EcoRI digest or by comparing the Tth111I hybridization profiles with copy-number reconstructions. Thirty-five to 40 bands per diploid genome were present in all plants. When these same plants were tested for modification at the HinfI sites in their Mu elements, none was found. In one 1985 cross (85-6151-8/5151-3T), some class 3 and class 2 seeds were found in addition to the predominant class 4. DNAs from plants of classes 4, 3, and 2 seeds were digested with HinfI to determine if there was modification at the restriction sites for this enzyme in the Mu DNA. Little or no modification was observed in the plants from the class 4 seeds, while plants from class 3 and class 2 seeds had modification, with class 2 plants showing the most modification (Fig. 4).

DNA from plants of family 84-1119, which probably had lost germinal activity because germinal activity was missing in the previous generation, showed no modification at the HinfI sites of their Mu DNA, nor was HinfI site modification in Mu DNA found in plants from the 1981-82, 1982, 1982-83, 1983, and 1984 generations of this pedigree despite the high level of inbreeding that occurred. When the Mu DNAs from two plants with high somatic mutability and two with low somatic mutability from the 1985-86 crosses were tested for modification at the HinfI sites, no modification was found in either the "high" or "low" classes. When these same plants were tested for modification at the HpaII sites, however, the DNAs from the "low" class plants were found to be modified, but not the Mu DNAs from the "high" class plants. Among the "low" class plants with Mu DNAs modified at the HpaII sites, both germinally active and germinally inactive plants were observed. Among the "high" class plants with unmodified Mu DNAs, both germinally active and germinally inactive plants were also observed (Fig. 5).

Tab. 2. First generation tests for the correlation of somatic mutability and germinal activity of three al-Mum alleles, crossed onto al sh2 silks.

Family number of the outcross plants	Allele	Average mutability score of the outcross ear	Total number of plants selfed	Number of plants segregating for mutants	Percent of plants segregating new mutants
Tests of lines with low mutability					
85-3154	al-Mum1	1.78	46	0	0.0
85-3155	"	1.86	43	0	0.0
85-3156	"	1.40	44	0	0.0
85-3157	"	1.08	44	0	0.0
85-3158	"	1.04	45	0	0.0
85-3159	al-Mum1	1.05	49	0	0.0
85-3168	"	1.99	49	0	0.0
85-3173	al-Mum3	1.08	37	0	0.0
Tests of lines with medium mutability					
85-3160	al-Mum2	2.75	47	1	2.13
85-3161	"	2.85	45	2	4.44
85-3162	"	2.46	48	0	0.0
85-3164	"	2.53	47	0	0.0
85-3170	"	2.35	43	0	0.0
85-3171	al-Mum3	2.89	45	0	0.0
85-3174	"	2.28	46	0	0.0
85-3175	"	2.29	50	1	2.00
Tests of lines with high mutability					
85-3163	al-Mum2	3.36	119	0	0.0
85-3166	"	3.90	43	2	4.65
85-3169	"	3.88	47	3	6.38
85-3172	al-Mum3	3.40	47	0	0.0

Ears with an average mutability score of 2.24 or less were classified as low; those with scores of 2.25 to 3.24 were classified as medium; and those with score 3.25 and greater were classified as high mutability.

Tab. 3. Second generation tests for the correlation of somatic mutability and germinal activity of three al-Mum alleles crossed onto al sh2 silks.[a]

Family number of the outcross plants	Allele	Average mutability score of the outcross	Family number of the first generation tests	Total number of plants selfed	Number of plants segre-gating mutants	Percent of plants segre-gating mutants	Percent of first generation segre-gating mutants
Tests of lines with low mutability							
85-86-3501	al-Mum1	1.31	3154	35	0	0.0	0.0
86-1201	"	1.40	3154	41	0	0.0	0.0
85-86-3502	"	1.51	3155	26	0	0.0	0.0
86-1202	"	1.33	3155	47	0	0.0	0.0
86-1203	"	1.52	3156	44	0	0.0	0.0
85-86-3503	al-Mum2	1.10	3159	14	0	0.0	0.0
85-86-3505	"	2.10	3168	22	0	0.0	0.0
85-86-3508	al-Mum3	1.68	3173	20	0	0.0	0.0
85-86-3509	"	1.20	3173	27	0	0.0	0.0
86-1210	"	1.23	3173	40	0	0.0	0.0
86-1211	"	1.37	3173	42	0	0.0	0.0
86-1447	"	1.26	--	37	0	0.0	0.0
Tests of lines with medium mutability							
85-86-3512	al-Mum2	1.13[b]	3160	28	1	3.57	2.13
86-1207	"	3.11	3160	56	0	0.0	2.13
85-86-3507	"	2.70	3161	25	0	0.0	4.44
85-86-3515	"	1.85[c]	3161	28	0	0.0	4.44
85-86-1208	"	2.08[c]	3161	44	0	0.0	4.44
86-1204	"	3.18	3162	40	0	0.0	0.0
85-86-3504	"	2.77	3164	12	0	0.0	0.0
86-1205	"	3.21	3164	39	0	0.0	0.0
86-1206	"	3.32	3164	42	3	7.14	0.0
85-86-3510	al-Mum3	2.30	3171	33	0	0.0	0.0
85-1212	"	2.55	3171	37	0	0.0	0.0
85-1215	"	2.68	3171	42	0	0.0	0.0
85-86-3511	"	2.30	3172	63	(1)	(1.59)[e]	0.0

Family number of the outcross plants	Allele	Average mutability score of the outcross	Family number of the first generation tests	Total number of plants selfed	Number of plants segre-gating mutants	Percent of plants segre-gating mutants	Percent of first generation segre-gating mutants
Tests of lines with medium mutability (continued)							
86-1213	al-Mum3	2.45	3172	44	0	0.0	0.0
85-86-3514	"	2.56	3174	30	0	0.0	0.0
Tests of lines with high mutability							
85-86-3506	al-Mum2	3.86	3169	20	0	0.0	6.38
86-1209	"	3.89	3169	46	0	0.0	6.38
86-1450	"	1.01[d]	3169	35	1	2.86	6.38
86-1448	al-Mum3[f]	3.19	--	34	0	0.0	--
86-1449	" [f]	3.10	--	37	1	2.70	--

[a]In the second and third generations, plants were assigned to somatic mutability classes on the basis of the classification in the first generation. There are occasions in later generations where somatic mutability has diminished or increased. Hence in generations 2 and 3, an occasional individual will be found with mutability scores that do not agree with the class they are listed under in the table.

[b]In the previous generation, this line had medium somatic mutability with germinal Mu activity. This plant now has low somatic mutability but retains germinal activity.

[c]In the previous generation, these lines had medium somatic mutability with germinal Mu activity. These plants now have low somatic mutability and possibly have lost germinal activity.

[d]In the previous generation, this line had high somatic mutability with germinal Mu activity. This plant now has low somatic mutability but has retained germinal activity.

[e]There was one doubtful mutant present in this test.

[f]These plants were not tested for germinal activity in the first generation.

Tab. 4. Third generation tests for the correlation of somatic mutability and germinal activity of three a1-Mum alleles crossed onto a1 sh2 silks.

Family number of the outcross plants	Allele	Average mutability of the outcross ears	Family number of the first generation test	Family number of the second generation test	Total number of plants selfed	Number of plants segregating mutants	Percent of plants segregating mutants	Percent of first generation plants segregating mutants	Percent of second generation plants segregating mutants
Tests of lines with low mutability									
86-1479	a1-Mum1	1.58	3154	3501	36	0	0.0	0.0	0.0
86-1480	"	1.54	3155	3502	37	0	0.0	0.0	0.0
86-1481	a1-Mum2	1.25	3159	3503	41	0	0.0	0.0	0.0
86-1483	"	2.63[a]	3168	3505	37	0	0.0	0.0	0.0
86-1487	a1-Mum3	1.50	3173	3508	34	0	0.0	0.0	0.0
86-1488	"	1.18	3173	3509	33	0	0.0	0.0	0.0
Tests of lines with medium mutability									
86-1482	a1-Mum2	1.68[b]	3164	3504	35	0	0.0	0.0	0.0
86-1485	"	3.91[c]	3161	3507	32	4	12.5	4.44	0.0
86-1505	"	1.07	3161	3507	34	0	0.0	4.44	0.0
86-1508	"	1.03	3161	3507	33	0	0.0	4.44	0.0

Tests of lines with medium mutability (continued)									
85-1489	al-Mum3	2.49	3171	3510	41	0	0.0	0.0	0.0
85-1490	"	1.09[b]	3171	3510	37	0	0.0	0.0	0.0
85-1491	"	1.83	3172	3511	41	0	0.0	0.0	(1.59)
85-1501	"	3.36[d]	3174	3514	38	4	10.53	0.0	0.0
Tests of lines with high mutability									
86-1484	al-Mum2	3.91	3169	3506	24	0	0.0	6.38	0.0

[a]In the previous generations, this line had low somatic mutability. This plant has borderline medium somatic mutability. There has been no change in germinal activity (=0.0).

[b]In the previous generations, these lines had medium somatic mutability. These plants now have low somatic mutability. Germinal activity has not changed (=0.0).

[c]In the previous generations, this line had been variable in somatic mutability. In the first generation, it was medium. In the second generation, three plants were tested; two had acquired low mutability, while one (85-86-3507) had retained borderline medium mutability. It was a sibling plant to those tested under 85-86-3507 that was crossed to give the ear from which the seed for 86-1485 came. This ear now has high mutability. Germinal activity has been variable in this line also. In generation 1, there was germinal activity although numbers were low in two instances. In the third generation, with the increase in somatic mutability, germinal activity has been regained.

[d]In the previous two generations, this line had medium somatic mutability and never showed germinal activity. With this plant there has been a sudden increase in somatic mutability, from 2.70 to 3.91, and a concomitant acquisition of germinal activity.

Tab. 5. Tests for the correlation of somatic mutability and germinal activity involving outcross ears with approximately 3:1 or 1:1 ratios for classes 3 + 4 : 1 + 2.[a]

Family number of the outcross family	Allele	Average mutability score of the outcross ears from which the seeds came	Percent of class 3 + 4 seeds on the outcross ears from which the seeds came	Family number of the first generation	Family number of the second generation	Total number of plants selfed	Number of plants segregating mutants	Percent of plants segregating mutants	Percent of first generation plants segregating mutants	Percent of second generation plants segregating mutants
85-86-3514	<u>al-Mum3</u>	2.56	57.14%	3174	--	30[c]	0	0.0	0.00	0.0
85-86-3566	"	2.56	57.14%	3174	--	39[d]	2	5.13	0.00	0.0
86-1503	<u>al-Mum2</u>	3.10	73.41%	3161	3507	37[e]	0	0.0	4.44	0.0
86-1504	"	1.80	26.33%	3161	3507	33[f]	0	0.0	4.44	0.0
86-1506	"	1.81	24.43%	3161	3507	36[f]	0	0.0	4.44	0.0
86-1507	"	2.01	36.89%	3161	3507	36[f]	0	0.0	4.44	0.0
86-1529	"	2.58	51.20%	--	--	28	0	0.0	--	--
86-1598	<u>al-Mum3</u>	2.24	40.91%	3172	3511	36[g]	0	0.0	0.00	(1.59)
86-1599	"	2.16	38.37%	3171	3510	37[g]	1	2.70	0.00	0.0
86-1600[b]	"	2.73	54.60%	3174	3514 and 3566	37[g]	1	2.70	0.00	0.0

86-1531	al-Mum3	2.74	56.05%	--	--	43[g]	0	0.0	--	--
86-1489	"	2.49	45.20%	3171	3510	41[g]	0	0.0	0.00	0.0
86-1491	"	1.83	35.71%	3172	3511	41[f]	0	0.0	0.00	(1.59)

[a]Classes 3 and 4 are combined because they both show significant somatic mutability. Classes 1 and 2 are combined because, frequently, class 2 seeds will give rise to plants that have predominantly stable seeds in the next generation. Also, plants from stable seeds when self-pollinated and (or) outcrossed to al sh2 will frequently produce ears with predominantly stable seeds plus, sometimes, a few class 2 seeds as well.

[b]The ear giving the seeds for family 1600 was the result of an outcross to al sh2 of a plant from a class 4 seed from the ear that produced the seeds for families 3514 and 3566.

[c]Only class 1 seeds were planted. The seeds for families 3514 and 3566 were taken from the same ear.

[d]Only class 4 seeds planted.

[e]1/2 class 3, 1/2 class 4 seeds planted.

[f]1/2 class 3, 1/2 class 1 seeds planted.

[g]1/2 class 4, 1/2 class 1 seeds planted.

Tab. 6. The correlation of somatic mutability and germinal activity in _al-Mum_ stable X Pl. _Mu_ F_1's and F_2's in which somatic mutability was restored.

Number of F_1 or F_2 plants outcrossed	Allele	Test of Pl. _Mu_ parent for _Mu_ germinal activity	Recovered[a] somatic mutability for selfed ears of sibling F_1 plants tested in outcross	Recoverd somatic mutability for selfed ears of F_2 plants tested in outcross	Estimated average somatic mutability for all selfed ears in the outcross families of the F_1's and F_2's	Total selfed outcross ears	Number of outcross ears segregating mutants	Percent of plants segregating mutants
_F_1 tests_								
84-6192-1[b]	_al-Mum3_	+	3.68	N.T.	3.14	41	3	7.32
84-6192-2	"	+	3.68	N.T.	3.30	35	1	2.86
84-6192-3	"	+	3.68	N.T.	2.61	161	0	0.0
84-6192-9	"	+	3.68	N.T.	3.37	179	5	2.79
84-6182-1	"	+	3.31	N.T.	3.05	38	1	2.63
84-6182-2	"	+	3.31	N.T.	3.38	44	0	0.0
84-6182-9	"	+	3.31	N.T.	2.78	37	0	0.0
83-8192-4	_al-Mum2_	N.T.	N.T.	3.09	N.S.	37	5	13.51
83-8192-5	"	N.T.	N.T.	3.41	N.S.	49	0	0.0
83-8192-3	"	N.T.	N.T.	1.98	N.S.	44	0	0.0
83-8196-5	"	N.T.	N.T.	3.15	N.S.	37	5	13.51

F_2 tests								
84-6176-2	al-Mum3	+	3.31	N.T.	3.48	33	4	12.12
84-6176-8	"	+	3.31	N.T.	2.80	35	7	20.0
84-6176-9	"	+	3.31	N.T.	3.41	44	2	4.55

[a]Values from Tab. 2. in Robertson et al. (19).

[b]Sibling plants of family 6192 were tested for Mu DNA modification at HinfI sites and found to be unmodified (19). The remainder of the families were not tested for Mu DNA modification.

N.T. = no test

N.S. = not scored

DISCUSSION

The outcrossing series of experiments, designed to test the correlation of the level of Mutator somatic mutability activity with germinal Mu activity, provide significant insight into the relationship between these two aspects of Mu activity.

1. Generally, stocks that exhibited a low level of somatic mutability showed no germinal activity. There were, however, some exceptions to such a generalization (e.g., 85-86-3512 and 86-1450, Tab. 3). Both these exceptions represent a sudden (in one generation) shift from high mutability (classes 3 or 4) to low. Both had high germinal activity in the previous generation. Thus, in this case, germinal activity may represent a "carry-over" activity from the first generation. If so, the next generation should show a loss of Mu activity. Unfortunately, we do not have the next-generation results on these individuals.

A delayed effect in the expression of Mu-regulated somatic mutability has been observed by Walbot (26). The observations by Roth (20) represent another exception to the rule that the loss of somatic mutability is accompanied by the loss of germinal activity. At least three 1985-86 plants with low levels of somatic mutability still retained germinal activity. Again, these were plants from the first generation in which the low-level condition was observed. Until we know more about these exceptions, it probably is too early to conclude that loss of somatic mutability is always accompanied by loss of germinal activity.

2. There seems to be an unusually high frequency of the loss of Mu germinal activity in this series of crosses. (In previous experiments, germinal activity was lost in about 10% of the outcross progeny of Mutator plants, Ref. 14.) Sometimes this is accompanied by a loss of somatic mutability and sometimes not. This high loss could be due to the repeated crossing to a relative inbred tester stock, a1 sh2. Robertson (17) has shown that the loss of Mu germinal activity is common when active Mu systems are repeatedly backcrossed to inbred lines. Such a phenomenon may be occurring in the crosses to the a1 sh2 stock.

3. Accompanying the loss of germinal activity on outcrossing is the appearance of crosses in which ears with 1:1 ratios of mutable to stable seeds are observed. This suggests that, upon outcrossing, we may be diluting out copies of a regulatory factor(s) of somatic mutability that was (were) present in the original stocks. These 1:1 lines, in some instances, have been shown to possess a dual-element control of somatic mutability (18). Although most of these stocks had no germinal activity, they did have many class 3 and class 4 seeds.

Thus, if a single regulator controls somatic mutability, it does not seem to be sufficient to induce germinal activity. Some of these stocks, however, are observed to have germinal activity. Thus, it seems that something more than the single factor regulating somatic mutability is required for germinal activity. This "something more" may be a sufficient number of Mu1-like elements transposing to produce enough mutations so that some would be detected in the small populations used for determining germinal Mu activity. If the foregoing is true, this would obviate the need for the theory that a threshold number of Mu elements is needed to produce sufficient transposase for transposition to occur.

4. In this outcrossing series of experiments, there are frequently found lines that have high somatic mutability but no germinal activity. Such lines are found in every generation. We conclude that high somatic mutability cannot, therefore, be depended upon as an indicator of germinal activity. The results from families 84-86-3514 and 85-86-3566, however, might suggest otherwise. These two families consist of plants from seeds of the same 1:1 outcross ear, in which highly mutable seeds were planted in 3566, and stable seeds in 3514. The presence of mutants in the plants from the highly mutable seeds might suggest an association between germinal Mu activity and high somatic mutability. Because this is a 1:1 ratio family, these results might further suggest a relationship between the putative regulator of somatic mutability and germinal activity (i.e., the somatic mutability regulator also regulates germinal activity). Such conclusions are unlikely, however, for at least three reasons.

First, Robertson (15) has shown that at least some germinal Mu activity occurs premeiotically, and hence, new mutants induced in the 1:1 parent plant would be distributed at random with respect to the somatic mutability of the seeds in the outcross. This would not be true if all new mutants induced were linked to the regulator of somatic mutability; from what we know about the nature of the Mu system, this linkage is an unlikely event. Second, in the outcross family 86-1503, which has a putative 3:1 outcross ratio and in which only class 3 and class 4 seeds were planted, no germinal activity was observed. Because the outcross parent in this cross would have two putative regulatory elements of somatic mutability, new mutants would be expected in this outcross progeny if the somatic regulators were involved in germinal activity. Third, germinal activity was found in only three of the 12 outcross progenies of Tab. 5. If there was a relationship between somatic mutability and germinal activity, more of the outcrosses would have been expected to have plants with mutants present, because at least half of the seeds in all these progenies had significant somatic mutability. Thus, it seems that our conclusion that somatic mutability is a poor indicator of germinal activity is still valid.

The experiments involving the reactivation of stable derivatives of a1-Mum mutable alleles by crossing to Pl. Mu lines further support the conclusion that high somatic mutability is not indicative of the presence of an active germline Mu system.

Chandler and Walbot (8) found that loss of somatic mutability in a Mu-induced bz2 mutant was associated with DNA modification of the Mu elements. Enzymes such as MspI, HpaII, EcoRII, BglI, and AvaII, which do not cleave DNA containing 5-methylcytosine at their target sites, will not cut the Mu DNA sequences of plants from stable seeds. Robertson et al. (19), however, found that some stable derivatives of a1-Mum3 alleles had unmodified Mu1-like sequences and that these derivatives could recover somatic mutability if crossed with Pl. Mu stocks. The results of testing the somatic mutability and Mu germinal activity of one of these a1-Mum3 stable x Pl. Mu F_1 populations are presented in Tab. 6 (crosses involving family 84-6192). In one instance (84-6192-3), somatic mutability was retained, but germinal activity was lost. Unfortunately, the state of modification, if any, is not known for plant 84-6192-3. We do know, however, that the stable line from which it was derived did not have its Mu DNA modified.

We have more information on modification of Mu DNA in the experiments involving plants from the inbred a1-Mum2 pedigree (Fig. 3.) By the 1985-86 generation, we are dealing with plants that have been inbred for five generations, which is one generation more than those inbred (interbred) plants analyzed by Bennetzen (5). The inbred material Bennetzen analyzed had been shown by Robertson (16) to have lost germinal Mu activity, and Bennetzen (5) established that these inbred plants had modified Mu sequences. The modification in these stocks involved sequences cut by EcoRII, MspI, and HpaII.

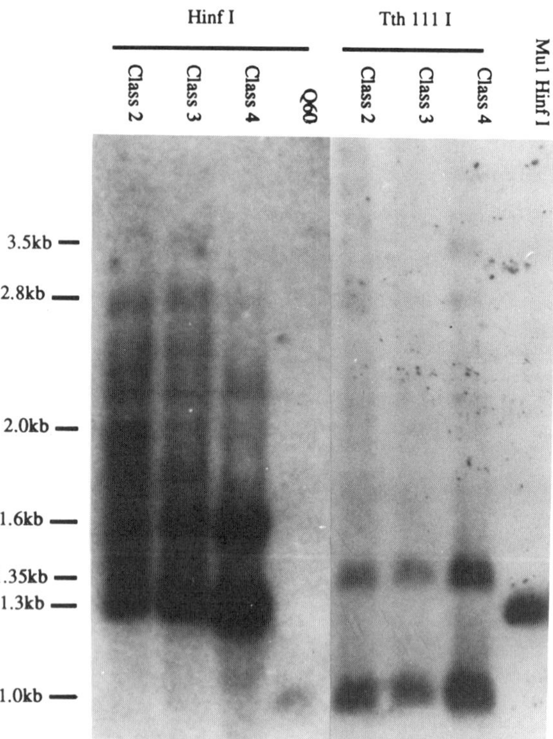

Fig. 4. Southern hybridization analysis of Mu elements in individual 1985 a1-Mum2 plants grown from class 2, 3, and 4 seeds. Maize DNAs were digested to completion with HinfI and Tth111I, fractionated on an 0.8% agarose gel, transferred to a Genatran sheet, and hybridized to a Mu1 probe. Lanes designated class 2, 3, and 4 contained DNA isolated from plants grown from seeds of that mutability level. Bands corresponding to Mu1-like and Mu1.7-like elements are seen in the Tth111I profiles (1.0 and 1.35 kb, respectively). This restriction endonuclease cuts both modified and unmodified Mu elements. Unmodified Mu1-like and Mu1.7-like elements are digested by HinfI to give bands 1.3 and 1.6 kb in size, respectively. Higher molecular weight bands seen in these profiles are derived from Mu elements which have been modified at the HinfI sites. Control lanes are DNA from standard line Q60, and plasmid pRB1 (which contains the whole of Mu1) digested with HinfI (Mu1 HinfI). Molecular sizes were estimated from lambda DNA fragments generated by PstI digestion and included on the same gel. [Figures here as in (20).]

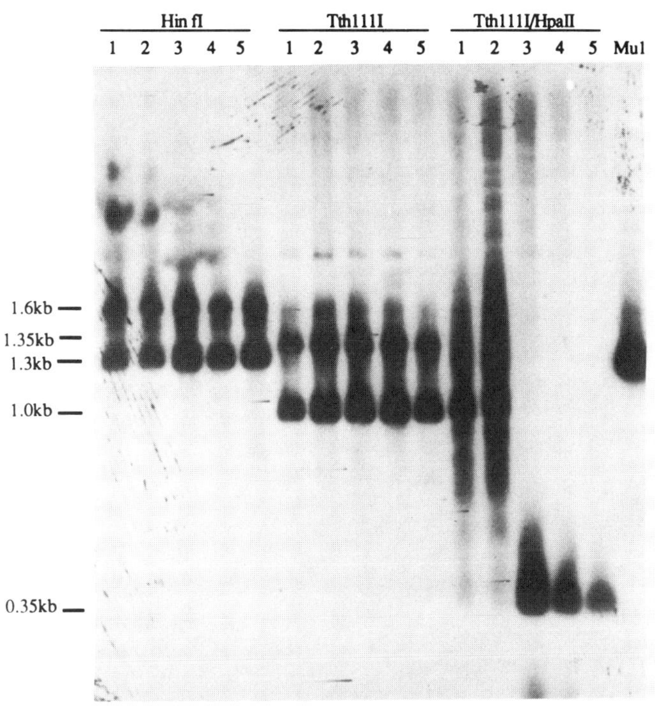

Fig. 5. Modification of Mu elements in plants whose progeny differ with respect to levels of somatic mutability and germinal activity. Maize genomic DNAs were digested to completion with HinfI, Tth111I, or Tth111I and HpaII in combination, fractionated by electrophoresis on an agarose gel, transferred to Genatran, and hybridized to a Mu1 DNA probe. Lanes: 1) 85-86-5558-1, low somatic mutability and no germinal activity; 2) 85-86-5558-3, low somatic mutability and germinal activity; 3) 85-86-6558-7, high somatic mutability and no germinal activity; 4) 85-86-6558-4, high somatic mutability and germinal activity; 5) 82-2154-5, a control demonstrating high somatic mutability, germinal activity, and no Mu element modification. Unmodified Mu elements are reduced by Tth111I and HpaII digestion to fragments 350 bp and less. If Mu elements are modified at some or all of the HpaII sites, the sizes of the bands increase above 350 bp, up to the sizes of the bands generated by Tth111I alone. Molecular sizes of fragments were estimated from lambda DNA fragments generated by PstI digestion included on the same gel. (Figures shown here as in Ref. 20.)

In our material (Fig. 3), by the fourth generation, all plants retained somatic mutability, and none had modified Mu element sequences. However, in the fifth generation, some plants shifted suddenly to low mutability, and these had their Mu DNA modified at their HpaII sites. Some of these low-somatic-mutability plants with modified Mu elements retained germinal activity while others did not. Highly modified lines which have germinal activity had not been seen before. Possibly, in some fifth-generation plants that grew up as class 4 plants with germinal activity, transposition occurred premeiotically, and modification occurred postmeiotically. If this had happened, the fifth-generation plants would have low somatic mutability, but some of their outcross progeny would carry mutants as a result of the premeiotic transpositions. Thus, these modified low-mutability plants with germinal activity should in the next generation have offspring all of which have modified sequences, and they also should lack germinal Mu activity. The results of tests to establish such a delayed effect are not yet available.

Until the fifth generation, all individuals of this pedigree (Fig. 3) had ears with predominantly class 4 seeds. Even in the fifth and sixth generations (not shown), many plants continued to have ears with highly mutable seeds. Many of these highly mutable plants, however, did not have germinal activity. These observations lend further support to the conclusion that high somatic mutability is not a good indicator of the presence of Mu germinal activity.

CONCLUSIONS

These studies establish that the level of somatic mutability is not a reliable indicator of germinal activity. This should not be an unexpected result when the obvious differences in the two phenomena are considered. Somatic mutability is measuring what happens to only one Mu element, the one at the locus (a1, in our tests). When dealing with germinal activity, many Mu elements are involved. Given the conditions by which germinal activity is determined, it will only be observed if transposition is occurring and if the number of Mu elements available is great enough so that sufficient new mutants will occur in the limited outcross populations utilized in these tests.

The degree of modification of the Mu elements in Mutator plants also may not be a reliable guide to the germinal activity in Mu plants, but more data are needed to confirm this. Information concerning the modification of the element at the aleurone locus is needed before it can be established whether somatic mutability, Mu element modification, and germinal Mu activity are functionally related. With this information, it may be possible to solve some of the apparent contradictions present in results reported from different laboratories. These seemingly contradictory results emphasize the need to know much more about the Mutator system before generalizations can be made about how it functions.

ACKNOWLEDGEMENTS

The authors are grateful to the National Science Foundation for the support of this research provided by grants PCM 79-23052, PCM 83-02214, and DCB 86-08188, and to the U.S. Department of Agriculture for the support provided by grant 84-CRCR-1-1392.

REFERENCES

1. Alleman, M., and M. Freeling (1986) The Mu transposable elements of maize: Evidence for transposition and copy number regulation during development. Genetics 112:107-119.
2. Barker, R.F., D.V. Thompson, D.R. Talbot, J.A. Swanson, and J.L. Bennetzen (1984) Nucleotide sequence of the maize transposable element Mu1. Nucl. Acids Res. 12:5955-5967.
3. Bennetzen, J.L. (1984) Transposable element Mu1 is found in multiple copies only in Robertson's Mutator maize lines. J. Molec. Appl. Genet. 2:519-524.
4. Bennetzen, J.L. (1985) The regulation of Mutator function and Mu1 transposition. In Plant Genetics, U.C.L.A. Symposium on Molecular and Cellular Biology, Vol. 35, M. Freeling, ed. Alan R. Liss, New York, pp. 317-331.
5. Bennetzen, J.L. (1987) Covalent DNA modification and the regulation of Mutator element transposition in maize. Molec. Gen. Genet. 208:454-451.
6. Bennetzen, J.L., J. Swanson, W.C. Taylor, and M. Freeling (1984) DNA insertion in the first intron of maize Adh1 affects levels: Cloning of progenitor and mutant Adh1 alleles. Proc. Natl. Acad. Sci., USA 18:4125-4128.
7. Bennetzen, J.L., R.P. Fracasso, D.W. Morris, D.S. Robertson, and M.J. Skogen-Hagenson (1987) Concomitant regulation of Mu1 transposition and Mutator activity in maize. Molec. Gen. Genet. 208:57-62.
8. Chandler, V.L., and V. Walbot (1986) DNA modification of a maize transposable element correlates with loss of activity. Proc. Natl. Acad. Sci., USA 83:1767-1771.
9. Chandler, V.L., C. Rivin, and V. Walbot (1986) Stable non-Mutator stocks of maize have sequences homologous to the Mu1 transposable element. Genetics 114:1007-1021.
10. Fedoroff, N.V. (1983) Controlling elements in maize. In Mobile Genetic Elements, J. Shapiro, ed. Academic Press, New York, pp. 1-63.
11. Feinberg, A.P., and B. Vogelstein (1983) A technique for radiolabeling DNA restriction endonuclease fragments to high specific activity. Analyt. Biochem. 132:6-13.
12. Maniatis, T., E.F. Fritsch, and J. Sambrook (1982) Molecular Cloning: A Laboratory Manual. Cold Spring Harbor Laboratory, Cold Spring Harbor, New York.
13. Rivin, C.J., E.A. Zimmer, and V. Walbot (1982) Isolation of DNA and RNA recombinants from maize. In Maize for Biological Research, W.F. Sheridan, ed. Plant Molecular Biology Association, Charlottesville, Virginia, pp. 161-164.
14. Robertson, D.S. (1978) Characterization of a mutator system in maize. Mutat. Res. 51:21-28.
15. Robertson, D.S. (1980) The timing of Mu activity in maize. Genetics 94:969-978.
16. Robertson, D.S. (1983) A possible dose-dependent inactivation of Mutator (Mu) in maize. Molec. Gen. Genet. 191:86-90.
17. Robertson, D.S. (1986) Loss of Mu mutator activity when active Mu systems are transferred to inbred lines. Maize Genet. Coop. Newsl. 60:10.
18. Robertson, D.S. (1987) Mendelian ratios in crosses of mutable Mutator-induced a1 mutants. Maize Genet. Coop. Newsl. 61:11-13.

19. Robertson, D.S., P.S. Stinard, J.G. Wheeler, and D.W. Morris (1985) Genetic and molecular studies on germinal and somatic mutability in Mutator-induced aleurone mutants of maize. In Plant Genetics, U.C.L.A. Symposium on Molecular and Cellular Biology, Vol. 35, M. Freeling, ed. Alan R. Liss, New York, pp. 317-331.

20. Roth, B.A. (1987) Robertson's Mutator system in maize: Studies on the regulation of activity and the prevalence of Mu1-homologous DNA sequences in diverse lines of maize. Ph.D. Dissertation, Iowa State University, Ames.

21. Southern, E.M. (1975) Detection of specific sequences among DNA fragments separated by gel electrophoresis. J. Molec. Biol. 98:503-518.

22. Strommer, J.N., S. Hake, J.L. Bennetzen, W.C. Taylor, and M. Freeling (1982) Regulatory mutants of the maize Adh1 gene caused by DNA insertions. Nature 300:542-544.

23. Taylor, L.P., and V. Walbot (1987) Isolation and characterization of a 1.7 kb transposable element from a Mutator line of maize. Genetics 117:297-307.

24. Taylor, L.P., V.L. Chandler, and V. Walbot (1986) Insertion of 1.4 kb Mu elements into the bronze1 gene of Zea mays L. Maydica 31:31-45.

25. Walbot, V. (1986) Inheritance of mutator activity in Zea mays as assayed by somatic instability of bz2-mu1 allele. Genetics 114:1293-1312.

26. Walbot, V., and C. Warren (1987) Regulation of Mu element copy number in maize lines with an active or inactive Mutator transposable element system. Molec. Gen. Genet. (in press).

THE MOBILE ELEMENT SYSTEMS IN MAIZE

Peter A. Peterson

Department of Agronomy
Iowa State University
Ames, Iowa 50011

ABSTRACT

Mobile element systems in maize have evolved from a distant curiosity of the 1950s to a prominent subject of study and research of the 1980s that has brought into the scientific arena a large number of investigators. This was largely stimulated by the findings associated with the IS elements, the Mu1 virus, and the Tn series in bacteria.

With the focus of McClintock's Ac-Ds system and its characteristic activities that include chromosome breakage (Ds-dissociation), it was initially difficult to relate the more ubiquitous P-VV behavior as the same type of phenomenon. More significantly, the mutability associated with Ac was not related to other instances of similar mutability phenomena such as Dt, En, and Spm. Thus, with these initial tests, it was clear that several forms of "control" of mutability could be identified, and these later became identified as "systems."

The isolation of unique systems such as Fcu, Uq, Bg, Mrh, Mut, Mu1, Cy, and M-st soon followed. There are others, such as MpI 1 at the c2m-3 allele. Other elements unrelated to variegation, such as Bs1, Tz, and the Cin elements, have been molecularly cloned.

The distribution of active elements varies whereby active Uq and Mrh are widely dispersed in corn breeding populations. The Cy element has long been a resident in our genetic materials, having first appeared in a 1966 a2m allele but not identified as such until the isolation of the rcy receptor element at the bz locus.

With the molecular characterization of these elements, their meaning and likely wide dispersal in corn populations will be revealed.

INTRODUCTION

Mobile elements in plants, first identified as regularized chimeric tissue of various plant parts (Fig. 1), were finally understood by the use of genetic analysis, and it was determined that the variegating genes were an expression of a series of interacting genetic elements that move (24). This understanding came from an initial set of observations by McClintock (19-24) on variegated kernels (Fig. 1B, 2A). In her early studies, McClintock determined that a weakened chromosome site (a Ds element) could be the site of chromosome breakage (Fig. 2B) induced by a second element, the regulatory element, Ac. This same two-unit Ac-Ds interaction, appearing as quite regularized events in view of the specific timing and frequency of mutation events (Fig. 1B, 2A), was soon seen to affect the functioning of a gene (the c-m1 allele). Because the genetic results indicated that this new instability at the c locus was under the control of the same Ac element, it was clearly evident that the origin of the c-m1 allele variegation was related to the same Ac-Ds phenomenon and, therefore, originated by the insertion (following transposition from a known site) of a Ds element. This conclusion was further strengthened by the consistent appearance of the same regularized pattern of Ac-Ds variegation. Additional mutants with the same type or a variant form of this variegation were subsequently isolated. These mutants were also controlled by Ac (21,22,24). From these observations, variegation could be related to genetic elements that move.

McClintock's observations were not alone. Similar phenomena were being unraveled by Dr. Brink's group at Wisconsin in their resurrection of the Emerson (7,8) P-VV studies in which unambiguous proof for transposition was present. Studies by Peterson were also under way at Urbana, Illinois, on the pale green mutable (pg-m, Fig. 1A) in the investigation of two newly arisen unstables originating from the Bikini A-bomb tests. The general concepts relating to transposable elements, those of instability, movement of elements, diversity of mobile element families, and insertion into genes, were confirmed and amplified.

A number of reviews written on mobile elements have included a full account of genetic and molecular findings of the last four decades (5,9,33) and of the past century (53). Thus, this information will not be repeated here. Yet a general summary is appropriate. Mobile elements are pervasive (4,53) and appear in a wide array of species (33,53). These elements can only be phenotypically seen when inserted into a gene, thereby altering the phenotype. The survival of these elements in these plant populations (51) (and in other forms also) emphasizes their functional role; this has been extensively discussed (64,72).

That the elements appear as systems of two-unit interaction (regulatory-receptor), though originally conceived as an exemplary means of gene control (controlling elements), now seems nonsignificant. These inserts, acting as receptor elements that are defective in mutator functions, frequently arise from fully functional elements. The insert, acting as a receptor element and causing the gene where it is inserted to be a null or partly affected allele, is now known to lack an excision-inducing capacity but itself can be induced to excise. Though seemingly nonfunctional and erroneously called defective, these elements acting as receptors can produce a product (Gierl et al, this Volume). The movement of an insert is induced by the transactive activity of a functional element.

The key features that provide the receptor element with the capacity to be mobile are the terminal inverted repeats (TIR) that are unique and significant for a particular system and represent the specificity whereby the transactive signals from a regulator recognize a specific receptor and thereby identify a system. The other specific feature associated with a system is the target site duplication (TSD) of the host chromosome that is generated upon insertion. The TSD can be any size, depending on the system, from the three-nucleotide sequence of the En group to over 200 of the Mu series (6). There is a consequence to the excision process. The TSD is generally altered after transposition (58), leading to alterations in the sequences of the gene, and this leads to altered proteins (64,72).

Considering that there are many DNA sites without a clear phenotype, it is evident that there is more mobility in the genome than can readily be observed. Indeed, inserts are commonplace and are found unexpectedly almost everywhere. Almost every DNA sequence that is investigated exposes a new and unexpected insert (63,65,67), and it is likely that there is a greater generality to these phenomena than was previously thought. One could say that the genomes (and especially individual genes) are adulterated with "foreign" DNA (at least foreign to that gene segment).

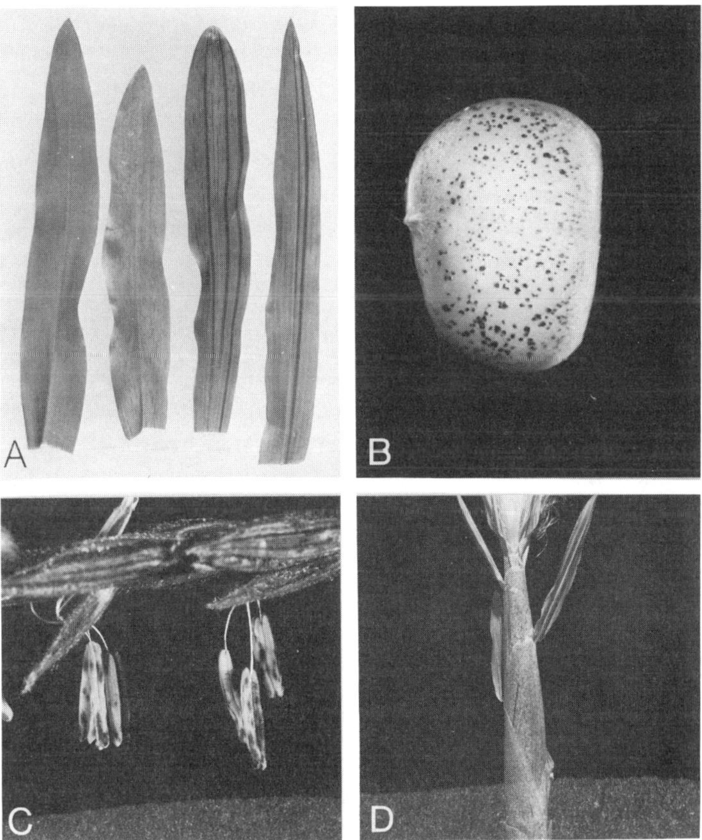

Fig. 1. Mutability in different tissues. A. Mutability on leaves of pg-m, the original source of the En insert (48). The lack of mutability indicates the absence of En. B. Kernel mutability of the a-m (En control) allele showing the uniform mutability. C. Mutability in anthers a-m(r) + En. D. Mutability in husk tissue (a-m dense).

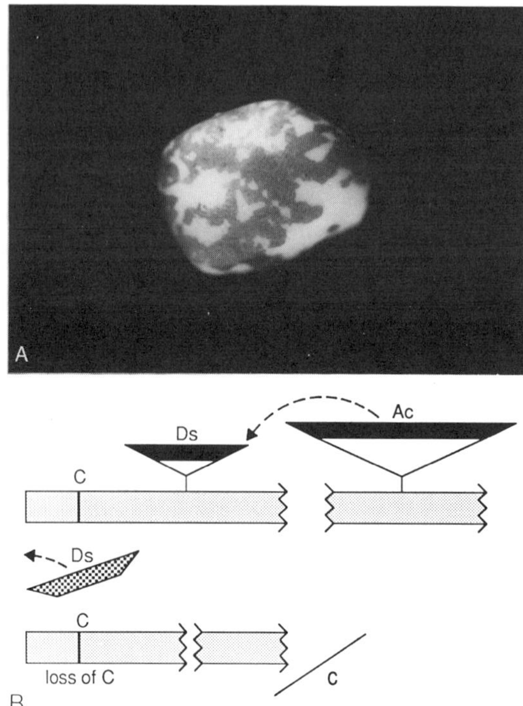

Fig. 2. A. Variegated kernel C Ds/c/c Ac. The loss of C uncovers c. B. Diagrammatic illustration of events in A, illustrating the action of Ac in causing Ds to excise, uncovering the c phenotype.

THE En-I SYSTEM AND ITS FOUR FUNCTIONS

The En-I(Spm) system has been well characterized (12,48,49) (En and Spm are "homologous," and term usage is dependent on whether En or Spm was used in the cross). It has been sequenced from the wx844 allele (wx844:En-1) (40,49), and the genes of this element have been identified (41). More extensive details of the two overlapping genes (gene 1 and gene 2) of En are given in this Volume by Gierl et al.

The Sp Function

At the genetic level, four functions of En (Spm) have been identified. The Sp (suppressor) activity can be visualized (Fig. 3A) with only certain alleles such as the a-m1 5719 allele (789 bp) (25-27) and the a-m11112 allele (945 bp) (71) that are effective substrates for the expression of this function. Here, a colored phenotype is suppressed, yielding a colorless background and colored spots owing to the M functioning, according to the original hypothesis by Schwarz-Sommer et al. (63). This is supported by protein binding studies by Gierl et al. (this Volume) that include steric hindrance of transcript development due to protein binding.

Tacke et al. (71) speculate on how the functional gene product (namely, the color produced by states 5719A-1 and 1112 of the a-m1 allele) could be synthesized. The two states, a-m15719A-1 and a-m11112, are derivative products of an original 2.2-kb element found at the wx-m-8 allele (63) and at the bzm13 (60) allele. Because this particular insertion (the a-m16078 and its derivatives) is one base pair from the exon-intron junction (Fig. 4), Tacke et al. further speculate that a modified exon is formed because of a putative intron donor site within the I element. Uniquely, the 2.2-kb insert of the a-m16078 allele at the A locus, though causing a near colorless kernel (Fig. 3B), does produce mature plant color indicating that a gene product is formed (Fig. 3C).

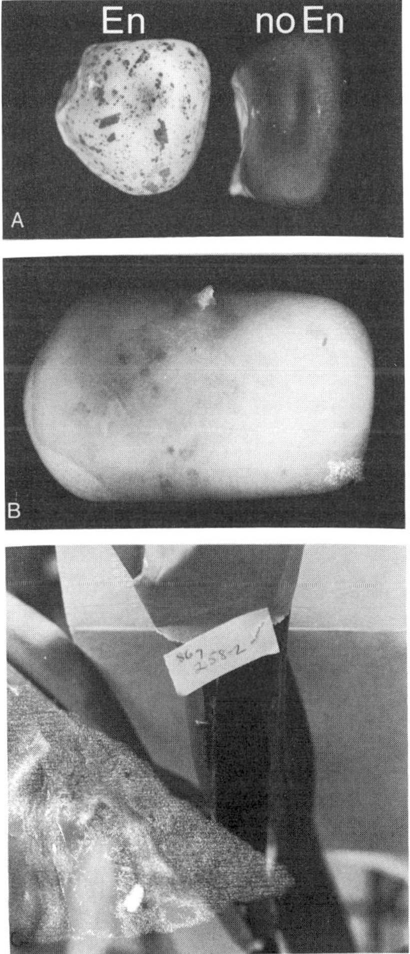

Fig. 3. A. The Sp function of En. The a-m1 1112 allele (55) with and without En from the cross a-m1 1112/a sh En x a sh/a sh. B. The a-m16078 allele. Slight coloration is expressed in kernels without En or Spm. C. Mature plant color of a-m16078 without En.

Thus, the Sp function was uncovered owing to the uniqueness of the site of insertion of this specific I element. With the site of placement (Fig. 4) of the I element (65) in a-m(r) (42), the Sp function cannot be expressed because, in this instance, the I is sited at the other end of this same exon, more than 20 bp in from the intron-exon junction at position 336 of exon 2 of the A gene; thus, the alternative splicing hypothesized by Tacke et al. (71) would not be an option. Other cases of inserts at genes that have colored phenotypes in the absence of En, such as c-m 41905 and c-m 41863 (47), would confirm the alternative splicing hypothesis (71). Whether it is the alternative splice site or the uniqueness of the insert, one base pair from the exon-intron junction is open to question.

There are other cases of gene product formed despite the presence of the insert. These include the c2m-2 allele (29,53), the "A2" state allele (28,30), and a number of other c-m alleles (47). Why the insert does not interfere with the transcript in these cases can be answered only by a further molecular analysis of these alleles.

That these inserts reside in alleles of native populations is exemplified by the A-Cuna allele. This allele, uncovered in studies with the Fcu system, expresses a full, dark-colored kernel phenotype. When crossed to a line containing En, there is a slight suppression of coloration, though not completely as in a-m1 (Fig. 3A), and spots are seen (Fig. 5A, 5B). With pr, the color suppression is even more striking (Fig. 5B).

The M Function

The M function of En/(Spm) is evidence for an active En. This is generally accompanied by the Sp function. This activity is expressed by the appearance of colored spots on kernels (Fig. 6A, 6B, 6C) or colored stripes on leaves (Fig. 1A), stems, or husks (Fig. 1D). The spots or stripes induced by M activity represent excision events. There is generally a correlation of early spotting pattern (Fig. 6A, 6B) and the recovery of germinal revertants. With the a-m(papu) allele, there is a high frequency of colorless germinal events, each of which is accompanied by the transposition of En from its donor site at A1 to another site (34,44). The large colorless areas frequently seen in a-m(papu) kernels (Fig. 6A, 6B, arrow) also represent M-induced excision events because they are correlated with the excision of transposed En's among test-cross progeny of a-m(papu). Alternatively, late events (Fig. 6C) do not produce germinal revertants because they occur in the later stages of endosperm formation, too late to be included in germinal tissue. Even with early-event alleles, the germinal revertants usually occur as single events (44) and are rarely included in an ear sector, suggesting a role of the meiotic cycle in generating transpositions of En.

The A Function

The A function was initially realized by genetic analysis from a mutant Spm in concert with a defective receptor of one allele at the a2 locus, namely, a2-m1 or A2-state (28,30). McClintock saw that the full color phenotype of the a2m-1 allele and, thus, the A2-state, because it has a full gene product, did not respond to the M signals of Spm but was responsive to the Sp function. Because this unique allele lacked excision potential but could be suppressed by Spm or En, the interaction of an active with an inactive Spm could be observed. This is possible because of

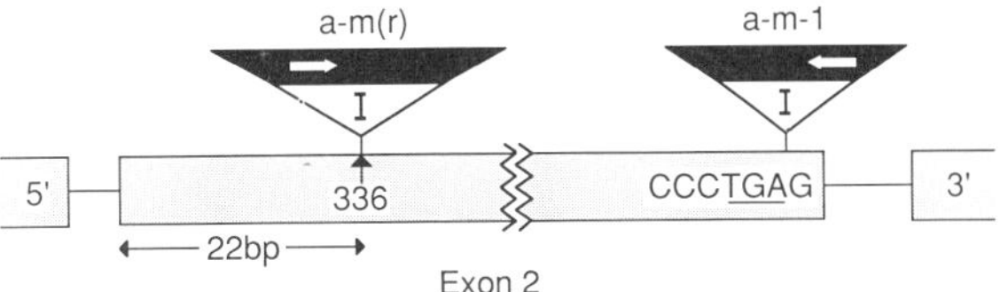

Fig. 4. The second exon of the A gene illustrating the insertion sites (not in scale) of the a-m(r) allele and the a-m1 allele. The a-m(r) allele is at position 336 of the A gene, 22 bp from the 5' junction of exon 2. Detailed view of the nucleotides in the 3' side of exon 2, modified after (71).

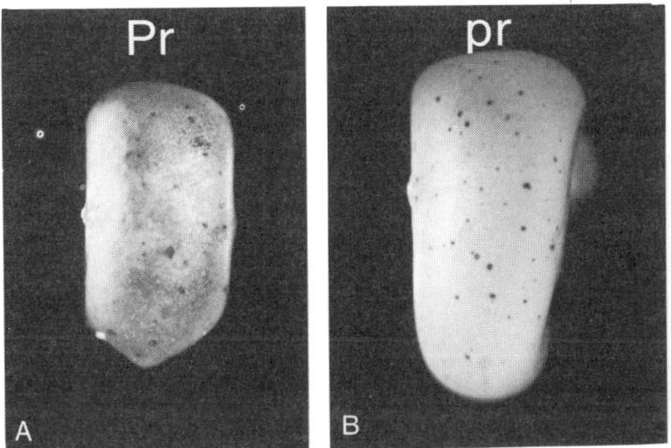

Fig. 5. The a-m(r)cu allele with En. A. Pr. B. pr.

Fig. 6. M function. A. The a2-m655169 allele. Two different mutability patterns. B. The a-m(papu) allele (11). The large colorless area (arrow) illustrates the lack of mutability and the excision of En from this autonomously mutable allele. C. The a2-m668144 allele, illustrating a uniform mutability pattern.

the four-dose option (0-4) of the maize endosperm (Fig. 7A, 7C). By virtue of expressing three doses of Spm activity where only one Spm was present, it was evident that the active Spm activated an inactive Spm by transactivating some feature of an inactive Spm, namely, an A function. Nevers and Saedler (32) hypothesized that the active Spm transactivated an inactive element. On the other hand, the active element might have demethylated an inactive element that persisted during the growth cycle of the endosperm. A weak En has only a modified effect on the a2-m1 allele (Fig. 7B).

The Coex Function

Another function realized from genetic crosses is also uncovered by the uniqueness of the site of an insert. The I element at the a-m2 allele (30,56), that is inserted at a unique receptor site located 15 bp upstream

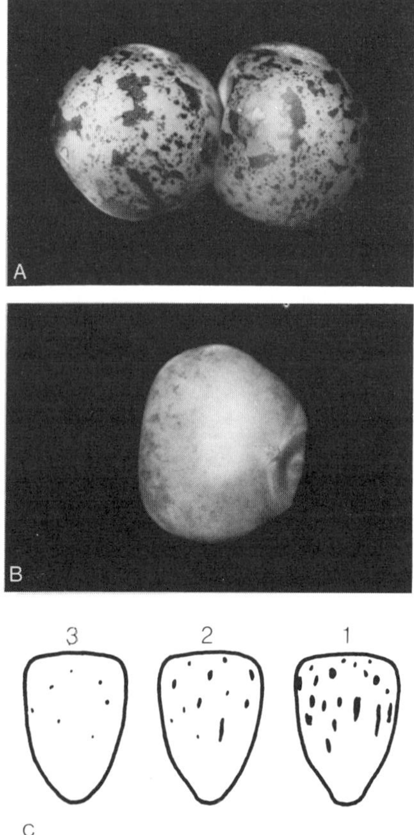

Fig. 7. A. The A2-state, the a2-m1 allele (28,30) with 1 En. B. The a2-m1 allele with a weakly acting En. C. Diagrammatic illustration of the effect of 3, 2, and 1 En on the a2-m1 allele.

of the CAAT box of the A1 gene promoter (65), expresses the Coex function. This function, Coex, induces an allele such as a1m4412 (55), that is colorless in the absence of En, to produce color and spots in the presence of En (Fig. 8A). Schwarz-Sommer et al. (65) speculate that the Spm(En)-mediated activation of transcription may be caused by the binding of an Spm(En)-encoded protein. As a consequence, this binding causes the replacement of normal A1 gene regulation already disturbed [colorless phenotype (Fig. 8A)] by the I insert that separates the general promoter by 1,080 bp from the upstream DNA regions; this region itself (the site of the I insert) might involve regulation of transcription.

Receptor Elements

Receptor elements such as I, ruq, rcy, etc., have been considered inactive elements. Molecular studies have confirmed (10,40,60,63,71) that these elements are deletion derivatives of a fully functional regulatory element. As a consequence, the term "defective," as in dSpm, has been applied to these elements.

Current findings suggest that these elements are defective in some capacities but functional in others. For example, they lack certain functions such as self-excision. This is usually tested by subjecting them to a receptor allele. In most cases, the results are negative, indicating a lack of the excision-inducing function. Yet, they do have gene products that can be expressed. This is evident for the a-m(r) allele (Fig. 8B) effect on the wx844 allele. Here, it can be seen that the full expressibility (Fig. 8B, a-m1) of the autonomously acting wx844 allele is modified [Fig. 8B, a-m(r)]. This is in contrast to the full mutability expression of the wx844 allele with the a-m1 allele (Fig. 8B). In other cases, it is very likely that the modifiers such as Rst (restrainer) (45) and En-Malt (54), or the complementary enhancing Mod (28), are similar elements with variously positioned deletions of the fully functional regulatory element.

There are other cases where modification of an autonomously mutable allele leads to an altered phenotype. In some cases, such as with the autonomously mutable a-m(Au) (42,46), the element suppressing this mutable expression (Fig. 8C) can be identified as En by standard tests. The same phenotype can be obtained with a segregating factor that does not show the En element by the standard reporter tests (Fig. 8C). Such elements have been identified as "factors." Now, it can be understood that the "mutant" derivatives of these regulatory elements are heterogeneous, and if they lacked the M function, such factors would not be identified as En. Now, these various elements will be tested with a-m1 alleles that will assay for the Sp function.

Relationship of an a-m(r) Protein to a Phenotypic Effect by a-m(r)

With the relationship of a protein to a-m(r), it is then possible to conclude that this En gene product is not the Sp function. Though the a-m(r) allele makes a gene product that delays mutability expression of the wx844 allele and, by analogy, similar interactions exist for other alleles, it is not the Sp product. This is based on the phenotype of the a-m(r)/am-1 heterozygote, the kernels of which are fully pale colored. If the a-m(r) allele expressed the Sp function, the heterozygous kernels would be colorless.

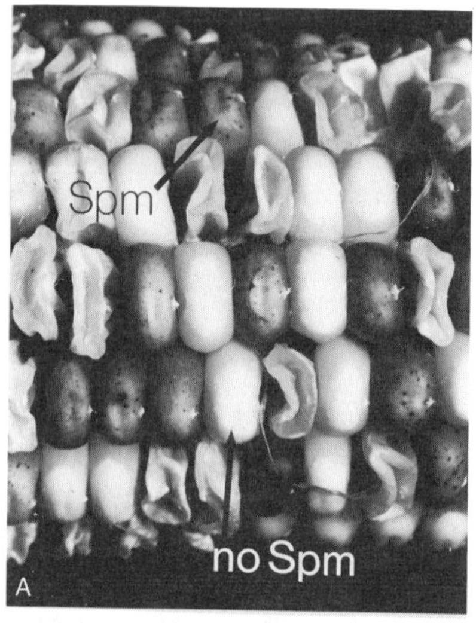
Spm
no Spm
A

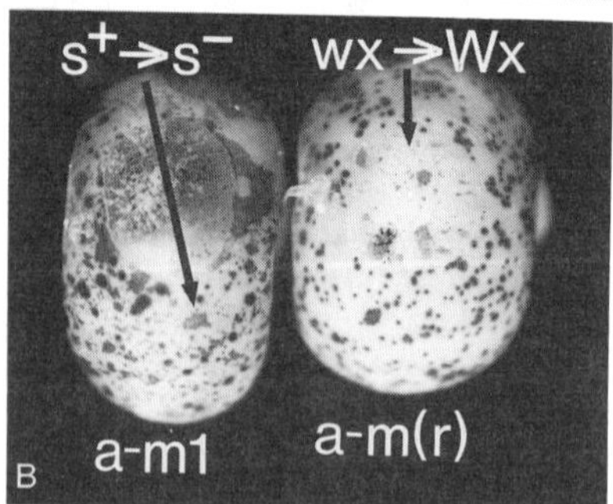
s+→s−
wx→Wx
a-m1
a-m(r)
B

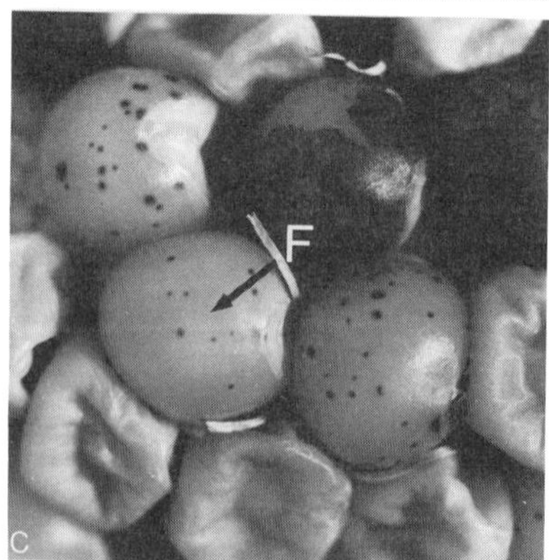
F
C

The Distribution of En

In terms of distribution, the En element is not as pervasive as the more ubiquitous Uq and Mrh elements. Many copies of En-homologous sequences exist, as determined by Southern analysis (63). Active elements, however, have been found only in a few populations (51,52). The unexpected homology of part of the open reading frame of En with part of the Tam1 element in Antirrhinum may reflect some form of evolutionary pattern or unusual transmission.

Uq

Uq was initially uncovered after crosses with virus-infected plants (39,50,68-70). It was determined (50) that an active element was initially present in the population (not necessarily the treated plant) and that a defective element inserted at the A locus, thus expressing instability. More importantly, an active Uq has been found in important breeding lines (51,52). Their persistence in the population despite a low initial gene frequency, followed by very strong selection for yield traits, does suggest a role of the element as a positive contributor to the population's fitness.

The ruq receptor element has been uncovered at the c locus by Oberthur and Peterson (35) in three different instances. Attempts to isolate the master Uq element are in progress.

In terms of distribution and persistence in a population, the development of the BSSS series (an Iowa Stiff Stalk Synthetic, an important contributor to breeding lines) breeding population is a good example. This population was initiated by the leader of the Iowa corn breeding project and advanced by Sprague and successors (16) in the project by intercrossing 16 inbred lines (for the early period, and each containing one or several desirable traits). The final composite became the initial BSSSCo population. After several cycles of testing and selection and many generations later, the Uq element was revealed in tests with a reporter allele (the a-ruq or c-ruq allele) in such a frequency, suggesting a positive selection. When the initial 16 lines are also examined, two of them (Ill Hy and Ia I159) are shown to have an active Uq (18). Thus, the elements survived in these populations despite an initial low frequency.

Fig. 8. A. Kernels on an ear from the cross am-2 8004/a Sh2/sh2 Spm x a/a sh2/sh2, illustrating the colorless round phenotype without Spm and the spotted-colored type with Spm. These same phenotypes are obtained with a standard En. The originating kernel giving rise to this ear was spotted-colored as illustrated in the progeny, confirming its heritability. B. Illustration of contrasting effects of the a-m(r) vs a-m1 alleles on the expressibility of the wx844 allele. The kernels originated from the cross a-o/a-o wx/wx x a-m(r)/a-m1 wx-844/Wx. The pale coloration arising from nonfunctioning En (S^+ to S^-) is indicated in the a-m1-containing kernel. C. The effect of a segregating factor (F) on the expression of the autonomously mutable a-m(Au) allele. The light-spotted kernels, when tested for heritability, again yield the a-m(Au) type (the dark kernels with the colorless areas) in the segregating test-cross progeny. When the shrunken kernels are tested against a reporter allele, no En can be detected.

In a survey of maize populations, Cormack and Peterson (3) uncovered the presence of Uq in a number of widely dispersed and unrelated populations. Uq was present in such populations as Hayes Golden, Iowa Long Ear, BSSS, BS11, Lancaster, and Illinois High Oil and Protein (IHOP) lines. Surprisingly, they appeared in the reverse selection phase of the IHOP lines but with the same sample size, but were not found in the original selection.

This is unlike the lack of the presence of active Cy (61) and of active Mu sequences (2). When populations are developed and maintained by random mating, it is evident that the frequency of some elements is not only maintained but increased. The lack of Cy or Mu elements is very likely due to their deleterious effects on maize populations. When the various elements in a genetic nursery are compared, the reason is very evident how frequently deleterious types arise in Cy-containing lines. It possibly is assignable to the long TIR, which makes precise excision difficult.

THE Fcu SYSTEM

The Fcu system was isolated as a two-unit system from a uniquely ornamental tribal corn (Fig. 9A) grown by the Cuna Indians in Northwestern Colombia (13). The variegation results from the interaction of a receptor at the r locus, identified as r-cu (receptor-Cuna containing an Icu element), and an independently segregating factor, Fcu (Factor-Cuna). In the absence of Fcu, r-cu expresses a variable kernel pigment phenotype (Fig. 9A) which expresses a graded series from near colorless, through a range of dilutes, to full colored. The expression of this graded series is an intrinsic property of the allele itself and not due to segregating modifiers. Seemingly, the pigment potential for an individual ear is "set" during plant development and then reset in the next generation, as indicated by heritability tests. Near full-colored types arising from test crosses such as:

Fcu / + r-cu / r-r x w22 r-g / r-g

are not revertants (R') but, rather, represent one extreme in the expression of r-cu pigmentation.

As with Dt (57) and Uq (11), the dosage effect is limited to the regulatory element, Fcu, and not expressed by varying the receptor element. Two doses of Fcu show more variegation than one dose, irrespective of the r-cu condition. In current terms, the Fcu transposase would have more protein and, therefore, more binding to the receptor site.

One interesting allele arising from the Fcu-rcu interaction is the R-mo(cu) allele that conditions a mottled phenotype (Fig. 9B). What makes this unique is that the mottling is expressed irrespective of its origin from the maternal or paternal parent. It was hypothesized (14) that the Icu element transposed to a part of the R locus that conditions mottling.

A partial homology was found between the Fcu and Spf (spotted-dilute) elements (14,59). Here, the Fcu element can activate the receptor for Spf R-r#2, but the Spf element cannot trigger the r-cu allele (Fig. 9C). It was hypothesized that Fcu and Spf differ in functional capacity, though consideration should be given to the differential capacities of the receptor alleles. With the current findings of transactive effects and differing gene products, it is more likely the former effect because, thus far, the receptors are not likely to possess different specificities, though they could include greater or less DNA for binding capacity (60,63).

The Fcu system is unique in being maintained as a tribal symbol for a remotely situated Indian tribe. Interestingly, the same tribal corn has harbored two other regulatory elements, En and Dt, both of them active. But, very likely, to retain the unique tribal pattern, strong pressure must have been applied to the Fcu system, likely to retain tribal identity (Fig. 9A). The A-Cuna allele (Fig. 5) as part of this Cuna population was previously mentioned.

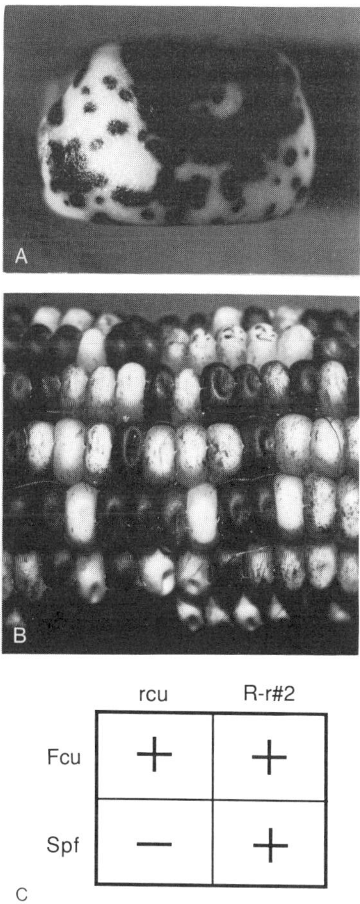

Fig. 9. A. The original pattern of the Cuna tribal corn: Fcu interacts with r-cu to produce this mutability. B. R-mo(cu) from the cross R-r/R-mo(cu) x W22 r-g/r-g. C. The interaction of the regulatory elements Fcu and Spf with receptors rcu and R-r#2. + indicates mutability, and - is the lack of mutability.

THE Cy SYSTEM

Cy (Cycler) was identified following the analysis of a newly originated bronze mutable allele, bzm805137, that arose in a TEL (transposon element laden) population (61). Four other elements were present in this population in addition to Cy.

The Cy system was found unrelated (functionally, nonhomologous) to all nonmutable stocks, but some relationship can be shown to the Mu element. The Cy system was established as a two-element system whereby an independently segregating Cy element conditions spotting (fully colored spots on a bronze background) (Fig. 10) at a bronze allele identified as bz-rcy (the receptor, rcy). In crosses, the Cy element segregates as a Mendelizing unit. In the absence of Cy, the bz-rcy allele expresses stable bronze phenotypes.

Like the En system, states arise both for the receptor allele bz-rcy and the regulatory element Cy. Whether, upon molecular analysis, these receptor-related states are alterations in the size of the receptor element, as has been identified for the En/Spm system (60,71), remains to be determined.

An important proof for the establishment of the Cy system is the reconstitution test. By crossing the reporter allele bz-rcy (nonspotted, bronze) to Cy-containing lines, it was possible to determine increases and decreases in the number of Cy elements. Starting from one-Cy lines, increases in the number of Cy elements have been determined, and the rate of increase is assignable to distinguishable groups of genotypes (62).

The question arises as to the origin of the population that gave rise to an active Cy element. The TEL population from which the bzm-805137

Fig. 10. Bronze-spotted kernels of the Cy system: Two different patterns from Cy bz-rcy/- x bz/bz.

allele originated was developed from crosses involving En-related a-mutable strains (61), beginning in the early 1950s, and never outcrossed except with a standard a sh tester or Dr. Brink's W22 color-reverted line. The population from which Cy originated was a TEL population developed in 1963. This TEL population is similar to the population that gave rise to a large number of a2 and c-m alleles, summarized in a 1975 Corn Breeding and Genetics meeting in Urbana (47). The frequency of recovery of mutable alleles confirmed with inserts ranged from zero to near 12 per 10^6 gametes. Some of the alleles have been cloned and studied molecularly by Paz-Ares et al. (37,38).

The a2-m668291 Allele

Among the unstable alleles uncovered in the attempt to move En into a2 and to c, several could not be assigned to a known mutable system in which the standard specificity tests were applied (11). With the available bz-rcy reporter allele, however, it was possible to test these for the Cy system. One such allele is the a2-m668291 allele; its progress to date is such that it can be presented, and its pedigree history is given in Tab. 1. As is evident, its origin traces back to the original En at the A allele (1952 561 x 370-1 cross) with similar parents as those of bz-m805137 (61), with the additional introduction of an abnormal 10 line in 1963. The population developed for this rescue of an a2-m allele was 1964 1472 x 828 (Tab. 6 in Ref. 47), which is a similar source for the En containing c-m alleles where En was identified both genetically (47,54) and molecularly (37).

The mutability of the a2-m668291 allele is characterized in the kernel by a high frequency of one-celled spots (Fig. 11). In the initial tests, no En could be detected, and in subsequent tests, En could not be detected in the a-m(r) reporter allele, though the a2-m allele was in the same parentage. The segregating ratios showed a frequency of more than one regulatory element segregating (Tab. 2, A, B, and C). In subsequent crosses, in which the multiplicity of regulatory elements would be diluted out, the ratios of spotted to colorless continued to indicate a large number of regulatory elements (Tab. 2C, 2D). The alternative that these colorless were due to a high incidence of germinal nonspotted types, as is found with the En transposition of the autonomously mutable a-m(papu) allele, is not applicable here because the mutability pattern associated with this a2-m is so late that germinals would not be expected.

In tests of the a2-m genotype with the Cy reporter allele bz-rcy, Cy was detected, and in subsequent tests of these heterozygotes the number of Cy present could be determined (Tab. 2E). It is clear that the number of Cy parallels the incidence of a2-m (Tab. 2D).

The confirmatory tests of relationship can be derived from the tests of segregating a2 spotting vs a2 nonspotting, as in the progeny in Tab. 2D, for the Cy element as determined with the bz-rcy reporter allele.

When the a2-spotted are separated from the a2-colorless, as in the progeny shown in Tab. 2D, tests can be made for the correlative presence of Cy as expressed by the bz-rcy allele in the tests. Similarly, the bz-spotted are separated from the bz-nonspotted (as shown in Tab. 2E), and these are tested for the presence of the mutability or nonmutability of the a2 allele. These results are given in Tab. 3A and 3B, illustrating this correlation of Cy presence and a2-mutability.

Tab. 1. The origin of a2m668291. Pedigree, originating from the initial origin of the insert of En into A2 (49).

Item	Year[5]	Plant number	Site[b]	Phenotype of kernel planted[7]	Cross: Relevant genotypes
1	1987	3133-3139	A	a2-sp and a2-non-sp	see Table 3A
	1987	3111-3114	A	bz-sp and bz-non-sp	see Table 3B
2	1986g	010/35	K	Cl	A2/a2m Bz/bz-rcy x bz/bz
3a	1986	5205Y-3/3923	A	mutable (7aa)	a2m/a2bt x bz-rcy/-
3b	1986	5204-4t/3924	A	(8a)	a2m/a2bt x bz-rcy/-
3c	1986	5206Y-9/3924	A	Cl	a2m/a2bt (5aa) x bz-rcy/-
4a	1970	6146-1/2955	A	(8aa)	A2/a2m x a2bt/a2bt
4b	1970	6144-1/2958	A	mutable	A2/a2m x a2bt/bt
4c	1976	4159-3/3103	A	(5aa)	a2m/a2bt x a2bt/a2bt
5	1969	1224-3/1508	A	mutable-coarse	a2m(r)/a2m[6]
6	1968	5310/5321-6	A	mutable-v. fine	a2m[6]/a2bt x a2m(r)/-
7	1967	2845/2846-9	A	mutable (7aa)	a2m/a2bt x a2bt/a2bt En
8	1966g	4B-11/19	H	mutable	a2m/a2bt x a2bt/2bt
9	1966	8291/7521	A	mutable (7aa)	a2m/a2bt no En x a2bt/2bt
10	1965	6017-92 x a2bt/s		Cl	A2A2 x a2m668291 arose on the ear as a single kernel exception[3]
11	1964	1472 x W22-line C	A	Cl	A2A2A1 a-m(papu) x Color line
12	1963	1115 x 1094[1]	A	Cl	a-m(papu) x Abnormal 10
13	1962	752-1 x	A	a-m(papu)	a-m(papu)/A
14	1961	824-6[2] x 1140-15	A	Cl	Line A x a-m(papu)

15	1960	973-3 x P	A	a-m(papu)-type	a-m(papu)/a sh x a sh/a sh
16	1959	606-2 x P	A	a-m(papu)-type	a-m(papu)/a sh x a sh/a sh
17	1958	431-11 x P	A	a-m(papu)-type	a-m(papu)/a sh x a sh/a sh
18	1957	213A x P	A	a-m(papu)-type	a-m(papu)/a sh x a sh/a sh, single kernel exception of a-m(papu)
19	1956	149-4 x 20-1	A	a-m(dense)-type	a-m(dense)/a sh x a sh/a source ear for a-m(papu) exception
sh					
20	1955	22-7 x 59-3	R	a-m(dense)-type	a-m(dense)/a-dl x a sh/a sh
21	1954	2-46 X(4)	R	mutable	a-m/- x
22	1953	37-5 x	R	mutable	a-m/a-dl x
23	1952	561 x 370-1	U	mutable	a-dl et Dt su x a-m/- 370-1- original source of En into A

a g = winter season.

b A = Ames, IA; H = Homestead, FL; K = Kauai, HI; R = Riverside, CA; U = Urbana, IL.

(1) 1953 1094 = Abn 10 line originated in 1958 from Emmerling 79:702A.2-12 x 702A.42/2-6 (g R^r/g R^r K pr pr).

(2) 1963 824 = A W22 converted line from Dr. R.A. Brink; homozygous purple aleurone.

(3) a2m668291 was isolated as a single exception on ear 1965 8017-92 as one kernel among 361,560 gametes (47). Six c-m mutants were isolated from this same 651472 population, and two (668613, 668655) were molecularly analyzed as En inserts (37,55).

(4) Pedigrees before the 1954 period are given in (61).

(5) Each year entry is derived from each preceding year entry. With double entries, 3a, 3b, etc., derived from 4a, 4b, etc.

(6) This phenotype was identified as very fine, clear, medium, high, which is later identified as 8aa (55).

(7) Cl = colored; mutable phenotype classification, 8a, 7aa as in (57); coarse or fine as described in photos. a-m(papu) - see Fig. 6B.

Tab. 2. Segregation pattern and progeny tests for a2-m668291 illustrating the frequency of nonspotted types.

Cross	Round			Brittle	
	Colored	Spotted 9aa	cl	Spotted	cl
A. Spotted progeny from the original ear a2-m668291 were again test crossed by a2 bt/a2 bt: a2-m/a2 bt x a2 bt/a2 bt:					
66g 4B-2/19	---	54	8	nc	nc
6gg 4B-11/19	---	58	18	nc	nc
B. Spotted progeny from 66g 4B-2 (see section A, above) were test crossed by a2-m/a2 bt x a2 bt/a2 bt:					
67 2841-21 x 1360	0	122	32	6	124
67 2841-22 x 1352	0	71	4	2	168
C. Colored progeny from cross of spotted progeny from 66g 4B-11 (see section A, above) to an A2 A2 parent were test crossed by A2/a2-m x a2 bt/a2 bt:					
67 2828-7 x 2918	136	110	16	ne	ne
67 2829-3 x 1359	190	173	21	ne	ne
67 2829-7 x 1360	155	131	36	ne	ne

D. Segregation of a2 spotted progeny following the outcross of a2-m x bz-rcy. a2-m/a2 bt Bz/Bz x A2/A2 bz-rcy-/- (see 3a in Tab. 1) are test crossed onto an a2 bt tester (86g 32); also tested for presence of Cy by crosses with a bz tester.

	Cl	a2-Spotted	a2-cl	Cy present	Approximate number of regulatory elements
86g 32 x 010-1	82	78	23	+	3
32 x 010-2	109	29	23	+	1
32 x 010-8	149	93	25	(na)*	3
32 x 010-9	67	49	19	(na)	2-3
32 x 010-11	99	81	40	+	2
32 x 010-11	90	76	38	+	2

E. Segregation of Cy in testcross progenies from the outcross (item 3c, Tab. 1) of a2-m spotted by bz-rcy: a2-m/a2 bt (5aa) x bz-rcy/--Colored (86g11) A2/a2-m Bz/bz-rcy x A2/A2 bz/bz--Colored, bz spotted and nonspotted selections.

	Cl	bz-Spotted	bz-Nonspotted	Approximate number of Cy
86g 11-22/105	nc	83	60	1
11-23/105	nc	85	18	3
11-24/105	nc	21	19	1
11-30/105	nc	21	19	1
11-32/105	nc	66	68	1
11-33-105	nc	91	40	2
11-34/105	nc	70	27	2-3

A2/am from outcrossed a2m/a2bt (66g 4B-11 in Tab. 1 and in section A, above).

ne = None expected (only those carrying the a2m allele are illustrated).

cl = Colorless; Rd = Round.

nc = Not counted.

*The original bz-rcy reporter allele plant was heterozygous for bz-rcy/bz-sh and therefore the bz would not report Cy presence. The presence of bz sh verifies the absence of the bz-rcy reporter allele.

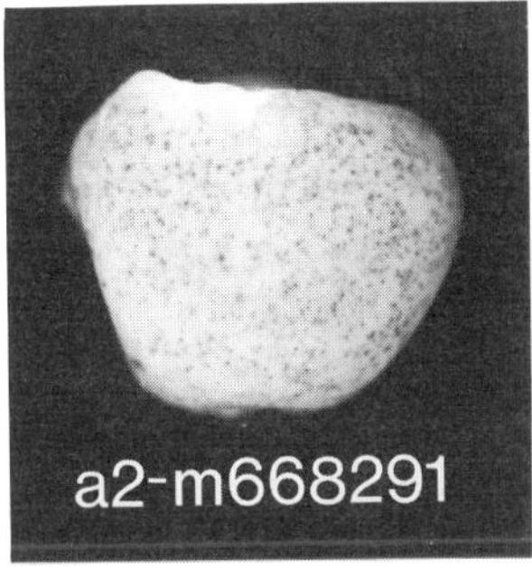

Fig. 11. The a2-m668291 allele. Kernels expressing this mutability contain Cy and lack En.

Tab. 3. Tests for the correlation of the presence of the Cy element as determined by the reporter allele bz-rcy and the presence of Cy among a2-spotted or nonspotted in a segregating population.

A. Tests of a2-spotted vs nonspotted selections from crosses itemized in Tab. 2D for the presence of Cy.

a2/a2 bt/bt x A2/a2-m Bz/bz-rcy Cy → a2-sp, a2-non-sp

a2-sp and a2-non-sp are tested by crosses with bz

Selection	Cl	bz-Spotted	bz-Nonspotted	Total	Approximate number of Cy
	Number of ear progenies segregating				
1987					
3133 a2-sp	6	7	0		
3134 a2-non-sp	5	1	5		
3138 a2-sp	4	6	0		
3139 a2-non-sp	8	0	5		

B. Tests for the occurrence of a2-mutability or nonmutability among Cy vs non-Cy selections from cross in 2E.

A2/a2-m Bz/bz-rcy Cy x bz/bz → bz-sp, bz-non-sp

bz-sp and bz-non-sp are tested by crosses with a2

Selection	Cl	a2-Spotted	a2-Nonspotted	Total	Approximate number of elements
	Number of ear progenies segregating				
1987					
3111 bz-sp	3	9	0	12	
3112 bz-non-sp	10	0	4	14	
3113 bz-sp	3	10	0	13	
3114 bz-non-sp	10	0	2	12	

Cl = Colored, only colored progeny, only the A2 is present.

a2-spotted: Among the noncolored, a2 spotting is segregating.

a2-nonspotted: Among the noncolored, only nonspotting is segregating.

It is clear that the a2-m668291 allele is Cy-responsive and was only related following the indication with the Cy-reporter allele. Thus, the a2-spotted are a2-rcy, Cy, and the a2-nonspotted lack Cy. This would support the contention that the Cy element was a resident very early in our TEL population.

Inserts Without Genetically Active Functions

In several investigations, inserts have been uncovered in genes, some unexpectedly, that cannot be related to genetically active units. The first corn insert discovered was uncovered accidentally in a search for the chalcone synthase gene (C2), utilizing a heterologous probe from *Antirrhinum* (67,73). This was the Cin1 (Cin-001) element, a 691-bp element located in an autonomous gene. With use of this original element as a probe, 14 additional Cin1 homologous sequences were isolated from the maize line W22, each with 80%-90% homology to each other as determined by heteroduplex analysis (15). These related sequences vary in size up to 700 bp.

Other elements have been described, and these are itemized in a listing in Döring and Starlinger's report (6). The BS1 element is a report as part of this Volume (Bennetzen et al.).

Certainly, a large number of other mobile elements exist, many of which are expressed as inserts in genes that are actively transposing. In many cases, the genes are unknown. For example, the V^mmp11817 unstable gene (43) affects the aleurone (Fig. 12A) when transmitted as a female and expresses a virescent phenotype late in the mature plant, where it expresses instability from virescent to green (Fig. 12B).

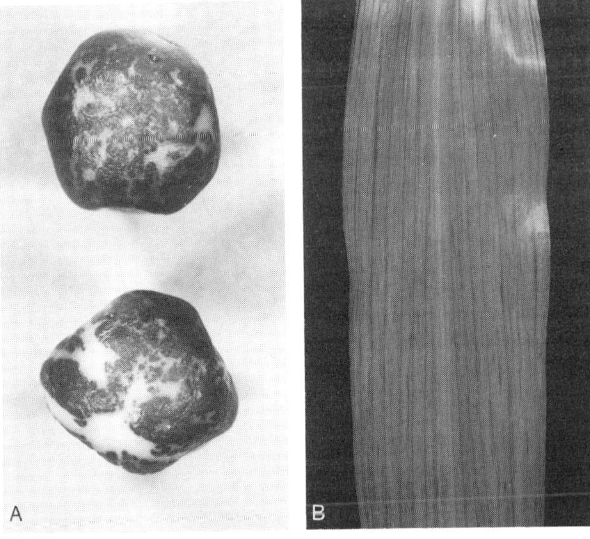

Fig. 12. The V^mmp-1817 allele (43). A. Etched kernels from the cross +/V^mmp1817 x +/+. B. Mutability on virescent-like leaves from plant originating from etched kernels.

DISCUSSION

With nearly five decades of active and consistent research on mobile elements, one can ask what has been learned about this transposing DNA. Originally, skeptics were not willing to concede that it was even inserted, despite the evidence of clear-cut genetic segregation ratios (1) as well as of recombination studies (31) that should have provided sufficient credibility for this concept. Certainly, as evidence accumulated from bacterial studies in the late 1960s (17,66), as well as in the later initial molecular studies in maize, it was well established that these transposing elements do indeed insert and excise.

There is a general characterization that can be applied to the elements currently under study. This has been given previously in this report. Also, much has been learned about the functions of these elements, and this is given in a report in this same Volume by Gierl et al.

More importantly, the greater impact of these elements has been on what they do to genes where they visit. This has also been detailed in numerous reports (64,72). But these elements have provided an access to genes for further study that would not have been previously possible. For example, the structure of a number of plant genes has been elucidated (16,36,37,74). The uncovering of unexpected inserts in almost every gene being studied certainly was not expected (63,65). It seems that unadulterated genes do not exist.

These mobile elements have also given investigators an opportunity to examine how genes have evolved. Access to a specific gene (such as the chalcone synthase gene) across a large number of species has provided a means to study this evolutionary process (H. Saedler, pers. comm.).

The utilization of these elements in tagging desirable genes is the beginning of new ventures in crop improvement. Any successes here would be ample reason for continuing studies with mobile elements.

ACKNOWLEDGEMENT

Journal paper No. J-12828 of the Iowa Agriculture and Home Economics Experiment Station, Ames, Iowa, Project No. 2381.

REFERENCES

1. Brink, R.A., and R.A. Nilan (1952) The relation between light variegated and medium variegated pericarp in maize. Genetics 37:519-544.
2. Chandler, V., C. Rivin, and V. Walbot (1986) Stable non-mutator stocks of maize have sequences homologous to the Mu1 transposable element. Genetics 114:1007-1021.
3. Cormack, J., and P.A. Peterson (1987) The Uq content of several breeding populations. Maize Genet. Coop. Newsl. 61:4.
4. Demerec, M. (1935) Unstable genes. Bot. Rev. 78:233.
5. Döring, H.-P., and P. Starlinger (1984) Barbara McClintock's controlling elements: Now at the DNA level. Cell 39:253-259.
6. Döring, H.-P., and P. Starlinger (1986) Molecular genetics of transposable elements in plants. An. Rev. Genet. 20:175-200.

7. Emerson, R.A. (1914) The inheritance of a recurring somatic variation in variegated ears of maize. Am. Nat. 48:87-115.
8. Emerson, R.A. (1917) Genetical studies of variegated pericarp in maize. Genetics 2:1-35.
9. Fedoroff, Nina (1983) Controlling elements in maize. In Mobile Genetic Elements, James A. Shapiro, ed. Academic Press, Inc., pp. 1-63.
10. Fedoroff, N., S. Wessler, and M. Shure (1983) Isolation of the transposable maize controlling elements Ac and Ds. Cell 35:235-242.
11. Friedemann, P., and P.A. Peterson (1982) The Uq controlling-element system in maize. Molec. Gen. Genet. 187:19-29.
12. Gierl, A., Zs. Schwarz-Sommer, A. Pereira, P.A. Peterson, and H. Saedler (1985) Molecular analysis of the En/Spm transposable element system of Zea mays. In Plant Genetics, M. Freeling, ed. UCLA Symposium on Molecular and Cellular Biology. New Series, Volume 35, Alan R. Liss, Inc., New York, pp. 383-389.
13. Gonella, J.A., and P.A. Peterson (1977) Controlling elements in a tribal maize from Colombia: Fcu, a two-unit system. Genetics 85:629-645.
14. Gonella, J.A., and P.A. Peterson (1978) The Fcu controlling-element system in maize. II. On the possible heterogeneity of controlling elements. Molec. Gen. Genet. 167:29-36.
15. Gupta, M., I. Bertram, N.S. Shepherd, and H. Saedler (1983) Cin1, a family of dispersed repetitive elements in Zea mays. Molec. Gen. Genet. 192:373-377.
16. Hallauer, A.R., W.A. Russell, and O.S. Smith (1983) Quantitative analysis of Iowa stiff stalk synthetic. Stadler Genet. Symp. 15:83-104.
17. Jordan, E., H. Saedler, and P. Starlinger (1986) 0° and strong-polar mutations in the gal operon are insertions. Molec. Gen. Genet. 102:353-363.
18. Karasawa, M., and P.A. Peterson (1987) Test for presence of controlling elements in populations. Maize Genet. Coop. Newsl. 61:4.
19. McClintock, B. (1946) Maize genetics. In Carnegie Institution of Washington Yearbook 45:176-186.
20. McClintock, B. (1947) Cytogenetic studies of maize and neurospora. In Carnegie Institution of Washington Yearbook 46:146-152.
21. McClintock, B. (1948) Mutable loci in maize. In Carnegie Institution of Washington Yearbook 47:155-169.
22. McClintock, B. (1949) Mutable loci in maize. In Carnegie Institution of Washington Yearbook 48:142-154.
23. McClintock, B. (1950) Mutable loci in maize. In Carnegie Institution of Washington Yearbook 49:157.
24. McClintock, B. (1951) Chromosome organization and genic expression. Cold Spring Harbor Symp. Quant. Biol. 16:13-47.
25. McClintock, B. (1952) Mutable loci in maize. In Carnegie Institution of Washington Yearbook 51:212-219.
26. McClintock, B. (1955) Controlled mutation in maize. In Carnegie Institution of Washington Yearbook 54:245.
27. McClintock, B. (1964) Aspects of gene regulation in maize. In Carnegie Institution of Washington Yearbook 63:592-602.
28. McClintock, B. (1965) The control of gene action in maize. Brookhaven Symp. Biol. 18:162-184.
29. McClintock, B. (1967) Regulation of pattern of gene expression by controlling elements in maize. In Carnegie Institution of Washington Yearbook 65:568-578.

30. McClintock, B. (1968) The states of a gene locus in maize. In Carnegie Institution of Washington Yearbook 66:20-28.
31. Nelson, O.E. (1968) The Waxy locus in maize. II. The location of the controlling element alleles. Genetics 60:507-524.
32. Nevers, P., and H. Saedler (1977) Transposable genetic elements as agents of gene instability and chromosomal rearrangements. Nature (London) 268:109-115.
33. Nevers, P., N.S. Shepherd, and H. Saedler (1985) Plant transposable elements. Adv. Bot. Res. 12:104-203.
34. Nowick, E.M., and P.A. Peterson (1981) Transposition of the enhancer controlling element system in maize. Molec. Gen. Genet. 183:440-448.
35. Oberthur, E., and P.A. Peterson (1984) Uq controlled mutable allele at the c locus. Maize Genet. Coop. Newsl. 58:4-5.
36. O'Reilly, C., N.S. Shepherd, A. Pereira, Zs. Schwarz-Sommer, I. Bertram, P.A. Peterson, and H. Saedler (1985) Molecular cloning of the a1 locus of Zea mays using the transposable elements En and Mu1. EMBO J. 4:591-597.
37. Paz-Ares, J., U. Wienand, P.A. Peterson, and H. Saedler (1986) Molecular cloning of the c locus of Zea mays: A locus regulating the anthocyanin pathway. EMBO J. 5:829-833.
38. Paz-Ares, J., D. Ghosal, U. Wienand, P.A. Peterson, and H. Saedler (1987) The regulatory c1 locus of Zea mays encodes a protein with homology of myb proto-oncogene products and with structural similarities to transcriptional activators. EMBO J. 6:3553-3558.
39. Pereira, A., and P.A. Peterson (1985) Origin and diversity of mutants controlled by the Uq transposable element system in maize. Genet. Res. 46:219.
40. Pereira, A., Zs. Schwarz-Sommer, A. Gierl, I. Bertram, P.A. Peterson, and H. Saedler (1985) Genetic and molecular analysis of the Enhancer (En) transposable element system of Zea mays. EMBO J. 4:17-23.
41. Pereira, A., H. Cuypers, A. Gierl, Zs. Schwarz-Sommer, and H. Saedler (1986) Molecular analysis of the En/Spm transposable element system of Zea mays. EMBO J. 5:835-841.
42. Peterson, P.A. (1961) Mutable a1 of the En system in maize. Genetics 46:759-771.
43. Peterson, P.A. (1964) The dominant mutable V^{m}mp-1817. Maize Genet. Coop. Newsl. 38:80-81.
44. Peterson, P.A. (1970) The En mutable system in maize. III. Transposition associated with mutational events. Theor. Appl. Genet. 40:367-377.
45. Peterson, P.A. (1976) Basis for the diversity of states of controlling elements in maize. Molec. Gen. Genet. 149:5-21.
46. Peterson, P.A. (1978) A test of a molecular model of a controlling element transposon in maize. Proc. Int. Cong. Genet. 14:49.
47. Peterson, P.A. (1978) Controlling elements: The induction of mutability at the A2 and C loci in maize. In Maize Breeding and Genetics, D.B. Walden, ed. John Wiley and Sons, New York, pp. 601-635.
48. Peterson, P.A. (1985) The Enhancer (En) system: A maize mobile-element system. In Plant Genetics, M. Freeling, ed. Alan R. Liss Inc., New York, pp. 369-381.
49. Peterson, P.A. (1985) The isolation of En-1 in the wx-84-4 allele. Maize Genet. Coop. Newsl. 59:3.
50. Peterson, P.A. (1985) Virus-induced mutations in maize: On the nature of stress-induction of unstable loci. Genet. Res. 46:207-217.

51. Peterson, P.A. (1986) Mobile elements in maize: A force in evolutionary plant breeding processes. Stadler Genet. Symp. 17:47-78.
52. Peterson, P.A., and F. Salamini (1986) Distribution of active transposable elements among important corn-breeding populations. Maydica 31:163-172.
53. Peterson, P.A. (1987) Mobile elements in plants. Crit. Rev. of Plant Sciences 6:105-208.
54. Reddy, A.R., and P.A. Peterson (1983) Transposable elements of maize: Genetic basis of pattern differentiation of some mutable c alleles of the enhancer system. Molec. Gen. Genet. 192:21-31.
55. Reddy, L.V., and P.A. Peterson (1984) Enhancer transposable element induced changes at the A locus in maize: The a-m-1 6078 allele. Molec. Gen. Genet. 194:124-137.
56. Reddy, L.V., and P.A. Peterson (1985) Spm and I element changes with the a-m2 8004 allele in maize. Molec. Gen. Genet. 200:211-219.
57. Rhoades, M.M. (1938) Effect of the Dt gene on the mutability of the a_1 allele in maize. Genetics 23:377-395.
58. Saedler, H., and P. Nevers (1985) Transposition in plants: A molecular model. EMBO J. 4:585-590.
59. Sastry, G.R.K., and S.L. Kurmi (1970) Spotted-dilute and the instability of R-r. Maize Genet. Coop. Newsl. 44:101.
60. Schiefelbein, J.W., V. Raboy, N.V. Fedoroff, and O.E. Nelson, Jr. (1985) Deletions within a defective Suppressor-mutator element in maize affect the frequency and developmental timing of its excision from the bronze locus. Proc. Natl. Acad. Sci., USA 82:4783-4787.
61. Schnable, P.S., and P.A. Peterson (1986) Distribution of genetically active Cy transposable elements among diverse maize lines. Maydica 31:59-81.
62. Schnable, P.S., and P.A. Peterson (1987) Molec. Gen. Genet. (in press).
63. Schwarz-Sommer, Zs., A. Gierl, R.B. Klösgen, U. Wienand, P.A. Peterson, and H. Saedler (1984) The Spm (En) transposable element controls the excision of a 2-kb DNA insert at the wx^{m-8} allele of Zea mays. EMBO J. 3:1021-1028.
64. Schwarz-Sommer, Zs., A. Gierl, H. Cuypers, P.A. Peterson, and H. Saedler (1985) Plant transposable elements generate the DNA sequence diversity needed in evolution. EMBO J. 4:591-597.
65. Schwarz-Sommer, Z., N. Shepherd, E. Tacke, A. Gierl, W. Rohde, L. Leclercq, M. Mattes, R. Berndtgen, P.A. Peterson, and H. Saedler (1987) Influence of transposable elements on the structure and function of the A1 gene of Zea mays. EMBO J. 6:287-294.
66. Shapiro, J.A. (1969) Mutations caused by the insertion of genetic material into the galactose operon of Escherichia coli. J. Molec. Biol. 40:93.
67. Shepherd, N.S., Zs. Schwarz-Sommer, U. Wienand, H. Sommer, B. Deumling, P.A. Peterson, and H. Saedler (1982) Cloning of a genomic fragment carrying the insertion element Cin1 of Zea mays. Molec. Gen. Genet. 188:266-271.
68. Sprague, G.F. (1986) Mutability in the a-ruq, Uq system in maize. Maydica 31:17.
69. Sprague, G.F., and H.H. McKinney (1966) Aberrant ratio: An anomaly in maize associated with virus infection. Genetics 54:1287-1296.
70. Sprague, G.F., and H.H. McKinney (1971) Further evidence of the genetic behaviour of AR in maize. Genetics 67:533-542.
71. Tacke, E., Zs. Schwarz-Sommer, P.A. Peterson, and H. Saedler (1986) Molecular analysis of states of the A1 locus of Zea mays. Maydica 31:83-91.

72. Wessler, S.R., G. Baran, M. Varagona, and S. Dellaporta (1986) Excision of Ds produces waxy proteins with a range of enzymatic activities. EMBO J. 5:2427-2432.
73. Wienand, U., H. Sommer, Z. Schwarz, N. Shepherd, H. Saedler, F. Kreuzaler, H. Ragg, E. Fautz, K. Hahlbrock, B. Harrison, and P.A. Peterson (1982) A general method to identify plant structural genes among genomic DNA clones using transposable element induced mutations. Molec. Gen. Genet. 187:195-201.
74. Wienand, U., U. Weydemann, U. Niesbach-Klösgen, P.A. Peterson, and H. Saedler (1986) Molecular cloning of the c2 locus of Zea mays: The gene coding for chalcone synthase. Molec. Gen. Genet. 203:202-207.

GENETIC AND MOLECULAR ANALYSIS OF TRANSPOSABLE ELEMENTS IN *ANTIRRHINUM MAJUS*

Rosemary Carpenter, Andrew Hudson, Tim Robbins, Jorge Almeida, Cathie Martin, and Enrico Coen

AFRC Institute of Plant Science Research
John Innes Institute
Norwich NR4 7UH, United Kingdom

ABSTRACT

Transposable element activity in *Antirrhinum majus* has been studied genetically for many years. More recently the genetic analysis has been combined with molecular techniques, leading to a much greater understanding of various properties of these transposons. We have shown that the frequency of transposition of specific transposable elements can be controlled by a number of different factors including the environmental conditions under which the plants are grown and the genetic background. Transposable elements are also able to alter gene expression by imprecise excision, deletions, inversions, and chromosomal rearrangements, thus giving rise to allelic series. The knowledge gained from these studies has enabled transposable elements to be used for gene isolation. In this paper we describe the main features of the behavior of transposable elements in *Antirrhinum* and how they may be used to study gene action.

PATHWAY TO ANTHOCYANIN PRODUCTION

In *Antirrhinum majus*, transposon activity has been most intensively studied in genes responsible for anthocyanin synthesis because mutations in these loci are easily recognized phenotypically. Several genetically defined blocks in the anthocyanin biosynthetic pathway have been described (Fig. 1). A block at the *nivea* (*niv*) gene, which encodes the enzyme chalcone synthase (20), interrupts an early step in the pathway and gives an albino flower. A block at the *incolorata* (*inc*) gene acts after the flavones and dihydroflavones have been produced and before the production of dihydroflavonol. A null mutation of this gene blocks the activity of the enzyme flavanone 3-hydroxylase (8), thus giving an ivory flower. The *pallida* (*pal*) gene acts at a later stage in the pathway and encodes the enzyme dihydroflavonol-4-reductase (E.S. Coen, J. Firmin, and R. Carpenter, unpubl. data), the null mutants giving ivory flowers. Mutations in a fourth gene, *delila* (*del*), block anthocyanin production in the corolla tube while the petals are fully pigmented. This gene appears to

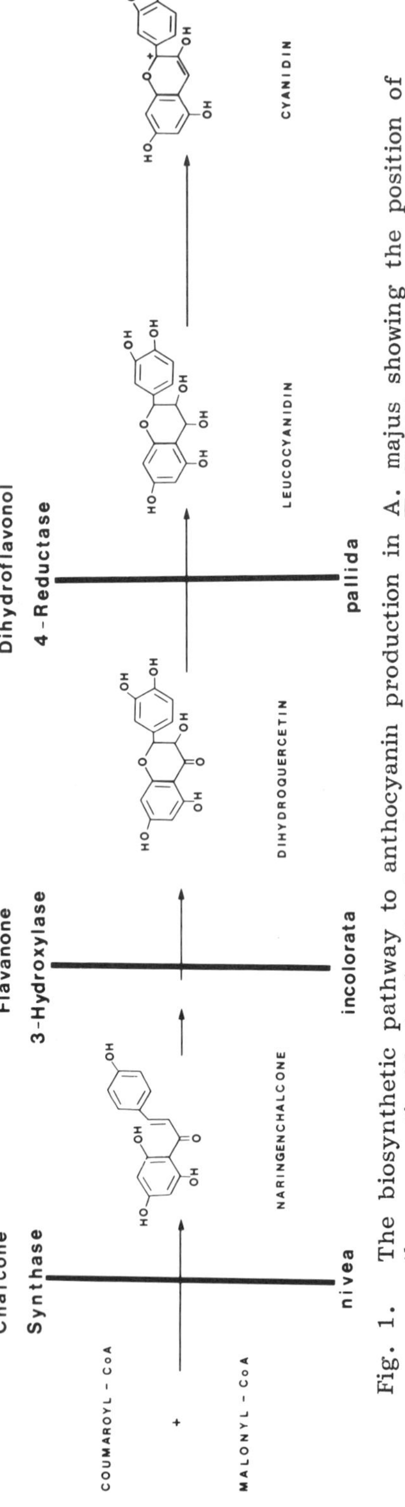

Fig. 1. The biosynthetic pathway to anthocyanin production in *A. majus* showing the position of three genetic blocks involved and the enzymes which they affect. A single arrow does not necessarily represent one enzymatic step.

act at a late stage in the pigment pathway (13). It may, however, be a regulatory gene, because evidence now indicates that it affects transcription of both the *niv* and *pal* genes (T. Robbins, R. Carpenter, and E.S. Coen, unpubl. data).

THE RECURRENS PHENOTYPES

Transposable elements inserted in genes involved in anthocyanin biosynthesis usually confer a variegated or "recurrens" phenotype. The unstable mutant *pallida*^{recurrens} (*pal*^{rec}) has ivory flowers with randomly distributed sites or sectors of red pigmentation. This mutation is now known to be due to the insertion of the transposable element *Tam3* in the promoter of *pal*, which excises and restores gene function, so allowing sites of pigmentation to be formed (18). The earlier in the ontogeny of the flower that this occurs, the larger the site of pigmentation. If excision occurs, gametically fully pigmented progeny are produced. *Tam3* is also present at the *nivea* locus in the *nivea*^{recurrens} (*niv*^{rec}) line 98. Flowers carrying this allele have a palely pigmented background, indicating that in this instance the insertion of the transposon does not completely block gene function. However, wild-type sites produced by *Tam3* excision can be seen superimposed on this background (5).

In contrast, when either the *Tam1* or *Tam2* element is inserted at the *niv* gene, the flower is albino, indicating that in these cases gene function is completely blocked. Randomly distributed sites of pigmentation on the albino background again indicate somatic excision of an active transposable element.

STRUCTURAL FEATURES OF TRANSPOSABLE ELEMENTS

Tam1 and *Tam2* have both been isolated at the *niv* locus. They have homologous termini and both produce a 3-bp duplication in the DNA into which they insert. *Tam1* is about 15 kb in length and is inserted in the promoter region (2,3) and confers a recurrens phenotype. *Tam2* is approximately 5 kb long and is inserted at the first exon/intron boundary. Although it has all the characteristics of a transposon (22), it normally confers a stable *nivea* phenotype; no pigmented sites are observed. The two elements show homology to the *En1* elements in *Zea mays* and the *Tgm1* element in *Glycine max*, suggesting that they belong to the same family of transposons (22).

Tam3 has been described at both the *niv* and *pal* loci and is 3.6 kb in length (6,19). In each case it is inserted in the promoter region and confers a recurrens phenotype. However, this element produces a larger target duplication (5 bp or 8 bp) and has different termini from those of *Tam1* or *Tam2*.

ACTIVATION AND REPRESSION OF TRANSPOSABLE ELEMENTS

The frequency with which a transposable element is excised from its site of insertion is variable and can depend on a number of factors. The structure of the transposon, the genetic background, and the environmental conditions under which the plants are grown all have a considerable effect on the rate of excision both somatically and gametically.

Environmental Effects

The frequency of Tam3 excision depends on the temperature at which the plants are grown. When plants of highly mutating lines pal^{rec}-2 and niv^{rec}-98 are grown at 25°C, only occasional full red sites (10-30 per flower) are seen. At 15°C a much higher frequency of sites is observed for both alleles (Fig. 2). This was shown to be 1,000-fold greater than at 25°C for both pal^{rec}-2 and niv^{rec}-98 lines (5,11). The 1,000-fold difference at the two temperatures is specific to Tam3. Although the unstable allele niv^{rec}-53, which is due to the insertion of Tam1, is temperature sensitive, only a six-fold difference in excision rate was recorded between plants grown at 15°C and those grown at 25°C (9).

A homozygous stock of an unstable niv^{rec} allele caused by Tam2 insertion (niv-568) has recently been isolated. Although no accurate figures are available, the effect of different temperatures on Tam2 excision rate would appear to be much nearer to the six-fold difference recorded for Tam1 than to the 1,000-fold shown by Tam3.

Genetic Control

The excision rate of Tam3 can be controlled genetically at both the pal and niv loci by an unlinked semidominant gene Stabilizer (St) (Fig. 3). At pal, St/St reduces the frequency of excision by about 1,000-fold from that seen in plants homozygous for st/st, with the heterozygous St/st showing an intermediate frequency (12). The effect of the St gene on Tam3 at niv was assessed by crossing niv^{rec}-98 to tester lines homozygous for St or st. In the F_2, three frequency categories were observed, with a

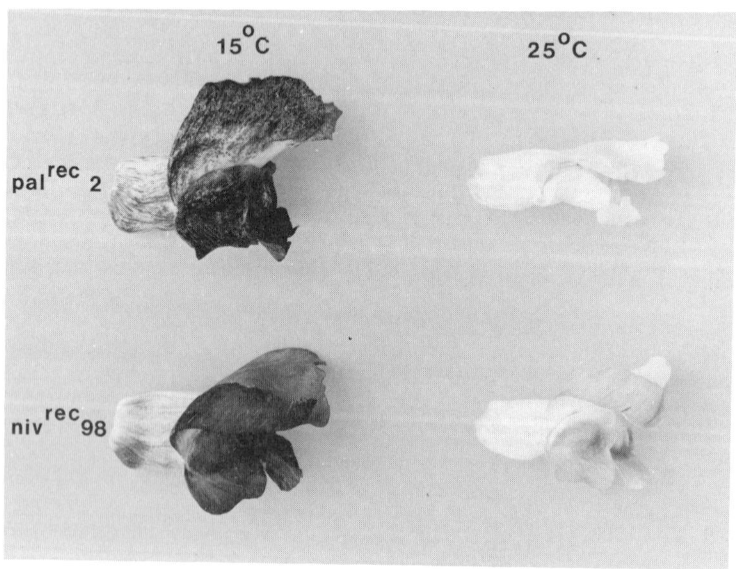

Fig. 2. Flowers from plants of pal^{rec}-2 and niv^{rec}-98 grown at 15°C and 25°C showing the effect of temperature on somatic excision rate of Tam3. In each case there was approximately a 1,000-fold difference between plants grown at the two temperatures.

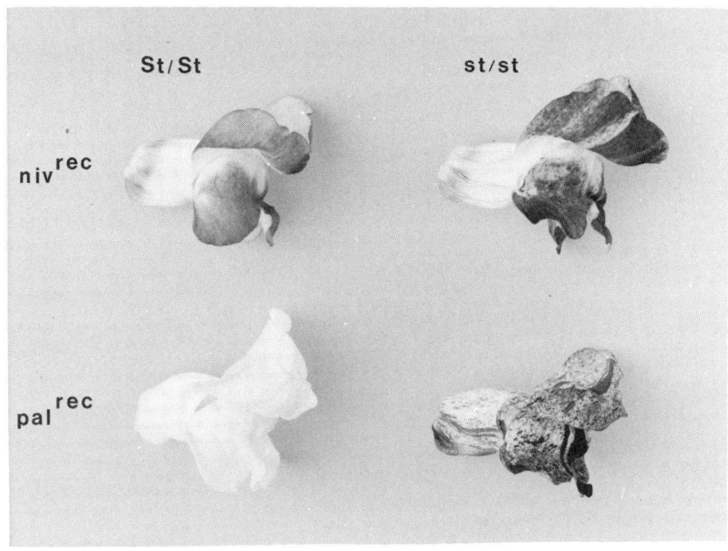

Fig. 3. Excision rate of Tam3 can be controlled genetically by the semi-dominant Stabilizer (St) gene. These flowers of plants carrying niv^{rec}::Tam3 or pal^{rec}::Tam3 are of the genotype St/St or st/st. In each case plants of St/St genotype have a much lower site frequency than those of st/st.

1,000-fold difference between plants of high and low frequency, indicating that the St gene also affects Tam3 at niv (5).

The St gene effect may well be element-specific, since it does not appear to have an effect upon the excision rate of Tam1 (9). As yet there are no data concerning its effect on the recently isolated homozygous niv^{rec}-568::Tam2.

Genetic factors which can modify Tam2 and Tam1 excision have also been described. The Tam2 transposon is inserted in the first exon/intron boundary of the niv gene and does not normally excise. In an attempt to activate Tam2, the line in which it is inserted, niv-44, was crossed to another line, niv-99, a stable white derivative of niv^{rec}-53::Tam1. Plants showing stable and unstable phenotypes were observed in the F_1, and in the F_2 wild-type revertants were recovered. Molecular analysis showed that one of these revertants arose by imprecise excision of Tam2 from the niv locus (15). In this case Tam2 is able to excise only when activated by another factor. Genetic analysis has shown that this factor is tightly linked to the niv-99 allele and may well be a derivative of Tam1. Interestingly, it has also been shown by Hehl et al. (14) that two other alleles derived from the line niv^{rec}-53 each carry a defective copy of Tam1 and are able to activate Tam2 in a similar way to that described above.

Earlier work by Harrison and Carpenter (9) showed that the line containing the niv-44 allele had a repressor effect upon niv^{rec}-53::Tam1. In F_1 plants obtained from crosses between the two lines there was a greater repression when niv^{rec}-53 was the male parent, and this repression increased with the aging of the plant. The repression was inheritable and in three subsequent generations did not revert to the highly unstable parental phenotype. It has been proposed that this repression of Tam1 transposition is due to the Tam2 element in niv-44 (22). However, genetic analysis of the repression suggests that it does not segregate with the niv-44 allele and may be due to as yet undefined genetic factors in the niv-44 line (A. Hudson, R. Carpenter, and E.S. Coen, unpubl. data).

ALLELIC DERIVATIVES

One of the main characteristics of unstable mutations caused by transposable elements is that they are able to give rise to gametic reversions. The most frequent event is reversion to wild type, but plants of different phenotypes may also be isolated which form an allelic series. One of the first and subsequently most extensively studied series to be described in Antirrhinum was the series at the pal locus (1,7,16). In 1984, a further series was produced by selfing of the pal^{rec}-2 line grown at 15°C, giving rise to plants that were isogenic except for small changes at the pal locus. These changes have now been shown, by molecular analysis, to be due to the activity of Tam3 (6). Since Tam3 is now present at both pallida and nivea and allelic series have been produced at both loci, we have been able to compare and contrast the two different sets of alleles.

The insertions of Tam3 at niv and pal show both similarities and differences. In both cases the transposon is inserted in the promoter region, confers a recurrens phenotype, and by its activity has given rise to allelic series. These allelic series can be divided into several different classes, as follows.

Allelic Series

Revertants. These are plants with fully pigmented flowers of wild-type intensity, e.g., Pal-501 and Niv-522 (Fig. 4). At both loci, this is the most frequent event observed and in each case has been shown to be due to imprecise excision of the transposon from its site of insertion.

Stable alleles. Stable alleles, with a reduced intensity of pigmentation compared to wild-type, have been described at both loci. In each case a continuous allelic series is produced from near full red to very pale intensity, e.g., pal-509 to pal-31 and niv-544 to niv-532 (Fig. 4). Many of these have been shown by molecular analysis to have arisen by imprecise excision of Tam3, which removes sequences required for efficient gene transcription.

Stable-patterned alleles. These have been obtained at both niv and pal. At pal, two such alleles, pal-32 and pal-33, show preferential pigmentation in the corolla tube. A third allele, pal-14, has flowers in which the pigment is restricted to a ring at the base of the corolla tube (Fig. 4). All three of these alleles have lost the transposon together with varying amounts of adjacent DNA sequences. In general, the larger the adjacent deletion, the more restricted the pigmentation (J. Almeida, R. Carpenter, and E.S. Coen, unpubl. data). Another patterned allele, pal-41, has pigment restricted to the center part of the corolla and the base of the tube. This allele has again lost Tam3, and linkage analysis suggests that an inversion of the DNA spanning about six chromosome map

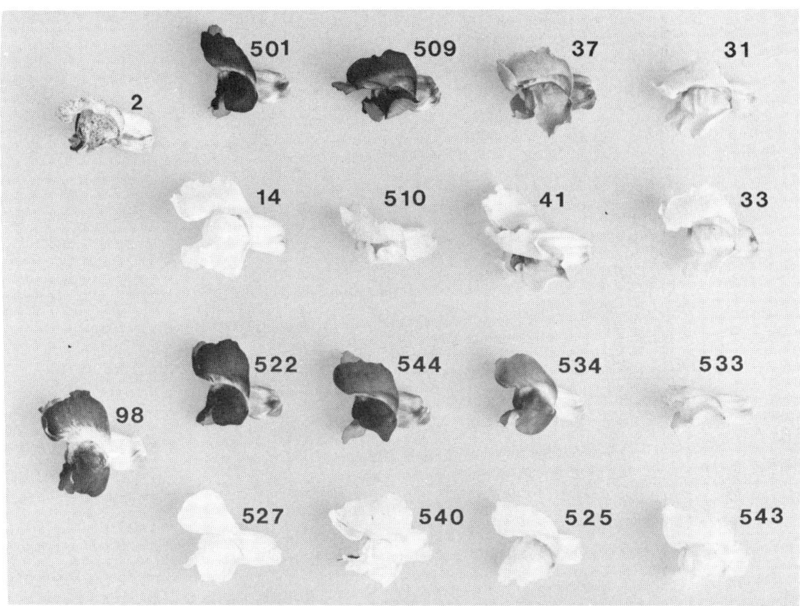

Fig. 4. Altered phenotypes generated by the activity of the transposable element Tam3. The top two rows of flowers are examples of those derived from pal^{rec}-2 and the bottom two rows were isolated from niv^{rec}-98. The progenitors are shown on the left of the photograph and stock numbers above each flower.

units has also taken place (T. Robbins, R. Carpenter, and E.S. Coen, unpubl. data). At niv, stable-patterned alleles have also been isolated, but these are quite distinct, phenotypically, from those derived by imprecise excision of Tam3 at pal.

Changed instability. New unstable alleles have been produced at both loci. The pal^{rec}-2 allele has given rise to a series of alleles conferring different frequencies, patterns, and intensities of sites. Of particular interest is pal-510 (Fig. 4), where instability is seen on an overall background of pale pigmentation, unlike pal^{rec}-2, where gene function is completely blocked by Tam3. The pal-510 allele was derived by a small deletion at one end of Tam3 which allows some pal expression to occur (A. Hudson, R. Carpenter, and E.S. Coen, unpubl. data). Another unstable allele of interest is pal-42, which has Tam3 present but has the same DNA inversion as that of the stable pal-41 allele already described (T. Robbins, R. Carpenter, and E.S. Coen, unpubl. data). Many of these alleles are under further analysis, but in all cases so far investigated Tam3 is present at the pal locus in the unstable alleles.

At niv, new unstable alleles have also been isolated. The niv-540 (Fig. 4) confers an albino background instead of the pigmented background of its progenitor, niv^{rec}-98. Molecular analysis indicates that in this case the Tam3 in niv^{rec}-98 has probably duplicated, excised, and reinserted at two different sites, one in the niv gene itself and the other some 3 kb upstream (A. Hudson, R. Carpenter, and E.S. Coen, unpubl. data). The upstream Tam3 appears to have no phenotypic effect on the flowers, while that inserted at the last exon of the niv gene blocks expression except where excision has occurred. The low frequency of excision sites presumably reflects the alteration in position of Tam3 from the promoter to the coding region.

Null mutants. So far, no complete null mutant has been obtained from the John Innes pal^{rec} stocks. However, several albino-flowered plants have been isolated from niv^{rec}-98, e.g., niv-527. In most cases this is due to deletion of part of the coding region of the gene, and is often accompanied by Tam3 excision (17). One possible explanation of this difference in ability of niv^{rec} and pal^{rec} to produce nulls is the opposite orientation of Tam3 with respect to the two genes (6).

Semidominant alleles. All mutant alleles so far described for both loci, other than those of group 1, are recessive, and, when crossed to their progenitor, randomly distributed wild-type excision sites can be seen in the F_1. However, an exceptional allele that has arisen by Tam3 activity is niv-525 (Fig. 4). In a cross of niv^{rec}-98 x pal-14, all F_1 plants should have been fully pigmented wild type, since the niv^{rec} line is homozygous for Pal^+, and the pal-14 line is homozygous for Niv+. However, one plant arose with a distinct patterned phenotype. To test whether this was a change at the niv or pal locus, flowers were fed with the precursors dihydroquercetin and naringenin. In each case additional synthesis of anthocyanin occurred, indicating a change at the niv locus. Mutations at niv would normally be recessive to Niv+, but so far all genetic analysis has shown niv-525 to be a semidominant allele. When backcrossed to its progenitor, only occasional red sites can be seen, restricted to the palely pigmented, patterned area. The niv-525 allele has been shown to carry an inverted duplication of 200 bp in the niv promoter, which is presumably responsible for its semidominance. No semidominant allele has ever been obtained from pal^{rec}-2. A similar semidominant allele, C_2ld_f, has, however, been described at the C_2 locus in maize (4) and may be due to a similar type of structural change.

Another allele at nivea, niv-543, also shows changed instability together with stable spatial patterning similar to that already described for niv-525. This allele is also semidominant, and it appears to be able to produce stable alleles of two different types. One is a fully pigmented, normal wild-type revertant, presumably caused by imprecise excision of Tam3. The second type, however, is phenotypically similar to niv-525 when homozygous, and again it is semidominant (Fig. 5). Plants showing further changes in instability have also been isolated from niv-543. It would appear that, in certain circumstances, a single transposition event can also act as a triggering mechanism for subsequent changes to occur at a relatively high frequency, immediately after the initial event.

Recently a series of semidominant alleles has been isolated from a direct selfing of the niv^{rec}-98 line at 15°C; they are therefore in an isogenic background. This series ranges from an allele which has very little effect on the wild-type gene to one that is almost albino even when heterozygous with Niv+. As yet, no molecular data are available, but this series should lead to a much grater understanding of the mechanism by which these semidominant alleles control gene expression.

The Tam1 transposon is inserted in the promoter region of the niv gene, and confers a recurrens phenotype. An allelic series has also been derived from the activity of this transposon, but of a more limited range

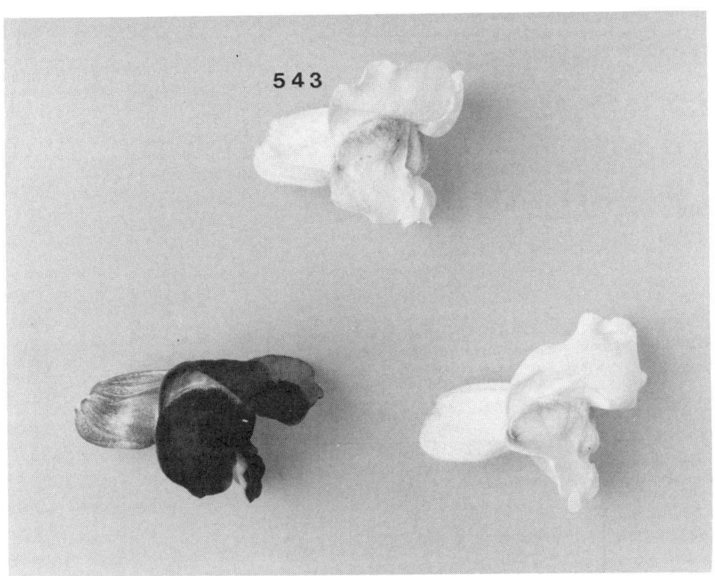

Fig. 5. The allele niv-543 was derived from niv^{rec}-98 and shows a changed recurrens phenotype together with stable spatial patterning. Two different phenotypes were isolated from this allele. On the left is the fully pigmented wild-type revertant presumably caused by imprecise excision of Tam3. The flower on the right shows only the stable pattern; no somatic excision sites have been observed.

than that produced by Tam3. Fully pigmented, wild-type revertants have been obtained and shown by molecular analysis to be due to Tam1 excision (3). Stable alleles of reduced pigmentation have also been produced by Tam1 activity and are allelic to the niv gene (9). It has been demonstrated, however, that at least two of these alleles are not due to Tam1 excision. It is thought likely that Tam1 has become inactive because of a deletion of 5 bp of the inverted repeat at the end of the Tam1 element, distal to the gene (14). Altered unstable phenotypes include one showing sites on the periphery of the flower with an albino background (9). Almost all sites were single-celled and therefore indicate a change in the timing of excision. A complete null has also been isolated, but so far all alleles have been recessive.

Tam2 at the niv gene is normally a stable null. However, after activation of the transposon, full red revertants were isolated and shown to be due to imprecise excision of Tam2. Recently some palely pigmented, stable phenotypes have also been obtained. These have not yet been tested genetically, but preliminary molecular analysis indicates that there may be other examples of deletions within the element conferring an altered phenotype.

Transposon Tagging

The technique known as transposon tagging has already been used successfully in A. majus. This has enabled the isolation and subsequent molecular analysis of the pal gene and its many alleles. The strategy employed in the isolation of *pal* illustrates several of the principles involved in tagging genes in plants.

The highly unstable line pal^{rec}-2 reverts to wild type at a frequency of about 20% when parental plants are grown at 15°C. Some of these revertants when homozygous show a high degree of randomly distributed sites on the full red background which are either palely pigmented or, very occasionally, white. It was postulated that in these plants a transposable element had excised from the *pal* gene but was reinserting into genes involved in pigment biosynthesis to give pale sites when only one allele was affected, and very occasionally white sites where reinsertion had occurred in both alleles. This line was designated *PalTR-75* and was used in an extensive genetic program, crossing it to inbred genetically defined stocks. These stocks were homozygous for a known anthocyanin mutation, i.e., *nivea*, *incolorata*, or *pallida* or various combinations of the three. F_1 plants were of full red intensity with randomly distributed white sites indicating the reinsertion of a transposable element. After screening approximately 10,000 plants, one new unstable plant was isolated (10). This was shown to be due to an insertion of *nivea* (19) and was designated the niv^{rec}-98 line. The transposable element *Tam3* was isolated from this line by H. Sommer and his colleagues in Cologne. This was then used as a molecular probe to isolate the *Tam3* element in pal^{rec}-2 (18), and subsequently to analyze the *pallida* alleles.

The principles of this technique are now being used in an attempt to isolate genes other than those involved in pigment synthesis. Several genes in *Antirrhinum* behave in a similar way to the unstable alleles described above. Plants carrying the unstable $cycloidea^{radialis}$ (cyc^{rad}) have an alteration in the shape of the flower from zygomorphic to radially symmetrical. The allele reverts readily to wild-type and an allelic series exists at the *cyc* locus. The allele $deficiens^{globifera}$ (def^{gli}) results in sepals growing in place of petals, while the anthers have been lost and their filaments fused to form a pollen-receptive ring. There is an unstable form of this which occasionally reverts to almost wild-type, and an allelic series also exists. The behavior of both genes could well be due to the insertion and subsequent excision of a transposable element. To analyze these genes at a molecular level, a known active transposon needs to be inserted into the genes.

In order to address this problem, two different large-scale experiments are in progress. The first is very similar to that used previously for the production of niv^{rec}-98. Lines containing highly active transposable elements have been grown at 15°C, then crossed to cyc^{rad} and def^{gli} plants and the F_1 progeny screened. So far, out of about 9,000 plants grown, one possible *cycloidea* mutation has been isolated, backcrossed, and grown onto the F_2. However, the difficulty with this approach is distinguishing unequivocally, in the F_2, the newly derived *cyc* mutation from that of the original.

To overcome this difficulty and also to enable us to screen for any phenotypically expressed mutation that may be due to transposon activity, plants of highly transposing lines and niv^{rec}-98 were grown at 15°C and self-pollinated. The subsequent seed capsules were collected separately and sown to give 6,500 plants in the glasshouse. As expected, these plants were mainly wild-type, and, since most mutations are likely to be recessive, the plants were self-pollinated to give rise to 20,000 plants in the next generation.

So far we have isolated a number of mutations, some of which could be due to transposon insertions. From the PalTR line there have been two new possible insertions into pigment genes. Two separate mutations changed the habit of the plants, one of which has a rosette of basal leaves while the other is similar to that described by Stubbe as cupliformis (21). Stubbe also described an allelic series for this gene, and, in certain lines, back mutation to wild-type was recorded. Perhaps one of the most interesting mutations obtained, so far, is one in which the leaves are unstable with dark-green sites of a clonal pattern on a pale yellow/green background. Maybe Tam3 has integrated into a gene involved in biosynthesis of photosynthetic pigments, partially blocking gene function. When excision occurs, gene function is again restored and dark-green sites observed.

CONCLUSION

Clearly, the activity of transposable elements can have many diverse effects on the structure of a gene and its expression. It is unlikely that these phenomena are confined to genes responsible for anthocyanin synthesis. The reason these genes have been so intensely studied is that they are the easiest to identify phenotypically; even a somatic event at the last stage of flower development can be observed as a single-celled pigmented site. Many of the allelic series described for other loci in Antirrhinum could well be the result of transposon activity. We hope that the transposon tagging experiments now in progress will enable us to isolate other genes and that subsequent molecular analysis will lead to a greater understanding of developmental processes in plants.

ACKNOWLEDGEMENTS

We wish to thank Prof. D. A. Hopwood and Dr. J. Bollmann for critical reading of the manuscript, Mr. P. Scott and Mr. A. Davis for photography, and Mrs. A. Williams for typing the manuscript. We are grateful to the Glasshouse and Field Services Department for growing the plants and to Miss K. Blewett for summer vacational assistance. This work was supported by the Agricultural and Food Research Council via a grant in aid to the Institute of Plant Science Research, John Innes Institute. E.C. and R.C. also acknowledge a generous grant from the Gatsby Foundation for the current transposon tagging experiments.

REFERENCES

1. Baur, E. (1924) Untersuchungen über das Wesen, die Entstehung und die Vererbung von Rassenunterschieden bei Antirrhinum majus. Bibliotheca Genetica 4:1-70.
2. Bonas, U., H. Sommer, and H. Saedler (1984) The 17 kb Tam1 element of Antirrhinum majus induces a 3 bp duplication upon integration into the chalcone synthase gene. EMBO J. 13:1015-1019.
3. Bonas, U., H. Sommer, B.J. Harrison, and H. Saedler (1984) The transposable element Tam1 of Antirrhinum majus is 17 kb long. Mol. Gen. Genet. 194:138-143.
4. Brink, R.A., and I.M. Greenblatt (1954) Diffuse, A pattern gene in Zea mays. Heredity 45:47-50.

5. Carpenter, R., C. Martin, and E.S. Coen (1987) Comparison of genetic behaviour of the transposable element Tam3 at two unlinked pigment loci in Antirrhinum majus. Mol. Gen. Genet. 207:82-89.
6. Coen, E.S., R. Carpenter, and C. Martin (1986) Transposable elements generate novel spatial patterns of gene expression in Antirrhinum majus. Cell 47:285-296.
7. Fincham, J.R.S., and B.J. Harrison (1967) Instability at the Pal locus in Antirrhinum majus. II. Multiple alleles produced by mutation of one original unstable allele. Heredity 22:211-227.
8. Forkmann, G., and G. Stotz (1981) Genetic control of flavanone 3-hydroxylase activity and flavonoid 3'-hydroxylase activity in Antirrhinum majus (snapdragon). Z. Naturforsch. 36c:411-416.
9. Harrison, B.J., and R. Carpenter (1973) A comparison of the instabilities at the nivea and pallida loci in Antirrhinum majus. Heredity 31:309-323.
10. Harrison, B.J., and R. Carpenter (1979) Resurgence of genetic instability in Antirrhinum majus. Mutat. Res. 63:47-66.
11. Harrison, B.J., and J.R.S. Fincham (1964) Instability at the Pal locus in Antirrhinum majus. I. Effects of environment on frequencies of somatic and germinal mutation. Heredity 19:237-258.
12. Harrison, B.J., and J.R.S. Fincham (1968) Instability at the Pal locus in Antirrhinum majus. III. A gene controlling mutation frequency. Heredity 23:67-72.
13. Harrison, B.J., and R.G. Stickland (1980) Precursors and the genetic control of pigmentation. V. Initiation of anthocyanin synthesis in Antirrhinum majus by Botrytis cinerea. Heredity 44:103-109.
14. Hehl, R., H. Sommer, and H. Saedler (1987) Interaction between the Tam1 and Tam2 transposable elements of Antirrhinum majus. Mol. Gen. Genet. 207:47-53.
15. Hudson, A., R. Carpenter, and E.S. Coen (1987) De novo activation of the transposable element Tam2 of Antirrhinum majus. Mol. Gen. Genet. 207:54-57.
16. Kuckuck, H. (1936) Über vier neue Serien multipler Allele bei Antirrhinum majus. Z.f. indukt. Abst.-u. Verebungsl. 71:429-440.
17. Martin, C., R. Carpenter, E.S. Coen, and T. Gerats (1987) The control of floral pigmentation in Antirrhinum majus. In Developmental Mutants in Higher Plants, H. Thomas and D. Grierson, eds. Cambridge University Press, pp. 19-52.
18. Martin, C., R. Carpenter, H. Sommer, H. Saedler, and E.S. Coen (1985) Molecular analysis of instability in flower pigmentation of Antirrhinum majus, following isolation of the pallida locus by transposon tagging. EMBO J. 4:1625-1630.
19. Sommer, H., R. Carpenter, B.J. Harrison, and H. Saedler (1985) The transposable element Tam3 of Antirrhinum majus generates a novel type of sequence alterations upon excision. Mol. Gen. Genet. 199:225-231.
20. Spiribille, P., and G. Forkmann (1982) Genetic control of chalcone synthase activity in flowers of Antirrhinum majus. Phytochemistry 21:763-776.
21. Stubbe, H. (1966) Genetik und Zytologie von Antirrhinum 1. sect Antirrhinum. VEB, Gastav Fischer Verlag, Jena, G.D.R.
22. Upadhyaya, K.C., H. Sommer, E. Krebbers, and H. Saedler (1985) The paramutagenic line niv-44 has a 5kb insert, Tam2, in the chalcone synthase gene of Antirrhinum majus. Mol. Gen. Genet. 199:201-207.

RECOMBINANT MUTABLE ALLELES OF THE MAIZE R GENE

Jerry L. Kermicle

Laboratory of Genetics
University of Wisconsin
Madison, Wisconsin 53706

ABSTRACT

Three insertions of Dissociation (Ds) into an R allele that confers strong seed color (Sc) have been transferred by crossing over into a plant color allele (P). The recombinant products show partial to complete inhibition of plant pigmentation in the absence of Activator (Ac). The rank of pigmentation levels in the sheaths of seedling leaves and in anthers is the same for the (P) recombinants as for seed color in the original (Sc) mutables. There is one reversal of rank order in coleoptiles. In the presence of Ac, intensity of sectoring in anthers parallels that in seeds of the respective (Sc) counterparts. The recombinant (P) mutables originate from (Sc)Ds/(P) heterozygotes as a reciprocal product of crossing over that restores normal (Sc) phenotype. For r-m3, both products were recovered together following meiotic nondisjunction. The transfer of mutability to (P) and the restoration of (Sc) pigmentation were used as criteria to investigate the relationship of (Sc) and (P) with other R genic elements. Transfer and restoration experiments involving (Sc) and (P) have succeeded in every combination tested. Although functional (P) derivations were obtained from (P)m-1 heterozygotes involving the R elements (S) and (D), mutable forms of these elements were not obtained. Neither transfer nor restoration occurred when (P)m-1 was combined with R-cherry, indicating a third class of R allele organization.

INTRODUCTION

Insertion mutants have been used in genetic fine-structure analysis of various maize genes. Based on the phenotype of stained pollen grains, Nelson (11) was able to map insertions in waxy relative to other wx mutant sites, some of which later proved to be cryptic insertions (14). By incorporating flanking markers, Dooner (4) not only mapped insertions in the bronze-1 gene but was able to orient the gene within chromosome 9 and to localize sites affecting the level and thermal stability of the bronze enzyme. A variation of such tests serves to map insertions relative to components conferring tissue-specific properties to different alleles of R. In this case, mutations in a seed-color element (Sc) resulting from insertion of Dissociation are paired with a nonmutant allele conferring only plant color (P) to produce functional (Sc) products (9). The experiments with wx, bz1, and R share a common feature: restoration of nonmutant phenotype through intragenic recombination.

A second product of recombination has been recognized with the R materials, namely, transfer of the insertion and its mutant effect from (Sc) to (P). Such events have been reported for (Sc) mutables m-1 and m-9, two of three Ds mutables established in an initial screening of the progenitor allele R-sc:124. Both inhibit (Sc) expression incompletely. Recombinant (P)m-1 and (P)m-9 mutables were identified as seedlings in which plant coloration was inhibited only partially. When P-vv, containing Modulator, a member of the Activator family of transposable elements, was reintroduced into the genome containing the putative (P)Ds derivatives, sectoring for strong coloration appeared in seedling and mature plant parts ordinarily pigmented by (P). Thus, Ds, rather than a coincidental mutation in (P), was responsible for inhibited plant color.

The implication of this finding concerning functional homology between (Sc) and (P) has been discussed separately (10). To be considered here are three additional questions concerning the (P)Ds recombinant mutable alleles. First, does the mutation of (P) by transfer of Ds from (Sc) Ds and the crossover restitution of (Sc) expression from the same heterozygote originate as reciprocal products of given crossover events? Second, what is the phenotype of the recombinant mutables? Is the phenotype of a (P)Ds mutable predictable based on its (Sc)Ds progenitor, or are there unexpected interactions between the insertion and different tissue-specific components of R? And third, can Ds be transferred from (P)Ds to seed-color alleles other than (Sc), indicating comparative homology?

RECOVERY OF COMPLEMENTARY PRODUCTS OF Ds TRANSFER IN SINGLE MEIOTIC CELLS

Two (P)Ds derivatives of mutable-9 and four of mutable-1 were reported previously as exceptional seedlings isolated from (Sc)Ds/(P) testcrosses (9,10). Each was borne on a recombinant chromosome, carrying proximally the allele of golden-1 from the (P) chromosome but distally the allele of M-st from the (Sc)Ds homolog. This combination of flanking markers is reciprocal to that associated with (Sc) restitution from these heterozygotes. The two derivative classes are also complementary in the sense that stability vs instability in response to Ac has been reversed relative to the parental alleles. It seemed plausible, therefore, that the two might originate as products of a given recombinational event. If so, it should be possible to recover both in circumstances where two of the four meiotic chromatids, rather than only one, are represented in single spores. In addition, the presence of restored (Sc), whose seed phenotype is dominant to both parental alleles, should facilitate recovery of the cryptic product (P)Ds.

TB-10a, a translocation involving the long arm of chromosome 10 and the maize B chromosome, was chosen for the half-tetrad study. About two-thirds of the long arm of chromosome 10, including R, is joined to the centromeric portion of B to constitute the B^{10} element. The centromeric portion of chromosome 10 together with the distal part of B comprise the second element, 10^{B}. In plants heterozygous for the translocation, 10^{B} usually disjoins from standard chromosome 10, whereas B^{10} assorts independently, giving approximately equal numbers of four classes: 10, 10^B B^{10}, 10 B^{10}, and 10^B. Lethality of the 10^B class accounts for the 25% sterility associated with such heterozygotes (12). Among the viable classes, 10 B^{10} is of interest in the present connection because the R locus is represented twice. Figure 1 illustrates translocation B-10a, the linked markers employed, an intragenic crossover of the sort postulated to account for (Sc) and (P)Ds products of (Sc)Ds/(P) heterozygotes, and meiotic nondisjunction of chromosomes 10 and B^{10}.

A single ear progeny of T(Sc)m-3/N(P) testcrossed by R-g:8-pale yielded 39,020 kernel offspring, including 24 that were intensely pigmented, thereby resembling the R-sc:124 progenitor of (Sc)m-3. Remarkably, all 21 of these 24 that were progeny tested successfully carried (Sc) on a B^{10} chromosome marked by the proximal marker G derived from (Sc)m-3 but the distal marker m-st from (P). The incidence of this recombinant class, expressed in terms of the two-thirds proportion of viable gametes expected to carry (Sc)m-3, is 9.2×10^{-4}. This estimate compares with 9.0×10^{-4} reported for (Sc)m-3/(P) using standard chromosomes (5).

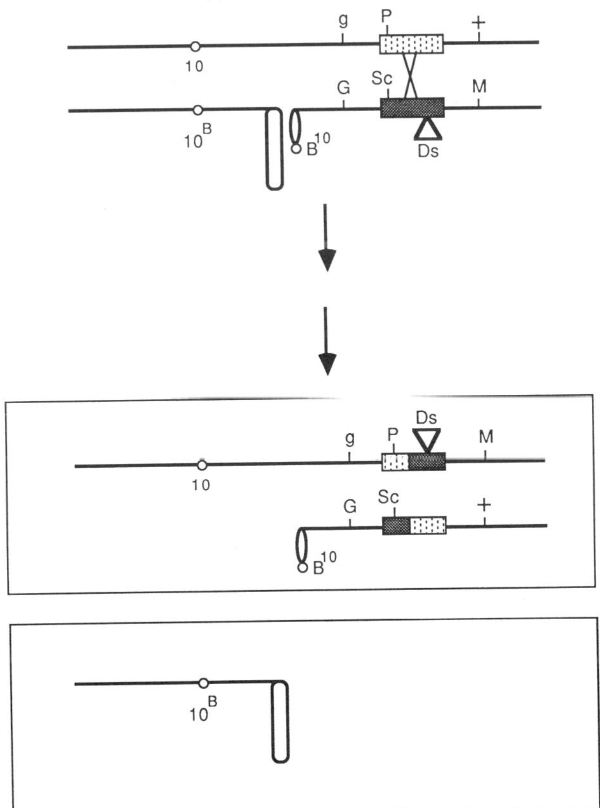

Fig. 1. Intragenic crossing over between (P) and (Sc)Ds with meiotic nondisjunction of chromosomes 10 and B^{10}. Complementary crossover products (Sc) and (P)Ds are delivered to the same meiotic cell. Only one of each pair of sister chromatids is shown.

Absence of an effect of the translocation on intragenic recombination is not unexpected, since the chromosome-10 breakpoint in TB-10a maps some 25 crossover units proximal to R (7).

Thirteen of the 21 exceptional (Sc) progeny were balanced TB-10a heterozygotes; eight were $10/10/B^{10}$ hyperploids, the class of potential interest. In five of these eight, the standard chromosome 10 derived from the test heterozygotes carried (P) and its parental combination of flanking markers, g m-st. As instances of recovery of a noncrossover chromatid in conjunction with crossover (Sc), this class also is not uniquely informative. The remaining three were g m-st recombinants, i.e., marked reciprocally relative to restituted (Sc). When crossed to an r-g P-vv stock, one ear segregated spotted kernels resembling the phenotype of parental (Sc)m-3. The remaining two gave colorless seed but frequent red sectors in seedling shoots and roots, as well as a vivid display of red mosaic anthers on mature plants, confirming the expectation for recombinant (P)Ds derivatives. Thus two positive outcomes occurred among eight eligible trials.

PHENOTYPE OF RECOMBINANT MUTABLE ALLELES

Recombinant (P)Ds strains, in comparison with parental (Sc)Ds, provide material for investigating whether insertions affect R expression differentially in various tissues. For a single insertion, the absolute level of pigmentation in vegetative tissues of recombinant (P)Ds plants cannot be predicted, of course, based on level of seed pigmentation in its (Sc)Ds counterpart. However, if the influence of Ds insertion on pigmentation in various plant parts is general, without unique interactions between given insertions and particular tissues, then the relative rank for a series of insertions and that of their (P)Ds recombinants should be the same.

Figure 2A pictures (Sc) and (P) counterparts of mutables m-1, m-3, and m-9 together with their progenitor alleles, designated as +, all strains lacking active Ac. Ranked from darkest to lightest, the order for seed color is + > m-9 > m-1 > m-3. The order is the same for the sheath of the first expanded seedling leaf. For anthers, the order is similar but the threshold of expression differs, with m-3 acyanic as well as m-1. Only in the coleoptile is there a reversal of rank order, with + > m-9, but m-3 > m-1. As a result, the sheath of m-1 seedlings is pigmented more darkly than the coleoptile, whereas the opposite is true of m-3.

A second set of recombinant (P) mutables representing m-1, m-3, and m-9 were examined to determine whether the unique rank among coleoptiles may have resulted from recombination within tissue-specific determiners of (Sc) and (P). Crossovers that result in loss or a different spectrum of tissue effects occur but are extremely rare (3,10). The rankings of coleoptiles and sheaths within the second set (not shown) were the same as those pictured in Fig. 2A. It is more likely, therefore, that Ds in insertion m-1 or m-3 influences expression in the coleoptile differently than in the other plant tissues studied.

In the presence of Ac (Fig. 2B), seed spotting of m-1 is light whereas m-3 is dense. Mutable-9 gives sectors of different intensities superimposed on a pale background. Sectoring in anthers is likewise light with m-1 and dense with m-3, whereas m-9 anthers show less well-defined sectors on a pale background. Particular characteristics of the insertions in

+

m 1

m 3

m 9

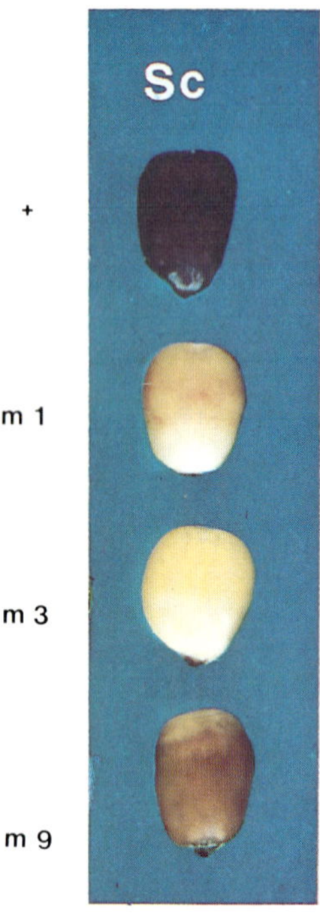

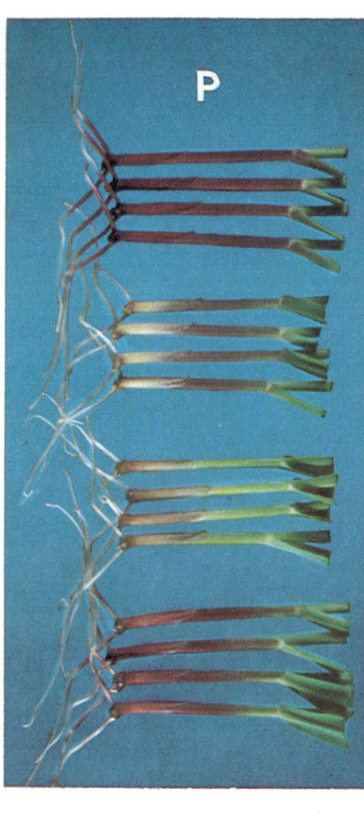

B

Ac present

+

m 1

m 3

m 9

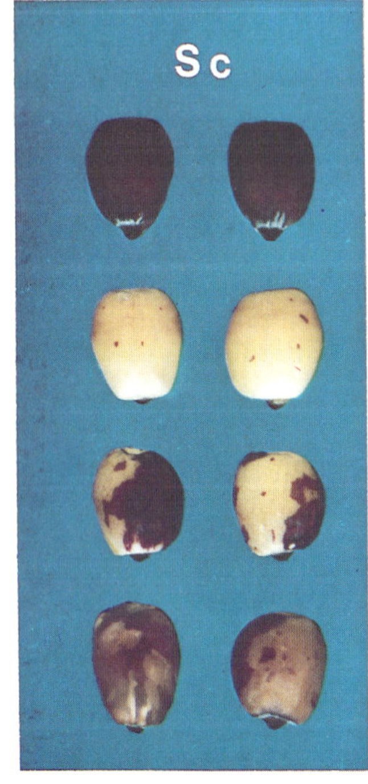

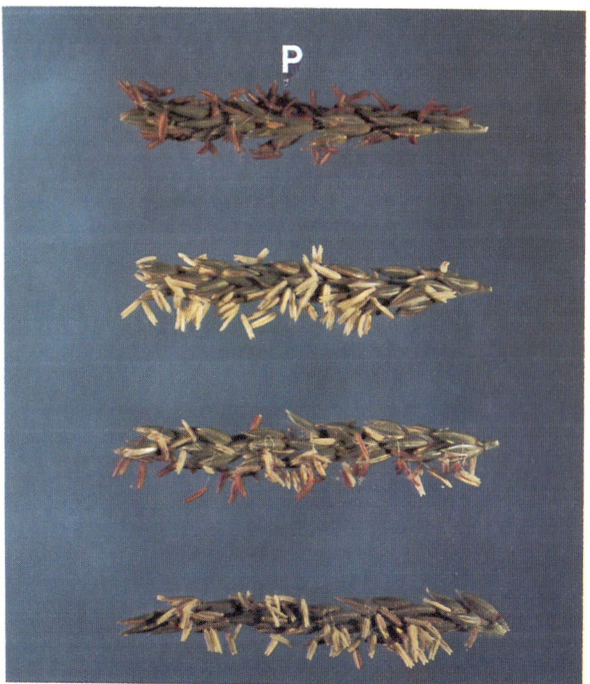

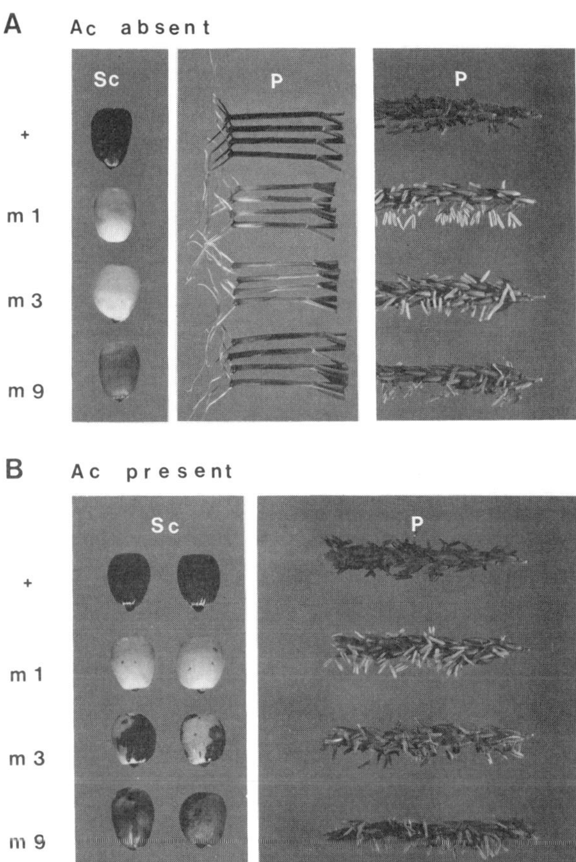

Fig. 2. Seed and anther pigmentation in progenitor (+) and r-mutable strains m-1, m-3, and m-9. The mutables were isolated as Ds variants of R-sc:124, and Ds was then transferred to (P) by crossing over. A: Stable phenotypes in the absence of Ac, showing different extents of partial expression (leakiness) in seeds, 11-day-old seedlings, and anthers. B: Mutable phenotypes in the presence of Ac. Mutable-1 and m-3, respectively, give sparse and dense sectoring in both seed and anthers. Mutable-9 gives spots of different intensities on a pale background. See facing.

this set are expressed in seed and anthers alike, as though reversion is not affected differentially in the two tissues. As was true for mutant phenotypes in the absence of Ac, the major source of variation is between insertions. Important determiners of phenotype presumably are type and orientation of Ds as well as site of insertion within R.

TRANSFERABILITY OF Ds AMONG R ALLELES

To what extent does the relationship between (Sc) and (P), indicated by the transferability of Ds between them, extend to other R elements?

Are other R elements organized such that Ds is transferred to them by crossing over, and is the effect of Ds parallel to that in (Sc) and (P)? Recombinant (P)Ds alleles are particularly suited for applying these questions to any colored seed allele. Consider the test heterozygote (P)Ds/R, where R represents generically any allele conferring seed color. Restitution of (P) function by the removal of Ds is recognized as a colored seedling produced in testcrosses conducted in the absence of Ac; transfer to R is detectable as a spotted kernel in testcrosses that incorporate Ac. The distinguishing phenotypes of R alleles that have been used for such tests are given in Tab. 1, listed together with (Sc) and (P).

Table 2 gives the incidence and linked marker composition of (P) restitution from (P)Ds when heterozygous with four R alleles. With (Sc), six red (P) seedlings, all proving to carry the distal marker of (Sc), occurred among a population of 5,050 seedlings grown from colorless kernels. This outcome represents resynthesis of one of the parental phenotypes used initially to derive (P)Ds. Similar frequencies of (P) restitution occurred in (P)Ds heterozygotes with (S) of R-g:1 and (D) of R-r:Catspaw. Again, all were borne on crossover chromosomes. No instance of functional (P) occurred, however, in parallel tests involving R-cherry.

In counterpart tests of Ds transfer from (P)Ds to R, 13 (Sc) mutables were established from a total kernel population of 18,130 (Tab. 3). Resynthesis of (Sc)Ds, the other parental allele used in (P)Ds derivation, provides a positive control for tests of (S), (D), and R-ch. Parallel tests involving these three alleles, however, gave altogether negative results. Thus, three different outcomes were observed with the four R alleles analyzed as (P)Ds heterozygotes: both (P) restitution and Ds transfer, involving (Sc); restitution of (P) but mutant transfer negative, involving (S) and (D); and neither restitution nor transfer, involving R-ch.

Positive outcomes to these tests suggest the following interpretation. With restitution, a segment from the R allele can substitute for (P) at the

Tab. 1. Source and principal distinguishing effects of certain R alleles.

R element or allele	Source	Pigment distribution: Kernel		Seedling		
		Aleurone (single dose)	Pericarp	Coleoptile	Roots	Anthers
(Sc)	R-stippled	Solid	-	+	-	-
(P)	R-r:standard	Colorless	-	+	+	+
(S)	R-r:standard	Mottled	-	-	-	-
(D)	R-r:Catspaw	Mottled	-	+	+	-
R-ch	via L.J. Stadler	Mottled	+	+	+	+

Tab. 2. Red plant exceptions isolated in the absence of Ac from colorless seed of the cross.

Test allele	Seedling population	Frequency (x 10^{-4})	Linked marker composition + +	g M	+ M	g +
(Sc) = R-sc:N1-571-1	5,050	11.8	0	0	0	6
(S) = R-g:1	3,580	11.2	0	0	0	4
(D) = R-r:Catspaw	8,330	8.4	0	0	0	7
R-ch:Stadler	9,250	nil	0	0	0	0

R represents any one of the four test alleles.

insertion site and distally in the gene, restoring (P) phenotype. With Ds transfer, conversely, a homologous segment including Ds effectively mutagenizes the R allele. Positive results imply functional equivalence, at the phenotypic level, of the segments concerned.

Negative results indicate differences in allelic organization and are, therefore, of particular interest in a comparative study. But because the outcome is negative, more than one interpretation usually will be possible. Consider first the situation where restitution succeeded but transfer failed, as for (P)Ds combinations with (S) and (D). Perhaps these alleles embody multiple copies of R organized such that genetic evidence of direct duplication is not evident (1). If so, Ds could be transferred from (P)Ds to one member without blocking expression of other copies. Alternatively, sites critical to the determination of (S) and (D) may lie on both sides of Ds in m-1 such that a single crossover would not incorporate Ds between them. Differential processing of the message in the region of Ds insertion in the tissues expressing (P) as compared with seed is another alternative. Still other possibilities pertain to R-ch, where neither crossover restitution

Tab. 3. Spotted kernels obtained following testcrosses with r-r r-r of the same R/(P)m-1 heterozygotes reported in Tab. 2. Here, Ac (as P-vv) was introduced with the r-r tester.

Test allele	Kernel population	Frequency (x 10^{-4})	Distal marker +	M
(SC) = R-sc:N1-571-1	18,130	7.1	0	13
(S) = R-g:1	15,170	nil	0	0
(D) = R-r:Catspaw	26,400	nil	0	0
R-ch:Stadler	28,310	nil	0	0

of (P) nor Ds transfer to R was observed. This would result if Ds were inserted in or between sites that are essential in determining (P) as opposed to the seed color effect of R-ch. Another possibility is that no recombination occurs in the segment between Ds and a site determining tissue-specific difference between (P) and R-ch. Or, if it does, it is possible that single crossovers are not recovered, such as if R-ch were inverted relative to (P).

The possibility of (Sc) restitution in R-ch heterozygotes was tested on a larger scale using the (Sc)Ds form of mutable-3. In order to distinguish solidly colored kernels due to (Sc) restitution from kernels of parental R-ch phenotype, a dominant factor (M-ch) that selectively inhibits R-ch pigmentation was incorporated into the testcross parent (13). Thus, parental classes of testcross kernels were either colorless or reduced to a low level by the inhibitor. Seven instances of (Sc) restitution occurred in a population of 65,800 kernels (Tab. 4). Not only is the incidence of (Sc) restitution reduced some four-fold, but the flanking marker distribution differs markedly from that obtained in combination with (P). Whereas all 12 (Sc)'s from (Sc)Ds/(P) were recombinant, carrying the distal marker from (P), all seven from (Sc)Ds/R-ch carry the distal marker of (Sc)Ds. Six of the seven also carry the proximal marker of (Sc)Ds, making them noncrossover type. Such cases could originate either as double crossovers or by gene conversion. Because instances of (Sc) restitution were not obtained in testcrosses of (Sc)m-3 homozygotes (9), origin of this class by transposition seems unlikely.

The present evidence based on Ds mutables of simplex alleles clarifies the nature of certain unstable recombinants isolated from heterozygotes involving the R-stippled complex. From R-st combinations with R-r: standard, a compound allele carrying separable plant and seed color functions, several investigators have recovered a colorless-seed but green-plant recombinant class that reverts to red (reviewed in Ref. 2).

Tab. 4. Self-colored (R-sc) kernels obtained following r-g r-g testcrosses of (Sc)m-3 heterozygotes with R-ch and (P). Tests made in the absence of active Ac.

Test heterozygote	Kernel population	R-sc frequency (x 10^{-4})	Linked marker composition 1a 2a	1b 2b	1a 2b	1b 2a
1a* R-ch 2a / 1b (Sc)m-3 2b	65,800	1.1	0	6	1	0
1a (P) 2a / 1b (Sc)m-3 2b	26,700**	4.5	0	0	0	12

* 1a and 1b designate alleles of the golden-1 locus; similarly, 2a and 2b indicate alleles of M-st.

** From Dooner and Kermicle (5).

Likewise, from combinations of R-st with R-navajo (R-nj), presumptive complementary crossover products were isolated and then combined to resynthesize parental R-st and R-nj (6,8). Subsequently, the instability factor (I-R) of R-stippled was demonstrated to be transposable (15). Analysis of I-R transfer between alleles by crossing over is complicated both by its autonomous transposition, giving revertants of the same phenotype as one recombination class, and by the fact that R-st carries more than one genic element.

ACKNOWLEDGEMENTS

Paper No. 2953 from the Laboratory of Genetics. I thank Charles Harrison, David Heller, Beverly Oashgar, and Nola Peterson for technical assistance, and Mary Alleman for stimulating discussions and for preparing Figure 1. Studies conducted at the Wisconsin Agricultural Experiment Station with support from the U.S. Department of Energy (FG02-86ER13539) and the National Science Foundation (PCM-8209036).

REFERENCES

1. Bray, R.A., and R.A. Brink (1966) Mutation and paramutation at the R locus in maize. Genetics 54:137-149.
2. Brink, R.A. (1973) Paramutation. Ann. Rev. Genet. 7:129-152.
3. Dooner, H.K. (1979) Identification of an R-locus region that controls the tissue specificity of anthocyanin formation in maize. Genetics 93:703-710.
4. Dooner, H.K. (1986) Genetic fine structure of the bronze locus in maize. Genetics 113:1021-1036.
5. Donner, H.K., and J.L. Kermicle (1986) The transposable element Ds affects the pattern of intragenic recombination at the bz and R loci in maize. Genetics 113:135-143.
6. Kermicle, J.L. (1970) Somatic and meiotic instability of R-stippled, an aleurone spotting factor in maize. Genetics 64:247-258.
7. Kermicle, J.L. (1970) Dependence of the R-mottled aleurone phenotype in maize on mode of sexual transmission. Genetics 66:69-85.
8. Kermicle, J.L. (1973) Organization of paramutational components of the R locus in maize. Brookhaven Symposium in Biology 25:262-280.
9. Kermicle, J.L. (1980) Probing the component structure of a maize gene with transposable elements. Science 208:1457-1459.
10. Kermicle, J.L. (1985) Alternative tests of allelism. In Plant Genetics, M. Freeling, ed. Alan R. Liss, Inc., New York, pp. 491-507.
11. Nelson, O.E. (1986) The waxy locus in maize. II. The location of the controlling element alleles. Genetics 60:507-524.
12. Roman, H. (1947) Mitotic nondisjunction in the case of interchanges involving the B-type chromosome in maize. Genetics 32:391-409.
13. Van der Walt, W.J. (1967) Paramutation of R alleles of diverse geographic origin. Ph.D. Thesis, University of Wisconsin, Madison.
14. Wessler, S., and M.J. Varagona (1985) Molecular basis of mutations at the waxy locus of maize. Correlation with the fine structure map. Proc. Natl. Acad. Sci., USA 82:4177-4181.
15. Williams, W.M., K.V. Satyanarayana, and J.L. Kermicle (1984) R-stippled maize as a transposable element system. Genetics 107:477-488.

STUDIES ON TRANSPOSABLE ELEMENT *Ac* OF *ZEA MAYS*

Peter Starlinger,[1] Barbara Baker,[2]
George Coupland,[1] Reinhard Kunze,[1]
Jürgen Laufs,[1] Jeff Schell,[3] Ursula Stochaj[4]

[1]Institut für Genetik
Universitat zu Köln
D-5000 Köln 41
Federal Republic of Germany

[2]United States Department of Agriculture
ARS Plant Gene Expression Center
Albany, California

[3]MPI für Züchtungsforschung
D-5000 Köln 30
Federal Republic of Germany

[4]Institut für Anatomie
Universität Marburg
D-3550 Marburg
Federal Republic of Germany

ABTRACT

Transposable element *Activator* (*Ac*) of *Zea mays* and several of its derivatives, called *Dissociation* (*Ds*), have been identified and studied both genetically and physiologically by B. McClintock (17,18). *Ac* has been cloned and sequenced (3,10,21,22), as have several *Ds* elements (4,5,11, 19,22,26).

For understanding transposition of *Ac*, both the transposition mechanism and its regulation must be studied. From the genetic experiments by McClintock (17), it is known that at least some of the functions necessary for transposition are encoded by *Ac*. Transposition of both *Ac* and *Ds* elements is linked to the presence of the *Ac* element, and no mutants in other loci are known to abolish this process. In addition, an *Ac*-dependent regulation of transposition is indicated by the fact that an increase in the dosage of *Ac* decreases the frequency of transposition events, and delays them to later times during endosperm development (15).

In order to study transposition and its regulation, we have investigated transcription of the Ac element and have begun studies of its translation (13). We have also introduced mutations into the cloned Ac element in order to test their influence on biological functions. Because efficiently reintroducing the Ac element into Zea mays is not yet possible, we made use of transferring Ac into tobacco via the Ti-plasmid of Agrobacterium tumefaciens. Baker et al. (1) have shown that Ac is able to transpose in tobacco. A phenotypic assay for the excision of Ac from its position in the T-DNA has been established by Baker et al. (Ref. 2, and this Volume), and was used for these experiments.

THE TRANCRIPTION OF Ac

The detection of Ac transcripts in northern experiments is complicated because all maize lines used carry more than 30 copies of sequences hybridizing to an Ac probe. Some of them are transcribed, as seen by subjecting poly(A) RNA from Ac-free maize lines to a northern analysis.

It was necessary, therefore, to distinguish Ac transcripts from the transcripts of Ds elements by identifying those that are present in all Ac-carrying strains and absent from all Ac-free strains. Only one transcript, approximately 3.5 kb in length, fulfilled these prerequisites. All other transcripts vary from strain to strain, and their appearance is not limited to Ac-carrying strains. In this part of the study, seven different maize lines carrying Ac were used. They carried Ac in at least four different positions on different chromosomes. We are confident, therefore, that only the 3.5-kb transcript is transcribed from Ac. The length of 3.5 kb shows that the major part of the 4.6-kb Ac element is spanned by the transcript. This is described in more detail below.

The amount of transcript was low and detection was easy when single-stranded DNA probes, labeled to a specific activity of more than 10^9 cpm/μg, were used. The fraction of Ac-RNA in the poly(A) RNA was estimated by comparing the signal strength of the RNA to that of appropriate dilutions of denatured plasmid DNA. We found the concentration of Ac-mRNA to be 1-3 x 10^{-7} of the poly(A) RNA preparations used, but have had to correct this slightly for the unknown amount of contaminating rRNA in our preparations.

Assuming that the amount of RNA in a maize cell is in the order of 0.16 pg, as found for Tradescantia palludosa (28), this amounts to far less than one Ac-mRNA molecule per cell. An approximate estimate, which is not too accurate, yields on the average one mRNA molecule per one in 20 to one in 100 cells. It is thus conceivable that the low frequency of transposition that occurs, in considerably less than every cell, is explained, in part, by the low rate of transcription. It would be interesting to know whether the amount of mRNA molecules is larger in those cells that are transcribed at all, or whether transcription is random, with Poissonian distribution of mRNA molecules per cell. Investigation of this point needs in situ hybridizations. We have not yet reached the necessary sensitivity for these experiments.

We have compared the amount of Ac-mRNA in different tissues, and found expression in all tissues investigated. The amounts were comparable and varied within a factor of approximately four. At this point, we are not sure whether the differences found are due to different rates of transcription or to experimental variations. We do not have a calibration probe for a gene which we know is equally well transcribed in all tissues.

Such a comparison should be possible if identical tissues of identical developmental stage are investigated. In this case, hybridization to a probe for a housekeeping gene (glyceraldehyde-1-phosphate dehydrogenase) could ascertain that the amount of RNA used was comparable. If this were done, we would be able to compare sporophytic tissues, carrying one or two copies of Ac, or endosperms carrying either two or three copies of Ac. One copy of Ac was not used under these conditions because: in the case of one dose of Ac brought in by the pollen, only the zygotic and endosperm tissue could contribute to the Ac transcript; while in the case of two or three doses, which have to come from the female side, remaining maternal tissues would equally contribute to the transcription of Ac.

When pairs of RNAs were compared, the procedure was not absolutely reproducible because the comparison relied upon the strength of hybridization signals in adjacent lanes. Therefore, these experiments were done repeatedly. In 11 of 13 experiments, the signal obtained from the RNA of the plant carrying the higher dose of Ac was stronger than the signal obtained from the RNA of the plant carrying the lower dose. We conclude that there is no negative dose effect at this level of transcription and that there seems to be an increase in the amount of transcription with Ac dose.

If the amount of mRNA is low, the Ac elements in most cells are not transcribed at a given time. If, on the other hand, a negative dosage effect is observed, both copies of the Ac element must be active in the cells carrying two of them. Why this should be so is unknown at present. McClintock observed that two Ds elements, inserted at different sites within the genome, are often excised from these sites simultaneously (16). This observation is another indication that several Ac-dependent events tend to occur in the same cell.

A HYPOTHESIS FOR THE NEGATIVE DOSAGE EFFECT

If the amount of RNA increases with the Ac dose, the amount of protein may also increase, though this has not yet been shown. How can this explain the negative dosage effect? As only one transcript has been identified so far, the idea of more than one protein is not easy to conceive. The negative dosage effect might be explained thusly: Transposition may be initiated by the binding of a single transposase molecule to both termini of the transposable element, which, by this binding, are brought together in a transposition complex. If the inverted repeats provide identical binding sites for the transposase at both termini, the active transposase molecule may be a homodimer. An increase in the concentration of transposase dimers might lead to the binding of one dimer each to both ends of the transposable element. If these dimer molecules have no affinity to each other, they might not bring the transposable elements into close vicinity. This process may thus be inhibitory to transposition.

THE STRUCTURE OF THE Ac GENE

The structure of the Ac gene and the sequences of Ac transcribed into RNA were determined both by isolating and sequencing of cDNA clones, derived from the mRNA, and by S1 protection experiments and DNA-RNA hybridization experiments with small probes in order to determine the 5'-end. Three overlapping cDNA clones were isolated from a library of 1.5 x 10^6 insert-carrying lambda phages. It had to be shown that these were derived from the 3.5-kb mRNA and not from other transcripts present in the cells. As the transcripts detected in a northern experiment (besides the 3.5-kb Ac transcripts) have a size of 1.2 kb or smaller, only those cDNA clones that carried an insert larger than 1.2 kb were used for further study. Three of these clones were sequenced and overlapped considerably. This made it likely that they were derived from the same messenger molecule, presumably the most abundant detected by northern experiments. The sequence determination showed four introns present in the gene that were missing in the mRNA. When genomic subclones were hybridized to poly(A) RNA, they were protected from S1 digestion and, thus, were colinear with the major transcript except for the introns. We concluded, therefore, that they are derived from the 3.5-kb mRNA. The combined length of the introns is 654 bp.

The cDNA clones cover a region from a poly(A) addition site 265 bp inside one end of Ac to a position deduced to be approximately 300 bp downstream from the start of the transcript. As we were not able to isolate a cDNA clone extending to the 5'-end, the transcription start was estimated by hybridization of the RNA transcript to small single-stranded probes. By this method, the transcription start was located in the interval between base pairs 4,165 and 4,283. A more exact determination was done by S1 mapping. This showed not one but several transcription starts that were reproducibly obtained and spread over a distance of approximately 100 bp. The most prominent starting point was located at position 4,230.

Upstream of this region, no TATA box or CAAT box was seen in the DNA sequence. This is reminiscent of several housekeeping genes, including the gene for the mouse dihydrofolate reductase (9,24). Using various controls, we came to the conclusion that the ends fully reflect differences in length of the 5'-ends of the mRNA molecules, rather than different splice sites. Though we cannot exclude that the RNA molecules have only one start point and are digested to particular positions from the 5'-end, it is more straightforward to assume that transcription starts at several points.

RNA isolated from transformed tobacco plants (see below) showed the same pattern of transcription starts, but the relative intensities of these bands were different than in maize. We do not yet know the basis for this finding.

The mRNA thus found consists of a leader sequence approximately 650 bp in length, followed by an open reading frame encoding 807 amino acids. An untranslated 3'-end of 239 bases follows. The leader sequence is free of ATG-codons. The first ATG does not conform to the consensus sequences described by Kozak (12) and Lütcke et al. (14), as it has neither a G at position +4, nor a purine in position -3. We tried to determine whether this ATG is still used as a translation start. For this purpose, an internal fragment of the cDNA, containing the first ATG, was cloned behind an SP6 promoter and transcribed in vitro. The RNA obtained was translated in vitro in a wheat germ system. The largest and most strongly labeled band was eluted and radiosequenced after labeling with [^{35}S]methionine and [^{3}H]proline. The pattern of radioactivity obtained showed unequivocally that the methionine was cleaved off and that the following sequence was that of the N-terminus of the open reading frame, implying that the first ATG was being used.

EXPERIMENTS WITH MUTATIONS PRODUCED IN VITRO

The following experiments were done by producing mutations of the *Ac* sequence by in vitro methods (removal of restriction sites or restriction fragments, digestion with *Bal*31, or use of a cDNA clone behind a T-DNA promoter). To test the ability of *Ac* to excise from a position in the leader sequence of a chimeric NPTII gene, the assay described by Baker et al. (2) was used. The constructs were transferred into an *A. tumefaciens* Ti-plasmid (29) and subsequently into tobacco protoplasts. Kanamycin-resistant (kan^r) calli emerging from these co-cultivation experiments are an indication of the excision of *Ac*.

If passive transposability of an element was to be tested, the T-DNA carrying this construct in the leader sequence of the NPTII gene was introduced into protoplasts that were stably transformed by *Ac*. The activity of this *Ac* element could be shown by its ability to excise a transposable *Ds* element from the identical position within the NPTII gene.

In order to determine whether an *Ac* derivative was capable of expressing the transposase, a second modification of the assay was designed. A tobacco strain was constructed that was stably transformed with a construct containing a *Ds* element inserted within the NPTII gene. It was reasoned that, in the presence of the transposase, this element would be excised, resulting in a kan^r phenotype.

EXPERIMENTS WITH cDNA

As described above, only a single *Ac* transcript has been identified. Because several functions are encoded by *Ac*, the question posed is whether this transcript gives rise to only one protein. Different proteins created by differential splicing are conceivable. In order to investigate this possibility, a cDNA fragment containing the whole open reading frame and 420 bp of untranslated leader was inserted downstream of the plant promoter P2' (27) and upstream of a 3' polyadenylation signal in the binary vector pPCV720 (C. Koncz, unpubl. data). This plasmid was then transferred to tobacco protoplasts previously transformed with a *Ds* element inserted into the NPTII gene. These experiments are still in progress, but it is already clear that this construct provides a function necessary for transposition.

(a) Introduction of this construct into tobacco protoplasts yields kan^r clones. If the cDNA is cloned into the construct in the opposite orientation, no kan^r clones are observed. As the number of resistant clones is low, the cDNA has possibly reduced activity. But, this quantitative estimate is not considered to be definite. The amount of RNA must be measured because, due to the use of P2' instead of the *Ac* promoter, the amount of RNA should be more abundant than the proper *Ac* transcript. Does this decrease transposition frequency?

(b) In addition to the cDNA, the construct used carried a gene for hygromycin resistance. It was thus possible to isolate clones that received the construct on the basis of hygromycin resistance. Four of the six calli derived from hygromycin-resistant clones contained detectable NPTII activity, if the Ac-cDNA was in the correct orientation. No NPTII activity could be detected in six hygromycin-resistant clones derived from the construct carrying the cDNA of Ac in reverse orientation. Though these experiments are not yet complete with regard to the DNA characterization of the transformed clones, they are highly suggestive of a capability of our cDNA to give rise to the product(s) necessary for transposition.

This experiment narrows down the possibility for the production of several protein products. The possibility that a minority of primary transcripts are spliced further to remove one or part of the exons is not excluded. However, it will have to be excluded that a function missing from the cDNA clone is provided by complementation with the artificial Ds element used for the test. At least in the case of the Ds element in the bz-m2(DI) allele (8), the element is deficient both in transposition and in contribution to the dosage effect. Nor is it excluded that different functional proteins are formed by post-translational processing. The protein sequence derived from a cDNA clone carries a segment of ten repeats of the dipeptide pro-gln. Because not all of the codons encoding the gln residues are identical, it is conceivable that this sequence is under selective pressure. Similar sequences have been found in the hobo transposon in Drosophila (25) that also shares with Ac the duplication of 8 bp flanking the transposon and some of the 11-bp inverted repeat sequence. In this case, however, the pro-rich region is formed by the repetition of the tripeptide thr-pro-glu. Pro-glu repetitions are also found in the E1A protein of adenovirus 2 (20) and in a sporozoite protein of Trypanosoma brucei (23). Is it conceivable that these regions, which due to the abundance of proline are probably rather unstructured, are a point for the attack via protease? This possibly is presently under investigation.

Taken together, the present experiments point in the direction of only one translation product of Ac that may be multifunctional, either in its native conformation or after post-translational processing.

THE TRANS-ACTING FUNCTIONS OF THE Ac PROTEIN

Removal of the N-terminus, including the first two ATG codons, inactivates transposition. The same result is obtained when two amino acids are introduced into the open reading frame by filling in and religation of the AccI site at position 3,511. This strengthens the conclusion of the in vitro radiosequencing experiment: that is, the first ATG of the open reading frame is the N-terminus of the Ac protein.

Removal of approximately 600 bp of the 650-bp leader does not abolish transposition. This shows that the leader sequence is not necessary. Our experiments are not yet sensitive enough to exclude small changes in transposition frequency. The absence of ATG codons in all three reading frames from a leader of 650-bp length is suggestive of selection against the presence of ATG. The retention of the long leader may then indicate that it has at least some function.

EXPERIMENTS ON CIS-ACTING FUNCTIONS

Removal of one terminus extending from the last base pair (4,563) to base pair 4,471 abolishes transposability of Ac, though this construct is capable of expressing transposase. This was shown in a control where the recipient protoplasts were stably transformed by an NPT gene carrying a Ds element in the leader sequence. To our surprise, however, internal deletions, extending from base pair 4,518 to base pair 4,771 and from base pair 4,486 to 4,382, had the same result. These deletions start well inside the 11-bp repeat that is thought to play a role in transposition. It must be concluded that sequences located inside of the inverted repeats are necessary for the transposability of Ac elements.

This result is not easy to understand, if we try to envisage the interaction of a transposase molecule with the termini of the transposable element. It is commonly thought that the presence of inverted repeats at the termini of transposable elements indicates an equivalent binding of transposase molecules to these two termini in order to ensure an equivalent cleavage of DNA sequences right at the terminus, or more probably at a given distance from this terminus. If the binding to the inverted termini is the only interaction between a hypothetical transposase and a transposable element, no problems arise. If, however, a sequence inside of the inverted terminus is necessary for transposition, it is very likely that this sequence is somehow interacting with the transposase. As the sequences inside of the transposable element differ from each other, the transposase molecule may have difficulty interacting with different internal sequences. If the transposase molecules bind to the identical termini and face the internal sequences of Ds with the same residues, it is not clear how these could recognize different DNA sequences, unless parts of the transposase can be moved relative to each other. It could be argued that inverted repeats are present inside of the terminal inverted repeats. Such repeats were identified, though not in equivalent positions (21). However, the differences in the distances between the components of the internal inverted repeats at the two ends are too large to allow an equivalent binding of a protein molecule. Also, they are probably too small to allow the alignment of sequences in different places on one protein molecule by looping without denaturation of the DNA. Of course, we do not know whether the DNA is denatured in the process of transposition.

Alternatively, it might be possible that an inside sequence at the BamHI end of Ac is necessary for transposition, but that, at the other end, the presence of the inverted repeat is sufficient. This is being tested by the construction of internal deletions at this end. It is also being tested by the construction of a transposable element in which one terminus of the element carrying sequences inside of the inverted repeats is replaced by the other terminus. In the case of Ds elements, indirect evidence for the involvement of both a left and a right terminus of Ac or Ds in transposition is derived from the study of double Ds elements (6) or from derivatives thereof (7). In these cases, different structures carrying inverted repeats at their termini were present. Some of these carried a left and a right end as defined by the sequence of Ac. Others carried two left ends. In these cases, the transposition events detected were exclusively structures carrying a left and a right end. This is also compatible with the idea that sequences inside both of the left ends of the right terminus are necessary for transposition.

The situation is complicated by the fact that the Ds1 element (26) is efficiently transposed by the Ac element. The Ds1 element shares with Ac only 13 and 19 bp, respectively, at the two termini, but the internal sequences deviate strongly from Ac. The sequences present inside of the left end and inside of the right terminus of Ds1 are quite distinct from those of Ac.

We are presently doing experiments using chimeric transposable elements in order to substantiate these findings. At present, it is hard to conclude that transposase molecules interacting with both termini, in most probably identical binding sites, must interact with different binding sites at a short distance inside of the termini of the transposable element. It is hard to see how this would be possible without distortions of either the protein or the DNA or both. Such conformation changes may be necessary to initiate the transposition process.

ACKNOWLEDGEMENTS

This work was supported by Deutsche Forschungsgemeinschaft through SFB 74. G.C. received fellowships from the Royal Society and EMBO in different years.

REFERENCES

1. Baker, B., J. Schell, H. Lörz, and N. Fedoroff (1986) Transposition of the maize controlling element "Activator" in tobacco. Proc. Natl. Acad. Sci., USA 83:4844-4848.
2. Baker, B., G. Coupland, N. Fedoroff, P. Starlinger, and J. Schell (1987) Phenotypic assay for excision of the maize controlling element Ac in tobacco. EMBO J. 6:1547-1554.
3. Behrens, U., N. Fedoroff, A. Laird, M. Müller-Neumann, P. Starlinger, and J. Yoder (1984) Cloning of the Zea mays controlling element Ac from the wx-m7 allele. Molec. Gen. Genet. 194:346-347.
4. Courage-Tebbe, U., H.-P. Döring, N. Fedoroff, and P. Starlinger (1983) The controlling element Ds at the Shrunken locus in Zea mays: Structure of the unstable sh-m5933 allele and several revertants. Cell 34:383-393.
5. Döring, H.-P., E. Tillmann, and P. Starlinger (1984) DNA sequence of the maize transposable element Dissociation. Nature 307:127-130.
6. Döring, H.-P., M. Freeling, S. Hake, M.A. Johns, R. Kunze, A. Merckelbach, F. Salamini, and P. Starlinger (1984) A Ds mutation of the Adh1 gene in Zea mays L. Molec. Gen. Genet. 193:199-204.
7. Döring, H.-P., R. Garber, B. Nelsen, and E. Tillmann (1985) Transposable element Ds and chromosomal rearrangements. In Plant Genetics, M. Freeling, ed. Alan R. Liss, Inc., New York, pp. 355-367.
8. Dooner, H., J. English, E. Ralston, and E. Weck (1986) A single genetic unit specifies two transposition functions in the maize element Activator. Science 234:210-211.
9. Dynan, W.S. (1986) Promoters for housekeeping genes. Trends in Genetics (in press).
10. Fedoroff, N., S. Wessler, and M. Shure (1983) Isolation of the transposable maize controlling elements Ac and Ds. Cell 35:235-242.
11. Geiser, M., E. Weck, H.-P. Döring, W. Werr, U. Courage-Tebbe, E. Tillmann, and P. Starlinger (1982) Genomic clones of a wild-type

allele and a transposable element-induced mutant allele of the sucrose synthase gene of Zea mays L. EMBO J. 1:1455-1460.
12. Kozak, M. (1986) Point mutations define a sequence flanking the AUG initiator codon that modulates translation by eukaryotic ribosomes. Cell 44:283-292.
13. Kunze, R., U. Stochaj, J. Laufs, and P. Starlinger (1987) Transcription of transposable element Activator (Ac) of Zea mays L. EMBO J. 6:1555-1563.
14. Lütcke, H.A., K.C. Chow, F.S. Mickel, K.A. Moss, H.F. Kern, and G.A. Scheele (1987) Selection of AUG initiation codons differs in plants and animals. EMBO J. 6:43-48.
15. McClintock, B. (1948) Mutable loci in maize. Carnegie Institution of Washington Yearbook 46:146-152.
16. McClintock, B. (1949) Mutable loci in maize. Carnegie Institution of Washington Yearbook 48:142-154.
17. McClintock, B. (1951) Mutable loci in maize. Carnegie Institution of Washington Yearbook 50:174-181.
18. McClintock, B. (1965) The control of gene action in maize. Brookhaven Symposium on Biology 18:162-184.
19. Merckelbach, A., H.-P. Döring, and P. Starlinger (1986) The aberrant Ds element in the adhl-2F11::Ds2 allele. Maydica 31:109-122.
20. Moran, E., and M.B. Mathews (1987) Multiple functional domains in the Adenovirus E1A gene. Cell 48:177-178.
21. Müller-Neumann, M., J.I. Yoder, and P. Starlinger (1984) The DNA sequence of the transposable element Ac of Zea mays L. Molec. Gen. Genet. 198:19-24.
22. Pohlman, R.F., N. Fedoroff, and J. Messing (1984) The nucleotide sequence of the maize controlling element Activator. Cell 37:635-643.
23. Roditi, I., M. Carrington, and M. Turner (1987) Expression of a polypeptide containing a dipeptide repeat is confined to the insect stage of Trypanosoma brucei. Nature 325:272-274.
24. Sazer, S., and T.S. Schimke (1986) A re-examination of the 5' termini of mouse dihydrofolate reductase RNA. J. Biol. Chem. 261:4685-4690.
25. Streck, R.D., J.E. MacGaffey, and S.K. Beckendorf (1986) The structure of hobo transposable elements and their insertion sites. EMBO J. 5:3615-3623.
26. Sutton, W.D., W.L. Gerlach, D. Schwartz, and W.J. Peacock (1984) Molecular analysis of Ds controlling element mutations at the Adhl locus of maize. Science 223:1265-1268.
27. Velten, J., L. Velten, R. Hain, and J. Schell (1984) Isolation of dual plant promoter fragments from the Ti-plasmid of Agrobacterium tumefaciens. EMBO J. 3:2723-2730.
28. Willing, R.P., and J.P. Mascarenhas (1984) Analysis of the complexity and diversity of mRNAs from pollen and shoots of Tradescantia. Plant Physiol. 75:865-868.
29. Zambryski, P., H. Joos, C. Genetello, J. Leemans, M. von Montagu, and J. Schell (1983) Ti plasmid vector for the introduction of DNA into plant cells without alteration of their normal regeneration capacity. EMBO J. 2:2143-2150.

MAIZE TRANSPOSABLE ELEMENTS: STRUCTURE, FUNCTION, AND REGULATION

E.S. Dennis,[1] E.J. Finnegan,[1] B.H. Taylor,[1] T.A. Peterson,[1] A.R. Walker,[2] and W.J. Peacock[1]

[1]CSIRO Division of Plant Industry
[2]Biochemistry Department
Australian National University
Canberra, ACT, Australia

ABSTRACT

We have mapped a transcript of the maize transposable element Ac. Virtually the entire 4.5-kb Ac element is transcribed as a single RNA molecule from which four introns are spliced to yield a 3.5-kb mRNA. The Ac mRNA codes for an 807 amino acid protein, which is presumably the transposase. The Ac element can be reversibly inactivated by methylation of the promoter and 5' untranscribed leader region to convert the Ac element to a Ds-like element.

The Ac element can support the transposition both of itself and of a number of types of Ds elements. The Ds elements have the same terminal inverted repeats as does Ac, and also are associated with the characteristic 8-bp duplication of host DNA at the site of insertion. The Ds1 element, which was first isolated from the Adh1-Fm335 allele, is a 405-bp segment and is repeated 40-50 times in the genomes of all maize lines, and in the maize relatives Teosinte and Tripsacum. The Ds1 element generates an additional intron in the Adh1-Fm335 allele; all but 14 bp of the element are spliced from the Adh1 mRNA.

THE Ac/Ds FAMILIES OF ELEMENTS

The Ac/Ds controlling element system of maize has been extensively characterized at both genetic and physical levels (5,14). A number of unstable alleles caused by the insertion of Ac in or near the waxy gene (wx-m9, wx-m7) have been cloned and the Ac element completely sequenced from wx-m9 (24,25) and from wx-m7 (19). The Ac elements from wx-m9 and wx-m7 have 11-bp imperfect terminal inverted repeats and, upon insertion, cause 8-bp duplications of the flanking genomic DNA (24). An Ac element located in the P-VV allele which has been cloned and partially sequenced is not flanked by an 8-bp duplication (T.A. Peterson, unpubl. data).

While an Ac element catalyzes both its own transposition and that of Ds elements elsewhere in the genome, Ds elements cannot catalyze transposition. Ds elements are related to Ac by having similar terminal inverted repeats, and their insertion also generates an 8-bp flanking genomic duplication. Three types of Ds elements have been identified: deletion derivatives of Ac, typified by wx-m6 and wx-m9 Ds elements (5); the members of the Ds1 family which consist of a 405-bp element with an internal sequence completely unrelated to that of Ac (22); and other Ds elements, e.g., Ds2, which are deletion derivatives of Ac but have insertions of other sequences within the element (15). These different classes of Ds elements all share the 11-bp terminal inverted repeats preceded by a 2-bp direct repeat and a further 3-bp inverted repeat, suggesting that transposition requires this terminal repeat arrangement. It also raises the possibility of further families of Ds elements in the maize genome with similar termini but different internal sequences.

THE Ds1 FAMILY OF CONTROLLING ELEMENTS

The type member Ds1 was the first plant controlling element to be characterized at the molecular level (22,31). It was isolated from the Adh1-Fm335 allele in which the element is inserted in the 5' leader sequence of the Adh1 gene (21). Analysis of revertants to full ADH1 activity of this Ds1 allele provided the first molecular description of a reversion event (22). In these revertants, most of the 8-bp duplication is retained, although in one instance we isolated a revertant in which the Ds1 element had undergone a perfect excision event which restored the sequence to that of the progenitor allele (2). In general, most of the duplication is retained, while nucleotides directly adjacent to the site of insertion have been deleted or changed to the complementary nucleotides. Models to account for these events have been proposed (23,29).

Larger deletions which stretch from the site of Ds insertion are also associated with Ac/Ds excision and may involve recombination with related sequences located away from the site of insertion, or with another copy of the transposable element. We examined a null allele of Adh1 which was generated from the unstable Ds1-induced Adh1-Fm335 in the presence of Ac (2). The Ds1 element had excised and removed 77 bases of flanking DNA immediately adjacent to the position of Ds1 insertion. It appeared that the deletion could have arisen by a recombination event between the 8 bp of DNA which is duplicated and lies on the upstream side of the insertion and a similar sequence in the first exon of the Adh1 coding region (2).

We know that the 8-bp duplication is not necessary for excision because the Ac element located at the P-VV locus which does not have an 8-bp flanking duplication excises at high frequency (P.A. Peterson, unpubl. data).

FAMILIES OF Ds ELEMENTS IN MAIZE AND RELATED SPECIES

There are a number of Ac- and Ds-related segments in the maize genome; 40-50 bands of both Ds1 and the Ac-related Ds family are seen in

Southern hybridizations. Approximately the same number of segments which hybridize at high stringency to the Ds1 element have been found in Teosinte, a wild precursor of maize, and in Tripsacum dactyloides, a more distantly related species (7).

Eight of these related segments have been cloned (7). In all cases the cloned fragments contained a Ds1-like sequence of approximately 400 bp, with at least 90% sequence homology to Ds1. Seven of these segments have the 11-bp inverted repeat termini characteristic of Ds1 and are bounded by direct 8-bp repeats. The remaining element, Ds101, had a 10-bp inverted repeat at its termini and was flanked by a duplication of 6 bp rather than 8 bp, suggesting that the length of the terminal inverted repeat may control the length of the duplication. However, the Antirrhinum transposable element Tam3 generates different length genomic duplications at different insertion sites, indicating that other factors may be involved in determining the length of the genomic duplication.

Members of the Ds1 family which are associated with other mutant alleles have been cloned by Wessler et al. (33) from the wx-m1 allele, and from the bz-wm allele by J.W. Schiefelbein and O.E. Nelson (pers. comm.). There is no obvious consensus sequence for the integration site of Ds1 elements. An extensive stem loop structure can be drawn for the Ds1 sequence (Fig. 1) similar to that drawn for other plant transposable elements (20). The reason that members of the Ds1 family of elements are so conserved in both length and sequence may be that this secondary structure is required for transposition. The sequence changes which occur between members of the family (7) are small; the large internal deletion which occurs in sequence 101 relative to other members of the family is in a region which is not in one of the stem-loops drawn.

The elements from Tripsacum also have all the features of the maize sequences--for example, terminal inverted repeats and duplication of flanking genomic DNA as well as conservation of the length and sequence. Members of the maize family are only as similar to each other as they are to the Tripsacum elements, and it seems reasonable to assume that all Ds1 sequences trace back to a single element. We cannot distinguish between the alternatives of horizontal transfer of the elements--for example, by introgression among maize, Teosinte, and Tripsacum--and a vertical evolution from a common ancestor of the three genera. If we make the assumption that the nucleotide substitution rate in these elements is comparable to the neutral rates of nucleotide substitution which apply to animal pseudogenes of 5×10^{-9} substitutions per site per year (12), then the Ds1 elements duplicated and diverged from a common sequence between eight and 25 million years ago. These elements must have been resident in the Maydeae genomes for a long period of time.

TRANSCRIPTION OF Ac

The Ac element is 4.5 kb in length, and sequence analysis shows that it contains several large open reading frames (ORF) (24,25). Ds derivatives of Ac that have a deletion in either of the two largest ORFs do not complement in trans (4,5), suggesting that neither of these ORFs alone is sufficient for transposition and also raising the possibility that the two large ORFs form part of a single transcription unit.

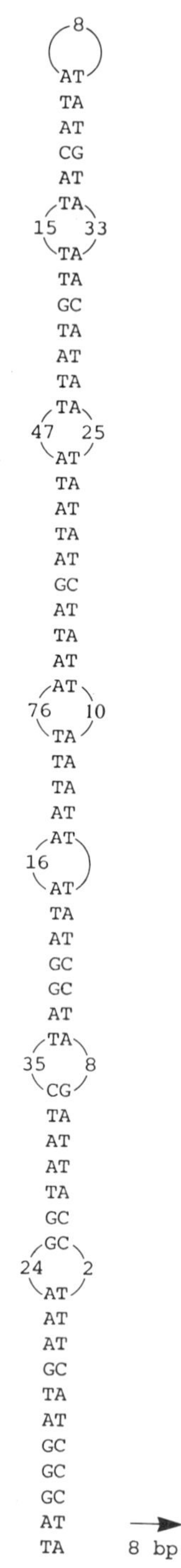

Fig. 1. Structure of the Ds1 element drawn to demonstrate the extensive stem loop structure possible. Numbers indicate the length of the loop, and 8 bp indicates the genomic duplication.

We assume that, like bacterial transposons and the P element in Drosophila, the Ac element encodes a transposase which catalyzes the transposition of both Ac and Ds elements. Proposed models of Ac transposition suggest that, during insertion, the Ac-encoded transposase makes single-stranded staggered cuts 8 bp apart in the genomic DNA, into which an Ac or Ds element inserts, and the single-stranded regions are "filled in" with a DNA polymerase. For excision, the same transposase could make single-stranded staggered cuts on either side of the element. The single-stranded regions in the element may then be digested away to form flush ends and the staggered cuts on the genomic DNA repaired (23,29).

In order to begin an analysis of the mechanism of Ac transposition, we have sequenced the transcript of Ac and predicted the amino acid sequence of the protein. RNA was isolated from Ac-containing seedlings and cDNA clones were prepared in λgt10; of 2 x 10^6 clones screened, 13 clones hybridized to a full-length Ac probe. The combined lengths of the cDNA clones sequenced covered 2,383 bases. Comparison of this sequence to the published genomic Ac sequence showed that three introns (of 71, 89, and 387 bases, respectively) had been spliced from the transcript during RNA processing. Removal of these intron sequences created a continuous ORF of 705 amino acids, with a 3' untranslated region of 239 bases (Fig. 2).

The 5' end of the transcript was determined by S1 analysis and showed two major initiation sites at positions 334 and 377 and a scatter of minor 5' ends over a 90-base region. The first ATG is 600-690 bases from the start of transcription and does not have the consensus sequence (10). A fourth intron of 107 b was detected by S1 analysis, located approximately 850 b downstream of the major transcription start sites. The published sequences of Ac-m9 (25) and Ac-m7 differ in the region 23 bp 3' to our proposed intron acceptor site; resequencing of this region confirmed the Ac-m7 sequence, which contains two additional bases at this position. Excision of the fourth intron extends the ORF by 102 amino acids to 807 amino acids (6) (Fig. 2). Our results essentially confirm those of Kunze et al. (11).

CHANGES IN ACTIVITY OF Ac

Results from experiments which correlate the inactivation of an Ac element with its modification by methylation are consistent with Ac encoding a single transcript which specifies a single protein. Schwartz and Dennis (30) have investigated the progenitor wx-m9 Ac allele in which Ac is fully active, an allele in which the Ac element had been converted to a Ds element (wx-m9 Ds-cy), and a number of revertants of this allele to full Ac activity (wx-m9 AcR). No differences between the alleles were seen when the three alleles were cloned and mapped with a number of restriction enzymes that cut frequently. The alleles were then mapped using Southern hybridization to genomic DNA, using the restriction enzymes HpaII and MspI, which both recognize the sequence CCGG. HpaII will not cut if the internal C is methylated, while MspI activity is not sensitive to methylation of this C residue. Sites at the 5' end of the gene (stretching from the promoter through the start of transcription to the ATG initiation codon) are variably methylated; in the active Ac element these sites are unmethylated, in the Ds-cy derivative all the sites are methylated, and in revertants to full Ac activity (AcR) some sites are no longer methylated (30) (Fig. 3). This analysis also showed that HpaII sites at the 3' end of

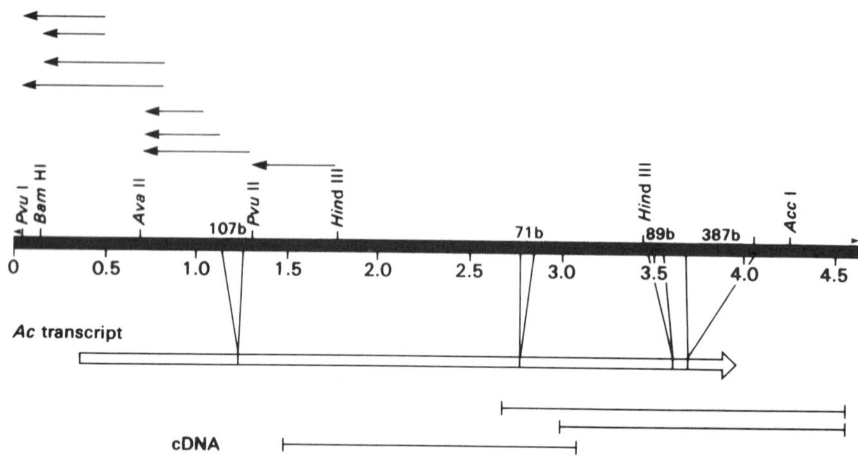

Fig. 2. Analysis of the transcript of *Ac*. The *Ac* element is drawn with the 5' end of the gene at the left, and numbering is from the first base in the inverted repeat. The four introns are indicated with their lengths; the cDNA clones we isolated are shown below the diagram, and S1 probes used in the definition of the 5' end are shown above.

the gene (Fig. 3), downstream of the transcription region, are methylated in both active and inactive *Ac* elements.

Chomet et al. (1) investigated a change in methylation associated with a change in phase of *wx-m7 Ac* from an active to an inactive form. They used the isoschizomers *EcoRII/BstNI*, both of which recognize the sequence CC(A or T)GG; but, while *EcoRII* does not cleave when the internal C is methylated, *BstNI* is insensitive to methylation. There are five *BstNI/-EcoRII* sites in the *Ac* element, one at the start of transcription and the other four in the untranslated leader region. These sites all became methylated when *Ac* cycled from an active to an inactive form. Methylation at *SstII* sites was also correlated with the change from an active to an inactive form. These *SstII* sites are again located in the 5' untranslated leader region (Fig. 3).

The data from *wx-m9 Ac* (30) and from *wx-m7 Ac* (1), together with our knowledge of the transcription pattern of *Ac*, suggest that the activity of the element is correlated with the degree of methylation of a region of the element covering the 5' 900 bp of the element that includes the promoter, transcription start site, and 5' untranslated leader (Fig. 3). Presumably, when *Ac* is fully methylated at the 5' region, transcription of the transposase is inhibited. The long 5' untranslated leader sequence is rich in GC dinucleotides and GXC trinucleotides and is an obvious target for control by methylation. One possible mode of action of DNA methylation is to alter the pattern of DNA-protein interaction over a reasonably long domain, leading to the inability of RNA polymerase to bind to the promoter (8).

The methylation of the *Ac* element does not affect its ability to transpose, since it can still function as a *Ds* element when in the fully methylated form (30). When the *Ac* element is methylated, only the element it-

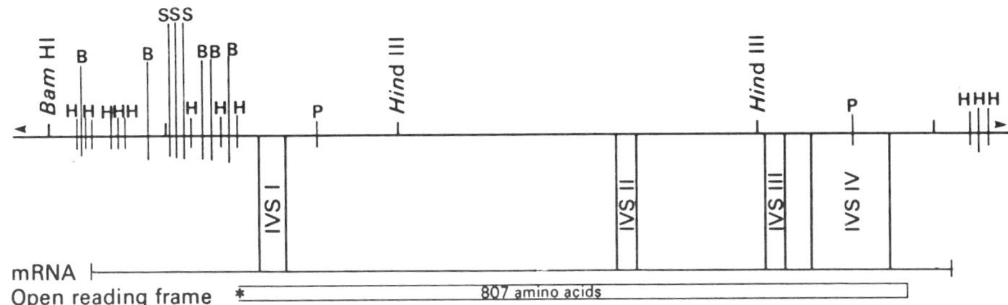

Fig. 3. Sites of methylation of Ac elements undergoing change of phase, for example, cycling from an active to inactive form (data from Ref. 1 and 30), shown relative to the structure of the Ac transcript determined by cDNA cloning and S1 analysis. H, HpaII sites; S, SstII sites; B, BstN/EcoRII sites.

self is methylated and not the flanking DNA; this is true for both the wx-m9 and wx-m7 cases. When Ac is inactive and methylated, the Ds elements which are deletion derivatives of Ac are not extensively methylated, indicating that methylation is specific to the Ac element. The change from wx-m9 Ac to wx-m9 Ds-cy occurred in the sporophyte or embryo but not in the endosperm, and affected both wx-m9 Ac alleles.

The mechanism of control of methylation of these Ac sequences is not clear, but the transition from Ac to Ds is a relatively rare event. Schwartz detected only one case of the wx-m9 Ac being converted to the wx-m9 Ds-cy (30). The correlation of Ac activity with lack of methylation of the leader/promoter region suggests that any mechanism which promotes demethylation can activate Ac elements. Methods such as chromosome breakage in bridge-breakage-fusion cycles, in high-knob/high-loss lines, or in irradiated lines, induce repair replication, which leads to regions deficient in methylation. Southern hybridization indicates the presence of cryptic Ac elements in the maize genome even in lines which do not contain an active Ac element (5); demethylation may activate these copies of Ac.

CHANGES OF STATE OF Ac

When transposable elements cycle from an active to an inactive state, as in the wx-m9 Ac and wx-m7 Ac cases described above, they are said to have undergone changes in phase. The effect of Ac on gene expression can also vary; such changes are termed "changes in state" and may be due to a change in the frequency or timing of excision events that might affect the expression of a target gene. This phenomenon may be caused by a change in the position of the transposable element within the gene or changes within the element itself.

We have examined a change in state of a P-VV allele (T.A. Peterson, unpubl. data). The P locus controls pigmentation of the pericarp and glumes of the cob. The P-VV allele, which specifies variegated pericarp and cob, comprises the transposable element Ac situated at the P locus. We have previously isolated a DNA clone from P-VV, and we are now using P DNA sequences to study the structure and expression of P.

We have isolated an allele termed P-OVOV (Orange Variegated) derived from P-VV. P-VV specifies colorless pericarp with red sectors, whereas P-OVOV specifies orange pericarp with many dark red sectors, and some colorless sectors. P-OVOV represents a change in state of the P-VV allele or of the Ac at the P locus.

Genetic tests showed that P-OVOV is a true allele of P, that Ac activity is tightly linked to P-OVOV, that P-OVOV is unstable, giving rise to revertant as well as other variant alleles, and that the trans-acting functions of Ac associated with P-OVOV are not different from the progenitor P-VV allele.

Southern analysis of DNA from the progenitor P-VV allele and the P-OVOV allele shows that the Ac element associated with P-OVOV remains within an 850-bp genomic DNA fragment at the P locus. However, the Ac element is inverted in P-OVOV relative to P-VV.

These results suggest that the change in state of P-VV to P-OVOV may be due to inversion of Ac at the same position in the gene. We are investigating how this inversion of Ac affects expression of the P locus. Perhaps the inverse orientation of the Ac element allows splicing of the element from the transcript (see later) and greater expression of the gene.

TRANSPOSABLE ELEMENTS CAN AFFECT GENE ACTIVITY

The insertion of a transposable element can have profound effects on the activity of a gene depending on where in the gene it is inserted, the orientation of the element, and the sequence of the flanking genomic DNA. If the insertion is located in a coding region, it will destroy the reading frame of the coding region unless it is spliced from the transcript. If it is located in a regulatory region, e.g., the promoter, it can affect expression either to increase or decrease it. Elements in an intron or untranslated region can affect transcription or cause aberrant splicing, which may result in an altered gene product.

Excision of elements can also affect the expression of a gene. Since additional bases remain following excision of an element, this can affect the amino acid sequence, resulting in a protein with altered properties--for example, a change in the level of enzymatic activity. Deletions extending from the point of insertion can produce stable null mutations.

The insertion of transposable elements can generate additional introns. One particular case we have examined is the Adh1-Fm335 allele, where the Adh1-Fm335 mutant phenotype is caused by the insertion of a Ds1 element into the transcribed leader sequence of the Adh1 gene (22, 31). In a homozygous Adh1-Fm335 stock, the amount of ADH1 enzyme is decreased to about 10% of the normal level but the specific activity (21) and temperature stability of the enzyme remain the same (M.M. Sachs and D. Schwartz, unpubl. data).

Northern hybridization analyses show that the length of the Adh1-specific mRNA in the mutant is approximately the same as it is in the progenitor allele (3); however, the amount of the mRNA is 100-fold lower. Under anaerobic conditions, the mutant plant shows a 20-fold increase in ADH1 enzyme activity and an increase in Adh1-specific RNA of 20- to 50-fold, similar to that seen in progenitor or revertant alleles. The timing of the induction of the Adh1 mRNA is similar to that in the progenitor Adh1-F plant.

The transcribed regions of Adh1-Fm335 and Adh1-F were compared directly in S1 mapping experiments, using probes derived from the 5' region of the gene (3,23). When the probe was prepared from the progenitor allele (Adh1-F), RNA from both the mutant and progenitor protected exactly the same length fragment, but with a much weaker signal in the mutant. This indicates that all the sequences present in the progenitor mRNA are also present in the Ds mutant mRNA and that the transcription start site is exactly the same as in the progenitor. When the mutant mRNA is used to protect a probe synthesized from the 5' region of the mutant gene, two fragments are seen. The first fragment extends 3' from the site of insertion of the Ds element to the first exon-intron boundary, and the second fragment, approximately 66 bp long, corresponds to a segment extending from the start of transcription to a point 14 bases inside the Ds1 element. These results indicate that only 14 bp of the Ds insertion are present in the mRNA of the mutant gene and that the remainder of the element is processed as an intron from the transcript (3). The intron donor sequence is 14 bp from the Ds1 5' terminus, and the intron acceptor site is at the junction between the 3' end of the Ds1 segment and the Adh1 leader sequence (Fig. 4).

The processing of the Ds1 element from the mRNA does not in itself explain the low level of messenger activity in the mutant. Run-on transcription experiments have shown that the mutant has approximately the same rate of transcription as the progenitor allele (L. Beach, pers. comm.), suggesting that the low steady-state level of mRNA is due to instability of the RNA in the mutant relative to the progenitor allele.

The behavior of the Ds element as an intron implies that sequences around the donor and acceptor splice sites should resemble the consensus sequences seen for splice sites in both plants and animals. They do; and, moreover, the sequence TCCTAAC occurs 30 bp before the 3' splice site. This sequence is identical to the consensus lariat acceptor sequence (27) shown to be present in introns and is located in the correct position.

We conclude that the Ds1 element in the Adh1-Fm335 gene has all the necessary sequence attributes to be spliced as an intron and that, although transcription occurs at a normal rate, the RNA is less stable. This provides an example of a transposable element introducing a new intron into a gene.

Recently, other cases of a controlling element being spliced from a transcript have been described. In two alleles of the maize waxy gene (wx-m9 and wx-B4) containing insertions of 4.3-kb and 1.5-kb Ds elements, respectively, the predominant transcripts are wild-type in length (33). Sequencing of cDNA clones of these transcripts for both alleles showed that the Ds element was spliced using donor sites in the Ds element and different 3' acceptor sites in the flanking wx gene. In the Adh1-2F11 allele, insertion of an Ac deletion derivative Ds element in the coding region of the Adh1 gene resulted in transcripts that were approximately wild type in length, indicating that the Ds element was spliced from the transcript (15).

The allele bz-m13cs9 of the bz-1 gene contains a 902-bp defective Spm transposable element in the second exon. However 40-50% of the wild-type enzymatic activity is present; analysis of cDNAs shows that this occurs through splicing, using the donor site of the normal bz-1 intron of the gene and an acceptor site within the inverted terminal repeat of the defective Spm element (9).

The finding that transposable elements can be spliced from transcripts allows for more subtlety in their effects on gene expression. Selection pressure for particular sequences (normally at the ends of transposable elements) capable of acting as splice junctions may be exerted. The orientation of an element relative to that of a gene will be important, as will the location of genomic DNA sequences which can act as the other splice junction.

LONG-TERM EFFECTS OF TRANSPOSABLE ELEMENTS

Once an element is inserted in a gene, the ends can mutate so that the element is no longer able to be excised. Many of the mutants and variants which occur are due to the insertion of transposable element-like sequences in genes.

Three examples from our laboratory show how transposable element-like sequences may become stably incorporated within genes. In the Adh1 locus, the naturally occurring 1F and 1S alleles show polymorphism in the 3' region; part of the polymorphism is due to the insertion of a transposable element-like sequence. This sequence is 332 bp long and has 13-bp inverted repeats with one mismatch and one extra base (Fig. 5). Flanking these repeats are 5- or 7-bp sequences which are not inverted repeats, and then a 4-bp direct duplication of the genomic DNA. This insertion provides four poly(A) addition signals for the Adh1F allele and results in different length mRNAs. Whether this insertion accounts for other differences between the 1F and 1S alleles (tissue-specific expression and lack of intragenic recombination between alleles) is not clear.

In wheat, an Adh gene has been isolated and mapped to chromosome 1A (16). A comparison of the sequence of this gene with that of the homologous Adh3 of barley (32) shows the wheat gene to have a 1.8-kb insertion in intron three. This insertion is flanked by an 8-bp duplication of host DNA and has 14-bp imperfect terminal repeats (11 bp of 14 bp are complementary) (Fig. 5). There are approximately ten copies of the insertion in the wheat genome. Analyses of the hexaploid, tetraploid, and diploid wheats show that this insertion has been present in the Adh gene on chromosome 1A for at least 10,000 years. It is located in a similar position in the Adh gene of all tetraploid and hexaploid wheats, but not in any of a number of A genome diploid wheats (17). Whether the autonomous controlling element has never been present to catalyze the element's transposition or whether it is incapable of moving is not clear. The same element is also present in the genomes of barley and rye but is not located in the Adh locus (18).

A third example of an effect of an insertion element on a locus is in the Bronze 1 gene of maize, where a 176-bp element is located in the first exon of a bronze 1 allele (bronze 1-mut). This is the responding allele of the Mut system described by Rhoades and Dempsey (26). This element has 8-bp perfect inverted terminal repeats, a 4- or 7-bp nonrepeated sequence, and then an 8-bp flanking duplication (Fig. 5). It is repeated

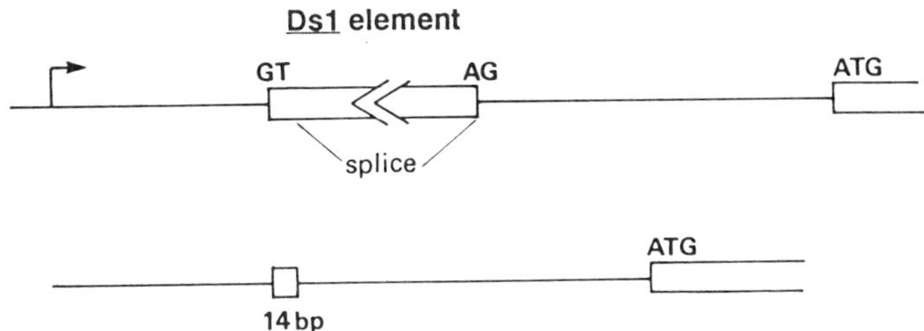

Fig. 4. Diagram of how the Ds1 element is spliced from the Adh1-Fm335 allele. The upper part of the diagram shows the location of the Ds1 element in the transcript. The lower part shows the mRNA containing 14 b of Ds1.

about 50 times in the genome. Revertants to full activity of this allele still contain the element in the coding region despite the element's containing numerous stop codons (A.R. Walker, unpubl. data). Perhaps the structure of this element makes it incapable of excision.

In each of these three cases it appears that an element has been resident in a gene for a long period. In one case, the element provides signals for polyadenylation; in the second, it has an unknown effect on gene expression; and in the third, it remains in the gene, although gene expression is changed radically in the presence of the autonomous element.

CONCLUSION

The Ac element can be thought of as a single gene which is transcribed and processed to produce an mRNA. This RNA can be translated to yield a protein (the transposase) which interacts with a specific sequence of 11 bp of characteristic structure (inverted repeats). These inverted repeat sequences can be variable distances apart and still be

Adh1-1F (Maize)

GTATtcatgTGTTTGTTTGCTT...293bp...AAGCCAACCAAACAtgcgcatGTAT

Adh1A (Wheat)

CTATGAAGCACTGGCGGAGGCC...1800bp...GTCCTCCGCCTCTGCTATGAAG

Bz1-R2 (Maize)

GCCGCCGCcatgGGGCCCAA...149bp...TTGGGCCCgcttctgGCCGCCGC

Fig. 5. Structure of the elements which are located in genes. The ends of the elements from the Adh1-1F allele, the Adh1A allele of wheat, and the Bz1-R2 revertant of the bz1-mut allele are shown. The genomic DNAs which are duplicated are underlined; mismatches in the terminal repeats are shown with dots; and bases which do not appear to be either inverted repeats or genomic DNA are in lower case.

recognized by the transposase. For example, they can be 4.5 kb apart when they flank Ac or as close as ∿400 bp when they flank the Ds1 element, and they can be located anywhere in the genome. The structure of the sequence between the terminal repeats may also be important. The activity of the transposase gene is subject to regulation by methylation of the promoter/5' untranscribed leader region. We are now in a position to manipulate the level of the transposase by altering the promoter or leader region of the gene so as to regulate expression.

The data available suggest that Ac and Ds elements can affect gene expression by a number of different mechanisms. The finding that controlling elements can be spliced from transcripts increases the possibilities for modulation of gene activity and introduces selection to ensure the retention of sequences both in the controlling element and in sites of insertion.

REFERENCES

1. Chomet, P.S., S. Wessler, and S.L. Dellaporta (1987) Inactivation of the maize transposable element Activator (Ac) is associated with its DNA modification. EMBO J. 6:295-302.
2. Dennis, E.S., W.L. Gerlach, W.J. Peacock, and D. Schwartz (1986) Excision of the Ds controlling element from the Adh1 gene of maize. Maydica 31:47-57.
3. Dennis, E.S., M.M. Sachs, W.L. Gerlach, L. Beach, and W.J. Peacock (1987) The Ds1 transposable element acts as an intron in the mutant allele Adh1-Fm335 (submitted for publication).
4. Dooner, H., J. English, E. Ralston, and E. Weck (1986) A single genetic unit specifies two transposition functions in the maize element Activator. Science 234:210-211.
5. Fedoroff, N., S. Wessler, and M. Shure (1983) Isolation of the transposable maize controlling elements Ac and Ds. Cell 35:235-242.
6. Finnegan, E.J., B.H. Taylor, E.S. Dennis, and W.J. Peacock (1988) Transcription of the maize transposable element Ac in maize seedlings and in transgenic tobacco. Molec. Gen. Genet. (in press).
7. Gerlach, W.L., E.S. Dennis, W.J. Peacock, and M.T. Clegg (1988) The Ds1 controlling element family in maize and its grass relative Tripsacum. J. Molec. Evo. (in press).
8. Keshet, I., E. Lieman-Hurwit, and H. Cedar (1988) DNA methylation affects the formation of active chromatin. Cell 44:535-543.
9. Kim, H.-Y., J.W. Schiefelbein, V. Raboy, D.G. Furtek, and O.E. Nelson (1987) RNA splicing permits expression of a maize gene with a defective Suppressor-mutator transposable element insertion in an exon. Proc. Natl. Acad. Sci., USA (in press).
10. Kozak, M. (1983) Comparison of initiation of protein synthesis in procaryotes, eucaryotes, and organelles. Microbiol. Rev. 47:1-45.
11. Kunze, R., U. Stochaj, J. Laup, and P. Starlinger (1987) Transcription of the transposable element activator (Ac) of Zea mays L. EMBO J. 6:1555-1563.
12. Li, W.-H. (1983) Evolution of duplicate genes and pseudo genes. In Evolution of Genes and Proteins M. Nei and R.K. Koehn, eds. Sinauer Associates, Sunderland, Mass., pp. 14-37.
13. McClintock, B. (1947) Cytogenetic studies of maize and Neurospora. In Carnegie Institution of Washington Yearbook 46:146-152.
14. McClintock, B. (1955) Controlled mutation in maize. In Carnegie Institution of Washington Yearbook 54:245-255.
15. Merckelbach, A., H.-P. Doring, and P. Starlinger (1986) The aberrant Ds element in the adh1 2F11 Ds2 allele. Maydica 31:109-122.

16. Mitchell, L.E., E.S. Dennis, and W.J. Peacock (1987) Molecular analysis of an Alcohol Dehydrogenase (Adh) gene from Chromosome 1 of wheat. (Manuscript in preparation)
17. Mitchell, L.E., E.S. Dennis, and W.J. Peacock (1987) Evidence for an insertion element in wheat. (Manuscript in preparation.)
18. Mitchell, L.E., E.S. Dennis, and W.J. Peacock (1987) The wheat Adh-1A gene as a molecular probe for analysis of the evolution of diploid and polyploid wheats. (Manuscript in preparation)
19. Mueller-Neumann, M., J.I. Yoder, and P. Starlinger (1984) The DNA sequence of the transposable element Ac of Zea mays L. Molec. Gen. Genet. 198:19-24.
20. Nevers, P., N.C. Shepherd, and H. Saedler (1986) Plant transposable elements. Adv. in Bot. Res. 12:103-203.
21. Osterman, J.L., and D. Schwartz (1981) Analysis of a controlling-element mutation at the Adh locus of maize. Genetics 99:267.
22. Peacock, W.J., E.S. Dennis, W.L. Gerlach, D. Llewellyn, H. Lörz, A.D. Pryor, M.M. Sachs, D. Schwartz, and W.D. Sutton (1983) Gene transfer in maize: Controlling elements and the alcohol dehydrogenase genes. In Proceedings of the Miami Winter Symposia, K. Downing et al., eds. Academic Press, New York, Vol. 20, p. 311.
23. Peacock, W.J., E.S. Dennis, W.L. Gerlach, M.M. Sachs, and D. Schwartz (1984) Insertion and excision of Ds controlling elements in maize. In Recombination at the DNA Level, Cold Spring Harbor Symposium on Quantitative Biology, Vol. 49, 347-354.
24. Pohlman, R.F., N.V. Fedoroff, and J. Messing (1984) The nucleotide sequence of the maize controlling element Activator. Cell 37:635-643.
25. Pohlman, R.F., N.V. Fedoroff, and J. Messing (1984) Correction: Nucleotide sequence of Ac. Cell 39:417.
26. Rhoades, M.M., and E. Dempsey (1982) The induction of mutable systems in plants with the high-loss mechanism. Maize Genet. Coop. Newsl. 56:21-26.
27. Ruskin, B., A.R. Krainer, T. Maniatis, and M.R. Green (1984) Excision of an intact intron as a novel lariat structure during pre-mRNA splicing in vitro. Cell 38:317-331.
28. Sachs, M.M., E.S. Dennis, W.L. Gerlach, and W.J. Peacock (1986) Two alleles of maize alcohol dehydrogenase 1 have 3' structural and poly(A) addition polymorphisms. Genetics 113:449-467.
29. Saedler, H., and P. Nevers (1985) Transposition in plants: A molecular model. EMBO J. 4:585-590.
30. Schwartz, D., and E.S. Dennis (1986) Transposase activity of the Ac controlling element in maize is regulated by its degree of methylation. Molec. Gen. Genet. 205:476-482.
31. Sutton, W.D., W.L. Gerlach, D. Schwartz, and W.J. Peacock (1984) Molecular analysis of Ds controlling element mutations at the Adh1 locus of maize. Science 223:1265-1268.
32. Trick, M., E.S. Dennis, K.J.R. Edwards, and W.J. Peacock (1988) Molecular analysis of the alcohol dehydrogenase gene family of barley. (Submitted for publication).
33. Wessler, S.R., G. Baran, and M. Varagona (1986) Transposable element Ds is spliced from RNA. Science 237:916-918.

STRUCTURE AND FUNCTION OF THE *En/Spm* TRANSPOSABLE ELEMENT SYSTEM OF *ZEA MAYS*: IDENTIFICATION OF THE SUPPRESSOR COMPONENT OF *En*

Alfons Gierl, Heinrich Cuypers, Stephanie Lütticke, Andy Pereira, Zsuzsanna Schwarz-Sommer, Sudhansu Dash,[1] Peter A. Peterson,[1] and Heinz Saedler

Max-Planck-Institut für Züchtungsforschung
D-5000 Köln 30
Federal Republic of Germany

[1]Department of Agronomy
Iowa State University
Ames, Iowa 50011

INTRODUCTION

The transposable element system Enhancer (*En*) of *Zea mays* was originally identified by Peterson (12) at the pale *green locus* as a mutable allele. Subsequently, the control of this mutability was shown to be homologous to the Suppressor-Mutator (*Spm*) system (5) both genetically (13) as well as molecularly (11). This *system* is comprised of two components, one of which (*En*, *Spm*) is capable of autonomous transposition and encodes all functions *associated* with this system. The second component, the non-autonomous Inhibitor (*I*) (12), is unable to promote transposition but can be trans-activated to *transpose* by an *En/Spm* element present in the same genome. Several *I* elements have been *isolated* (1,15-17,21,22) which bear the termini of *En/Spm* but carry internal deletions of varying extent.

The trans-active functions Mutator (M), Suppressor (S) (7), and Activator (A) (18) have been identified genetically for the autonomous element. M and S can be monitored simultaneously with certain alleles such as *a1-m1* in which the inserted *I* element does not lead to complete *inactivation, but* to an intermediate *level* of *A1* gene expression. In the absence of an *En/Spm* element, such a *mutation is* stable. However, in the presence of *the autonomous* element, the S-function abolishes the residual *A1* gene expression and the M-function mediates excision (transposition) of *the I* element. The Activator function has been proposed to explain the *observation* that an inactive element can be temporarily activated when a second fully active element is introduced into the genome (8,9).

Thus, the En/Spm system appears to be a complex genetic entity which can influence gene expression in various ways. In order to provide more insight into the structure and function of this system, the autonomous element En-1 was isolated in the wx-844 allele (14) and molecularly characterized (11).

STRUCTURE OF THE En-1 ELEMENT

The DNA sequence of the 8287-bp long En-1 element was determined and the structure of gene 1 which encodes the predominant 2.5-kb En transcript was established (3,10) by sequencing the corresponding cDNA.

The ends of En-1 are highly structured. There is a 13-bp terminal inverted repeat which is not repeated in the rest of the element. A 13-mer nucleotide motif is repeated eight times in the left and ten times in the right end in direct or inverse orientation (3). Dot plot computer analysis of the structure reveals that the homology between the two ends is restricted to about 200 bp of the left terminus and 300 bp of the right terminus. The sequence organization of the termini makes it possible to pair the left and right ends in a stem and loop structure as shown earlier (3,11,17). Alternatively, stem and loop structures can be formed within each end.

One striking feature of gene 1 is that it spans nearly the entire element (Fig. 1). The promoter of gene 1 is located within the highly structured left end of the element. The first nontranslated exon is followed by a 4434-bp long intron which precedes the cluster of ten exons on the right half of En-1. Gene 1 codes for a putative peptide of 621 amino acids (68 kD).

Another protein may be encoded, at least in part, by the large intron which contains two large open reading frames (ORF1 and ORF2). 1.3 kb of ORF1 share DNA sequence homology corresponding to 42% amino acid identity, to the similarly positioned open reading frame in the transposable element Tam1 (19). Not only the open reading frame but also the terminal 13-bp inverted repeats are conserved between these two plant transposable element systems (2,11). Since the transposase is thought to act upon the terminal inverted repeats, one may speculate that the region of conserved sequence homology encodes the element-specific transposase (M-function).

THE DELETION DERIVATIVE En-2 HAS RETAINED THE SUPPRESSOR ACTIVITY

In order to establish a structure and function relationship, we have cloned a mutant derivative of the wx-844::En-1 allele. This element, designated En-2, has retained the S-activity and abolishes the residual a1 gene expression from the suppressible a1-5719A allele, but it has only "weak" M-activity and shows a drastically reduced frequency of somatic excisions. In addition, it hardly trans-activates receptive elements to transpose. En-2 bears a 1,126-base pair internal deletion [position 3617 to 4742 is deleted; all positions refer to the En-1 sequence (15)]; ORF2 and 500-bp of ORF1 are deleted (Fig. 1).

An element similar to En-2, termed Spm-weak (Spm-w), was cloned (1) from the a1-m2 allele (6). Spm-w has a 1.6-kb deletion in the center

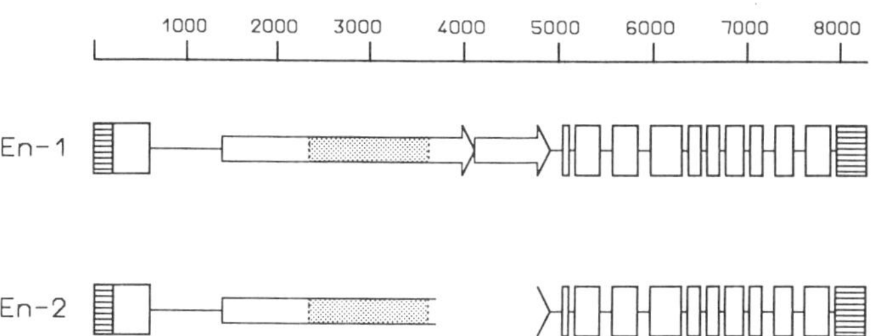

Fig. 1. Structural organization of En-1 and En-2. The upper part of the figure gives the size (nucleotides) of En-1. Hatched boxes indicate the left and right terminal repeats. The 11 exons of gene 1 are represented by open boxes. The TATA-box of gene 1 is indicated as P for promoter, and the translation start and stop codons of gene 1 are indicated as AUG and UGA, respectively. ORF1 and ORF2 are indicated by the arrows in the first intron of gene 1. The shaded box within ORF1 represents the region of conserved sequence homology with the Tam1 element.

of the element, the 3'-end of which is almost identical to that of En-2. Since the genetic behavior of both elements is also very similar, these data again indicate that the M-function may be encoded by ORF1 and ORF2. The residual M-activity observed in the presence of En-2 and Spm-w may be explained by the fact that cryptic, inactive elements are transiently or permanently activated by the mutant elements.

The deletion in En-2 and Spm-w does not interfere with the expression of gene 1. En-2 [as well as Spm-w (4)] produces a gene 1 transcript of original size, however, at a five-fold reduced rate. Therefore, one could speculate that the product of gene 1 is equivalent to the S-function, since both elements retain the suppressor activity.

THE PRODUCT OF GENE 1 BINDS TO THE ENDS OF THE En ELEMENT

In order to characterize the function of the product of gene 1, we have expressed the gene 1 cDNA in Escherichia coli using a bacteriophage T7 promoter expression system [pAR3039 (20)]. The gene 1 protein was partially purified by DEAE-chromotography from the bacterial crude extract. To test the binding behavior to the En element, restriction fragments of the element were incubated with the gene 1 protein preparation. Protein-DNA complexes were immunoprecipitated, the DNA was extracted from the precipitate by phenolization, and it was analyzed by gel electrophoresis. Only fragments carrying the ends of the element formed stable complexes with gene 1 protein. From the left end of the element, a fragment containing the terminal 270 bp interacted specifically with the gene 1 product. From the right end of the element, a fragment containing the terminal 350 bp was bound to the gene 1 protein. The complex formation showed a significant cooperativity, indicating that more than one binding site is present at the termini of En. This was corroborated by DMS-protection experiments for the left end of En, which indicated at least two specific binding sites around guanine residues at position 46 and 116, respectively. Both residues are at identical positions within the motif AAGAGTGTCGG, which contributes to the secondary structure of both ends of En.

Both binding sites are located upstream of the "TATA"-box (position 179) of gene 1. It is therefore possible that binding of gene 1 protein autoregulates the activity of gene 1 promoter.

The suppressible alleles, like a1-m1 (18) or bz-m13 (15), represent defective element insertions within the structural genes. We have speculated (3) that the suppressor function is exerted by binding of an element-encoded function to the ends of suppressible elements, thereby providing a steric block for transcription read-through. Our present findings agree very well with this view: The product of gene 1, which is probably the only product produced by the transposition-defective En-2 element, is a DNA-binding protein which recognizes specifically the ends of the En element and probably represents the suppressor component of the En/Spm system. It will be exciting to elucidate which role the gene 1 product plays in the transposition process: whether, for example, binding to all sites at the ends of the element results in the transition of a linear to a panholder-like structure by protein-protein interaction. Complex formation between gene 1 protein and En is probably the initial step in transposition. A second component (M-function?) may then bind to this complex and release the complex from the integrated state by cutting at the very ends of En.

REFERENCES

1. Banks, J., J. Kingsbury, V. Raboy, J.W. Schiefelbein, O. Nelson, and N. Fedoroff (1985) The Ac and Spm controlling element families in maize. Cold Spring Harbor Symposium on Quantitative Biology 50:307-311.
2. Bonas, U., H. Sommer, and H. Saedler (1984) The 17 kb Tam1 element of Antirrhinum majus induces a 3bp duplication upon integration into the chalcone synthase gene. EMBO J. 3:1015-1019.
3. Gierl, A., Zs. Schwarz-Sommer, and H. Saedler (1985) Molecular interactions between the components of the En-I transposable element system of Zea mays. EMBO J. 4:579-583.
4. Masson, P., R. Surosky, J.A. Kingsbury, and N.V. Fedoroff (1987) Genetic and molecular analysis of the Spm-dependent a-m2 alleles of the maize a locus. Genetics (in press).
5. McClintock, B. (1954) Mutations in maize and chromosomal aberrations in Neurospora. Carnegie Institute Washington Yearbook 53:254-260.
6. McClintock, B. (1963) Further studies of gene-control systems in maize. Carnegie Institute Washington Yearbook 62:486-493.
7. McClintock, B. (1965) The control of gene action in maize. Brookhaven Symposium on Biology 18:162-184.
8. McClintock, B. (1971) The contribution of one component of a control system to versatility of gene expression. Carnegie Institute Washington Yearbook 70:5-17.
9. Nevers, P., and H. Saedler (1977) Transposable genetic elements as agents of gene instability and chromosomal rearrangements. Nature 268:109-115.
10. Pereira, A., H. Cuypers, A. Gierl, Zs. Schwarz-Sommer, and H. Saedler (1986) Molecular analysis of the En/Spm transposable element system of Zea mays. EMBO J. 5:835-841.

11. Pereira, A., Zs. Schwarz-Sommer, A. Gierl, I. Bertram, P.A. Peterson, and H. Saedler (1985) Genetic and molecular analysis of the enhancer (En) transposable element system of Zea mays. EMBO J. 4:17-23.
12. Peterson, P.A. (1953) A mutable pale green locus in maize. Genetics 38:682-683.
13. Peterson, P.A. (1965) A relationship between the Spm and En control systems in maize. Am. Nat. 99:391-398.
14. Peterson, P.A. (1985) The isolation of En-1 in the wx-844 allele. Maize Genet. Coop. Newslett. 59:3.
15. Sommer, H., R. Hehl, E. Krebbers, R. Piotrowiak, W-E. Lönnig, and H. Saedler (1987) Transposable elements of Antirrhinum majus. (See this Volume.)
16. Schiefelbein, J.W., W. Raboy, N.V. Fedoroff, and O.V. Nelson (1985) Deletions within a defective Suppressor-Mutator element in maize affect the frequency and developmental timing of its excision from the bronze locus. Proc. Natl. Acad. Sci., USA 82:4783-4787.
17. Schwarz-Sommer, Zs., A. Gierl, R. Berndtgen, and H. Saedler (1985) Sequence comparison of "states" of a1-m1 suggests a model of Spm (En) action. EMBO J. 4:2439-2443.
18. Schwarz-Sommer, Zs., A. Gierl, R.B. Klösgen, U. Wienand, P.A. Peterson, and H. Saedler (1984) The Spm (En) transposable element controls the excision of a 2kb DNA insert at the wx-m8 allele of Zea mays. EMBO J. 3:1021-1028.
19. Studier, F.W., and B.A. Moffatt (1986) Use of bacteriophage T7 RNA polymerase to direct high-level expression of cloned genes. J. Molec. Biol. 189:113-130.br.i-4.
20. Tacke, E., A. Gierl, W. Rhode, L. Leclerq, M. Mattes, R. Berntgen, P.A. Peterson, and H. Saedler (1987) Influence of transposable elements on the structure and function of the A1 gene of Zea mays. EMBO J. 6:287-294.
21. Tacke, E., Zs. Schwarz-Sommer, P.A. Peterson, and H. Saedler (1986) Molecular analysis of states of the A1 locus of Zea mays. Maydica 31:83-91.
22. Wienand, U., U. Weydemann, U. Niesbach-Klösgen, P.A. Peterson, and H. Saedler (1986) Molecular cloning of the c2 locus of Zea mays, the gene coding for chalcone synthase. Molec. Gen. Genet. 203:202-207.

REGULATION OF MUTATOR ACTIVITIES IN MAIZE

Virginia Walbot, Anne Bagg Britt,
Kenneth Luehrsen, Margaret McLaughlin,
and Christine Warren

Department of Biological Sciences
Stanford University
Stanford, CA 94305

ABSTRACT

We discuss the properties of the Mutator (Mu) transposable element family of maize. We report the cloning of bz2-mu1, a mutable allele containing a 1.4-kb Mu element, using a combination of transposon tagging and tests for differential hybridization to northern and Southern blots. We report the sequence of this allele and the Mu element insertion, and propose a model for the structure of the Bz2 locus. We discuss the relationship between increased DNA modification of Mu elements and loss of somatic instability at bz2-mu1. To further explore this aspect of regulation of Mutator, we have used gene-specific probes to determine the level of modification at this locus in active and inactive Mutator lines. We have also utilized CsCl density gradients to estimate the overall level of DNA modification in active and inactive lines; we find that Mu elements in active lines are hypomethylated relative to other maize nuclear DNAs examined, and that in inactive lines the level of modification in Mu elements is similar to the genome as a whole. Utilizing γ-irradiation, we have demonstrated that inactive lines can be reactivated; this reactivation is first noted as restitution of the spotted kernel phenotype characteristic of bz2-mu1 in active Mutator lines. Hybridization analysis of DNA from reactivated plants demonstrates that the Mu elements in general, and specifically the Mu element at bz2-mu1, have the lower level of DNA modification characteristic of active lines. These results are discussed in terms of the role and timing of DNA modification in regulating Mutator activities.

INTRODUCTION

Discovery of the Mutator Transposable Element System

Approximately one decade ago, the Mutator phenomenon was described by Robertson (20) as a markedly increased mutation frequency in a particular stock. Because about one-third of the diverse types of mutants recovered were mutable, a transposable element system was suspected. The Mutator system, however, did not interact with other well-known transposable element systems of maize, so a new type of transposable element family was hypothesized (26). Subsequent characterization of an unstable mutant of Alcohol dehydrogenase-1 (Adh1) selected from a Mutator background verified the prediction that Mutator-induced mutations would result from a novel insertion sequence (6,31).

The 1.4-kb insertion in the first intron of Adh1 was termed Mu1. Sequence analysis demonstrated that Mu1 had ∿200-bp terminal inverted repetitive sequences (2) and that a 9-bp host sequence duplication was generated at the site of its insertion in the Adh1 gene (6). Hybridization of the Mu1 probe to Southern blots containing maize DNA digested with restriction enzymes that recognize sites in the terminal inverted repeat sequence demonstrated that Mutator stocks contain at least two size classes of elements: the 1.4-kb type and a slightly larger 1.7-kb element (3). Based on analysis of the sequences present in these two element types, the 1.4-kb type is postulated to be a deletion derivative of the 1.7-kb element (36).

Mu-homologous sequences are present in high-copy numbers in Mutator stocks (3). In addition, all non-Mutator lines examined thus far contain Mu element homology (8). There are more copies of Mu terminal inverted repeat sequences than of the internal portion of Mu1 (8).

As is noted with other transposable elements of maize, Mu elements are associated with frequent chromosomal deletion events. In one case, the sequence of a deletion has been obtained (35). This 74-bp deletion was contiguous to the Mu1 insertion site in the Adh1-S3034 allele; it eliminated the 5' donor site of the first intron and 2 bp of the first exon, rendering the locus nonfunctional. Genetic evidence for deletions has also been obtained by observing loss of markers on chromosome arm 9S (27).

That Mu elements--defined by the presence of terminal inverted repeat sequence homologous to Mu1--explain the high mutation frequency of Mutator stocks is well established. Only insertion elements related to Mu1 have been found in mutable alleles recovered from Mutator plants. Perhaps 20 unstable mutations in genes for which DNA hybridization probes exist have been analyzed at the level of Southern blots: these mutable alleles contain primarily 1.4-kb and 1.7-kb elements (see Bennetzen, this Volume; Ref. 18, 36). A 1.8-kb Mu element has also been described that shares little homology with the internal sequences of Mu1 (see Freeling, this Volume) and a 1.5-kb element in a mutable waxy allele (see Wessler, this Volume). As more genes are analyzed, other Mu size classes may be discovered.

To date, however, little is known about the molecular nature of the autonomous element programming excision and insertion of the (likely) nonautonomous elements that have been described thus far. Although the 1.4-kb Mu1-like group of elements are ubiquitous in Mutator stocks, they are considered unlikely candidates for the autonomous element. The Mu1.7 element can transpose (34) and is slightly larger than the Mu1-like elements, but it also lacks an open reading frame large enough to encode a typical transposase protein. Transcripts containing Mu sequences have been detected, but those analyzed in detail arise from transcription initiating in the gene into which a Mu element has inserted (28,36).

Of interest, therefore, is detection of a 5' consensus splice site in Mu1.7 in the appropriate orientation for splicing in bz1-mu2 transcripts. Based on fine structure mapping of the RNA transcripts using northern blots, this splice site is apparently utilized with a cryptic 3' acceptor site in the distal Bz1 coding region to produce a processed transcript containing part of a Mu element (36). More detailed studies analyzing the sequences of cDNA clones have demonstrated cryptic splice sites in Ds (a 5' donor sequence, Ref. 41) and dSpm (a 3' acceptor sequence, Ref. 15) elements. In these cases, splicing creates mRNAs encoding partially functional proteins.

Genetic Puzzles of the Mutator System

Despite considerable work in several laboratories, there are still major facets of the Mutator transposable element system that are poorly understood. Most important, unlike the autonomous elements Ac and Spm(En) that demonstrated simple segregation patterns critical to their identification as the autonomous (transposase-encoding) elements of their respective element families, the autonomous element(s) of the Mutator system generally fail to segregate in outcrosses to non-Mutator lines. Based on the infrequent loss of the Mutator phenomenon, Robertson (20) calculated that at least three independently segregating and active autonomous elements would be required. Despite years of outcrossing, stocks apparently segregating for an autonomous element have rarely been encountered, although a few are under scrutiny now (see Robertson, this Volume). And, as discussed above, no large elements have been discovered from the analysis of Mu element insertions into known genes.

A second, striking feature of the Mutator phenomenon is the late timing of Mu element activities in some tissues. Most somatic excisions in mutable genes occur late in development (21,22,24), resulting in tiny sectors (Fig. 1A). Similarly, new mutants are typically recovered as infrequent pollen events when the Mutator plant is used as male and as single mutable kernels when a Mutator plant is the ear parent (22,39). This bias for events late in the development of the aleurone and at the time of gamete production is unusual among transposable element systems of maize.

A third feature of the genetics of Mutator is that germinal revertants are rarely recovered. This creates a paradox, because both the forward mutation rate and the somatic reversion of individual elements are high. If somatic reversion usually results from element excision, then more germinal revertants would be expected if new insertions during meiosis result from transposition of existing elements from an original to a new site. Compelling evidence that new copies of Mu elements are generated comes from studies of the inheritance of element copy number through crosses. Progeny resulting from crosses of a Mutator and a non-Mutator plant have, on average, the same number of Mu elements as the Mutator parent (1), although the range of copy numbers is very wide (40). To maintain copy number through such crosses, the Mu element number must double. For a variety of reasons, this copy-number maintenance is proposed to occur late in plant development, perhaps during gamete development, and prior to fertilization (1,5,40).

The doubling of Mu element copy number does not, however, involve loss of existing Mu insertion sites. Alleman and Freeling (1) demonstrated that most Mu-containing restriction fragments are inherited in a simple Mendelian fashion through crosses, while new Mu-homologous fragments

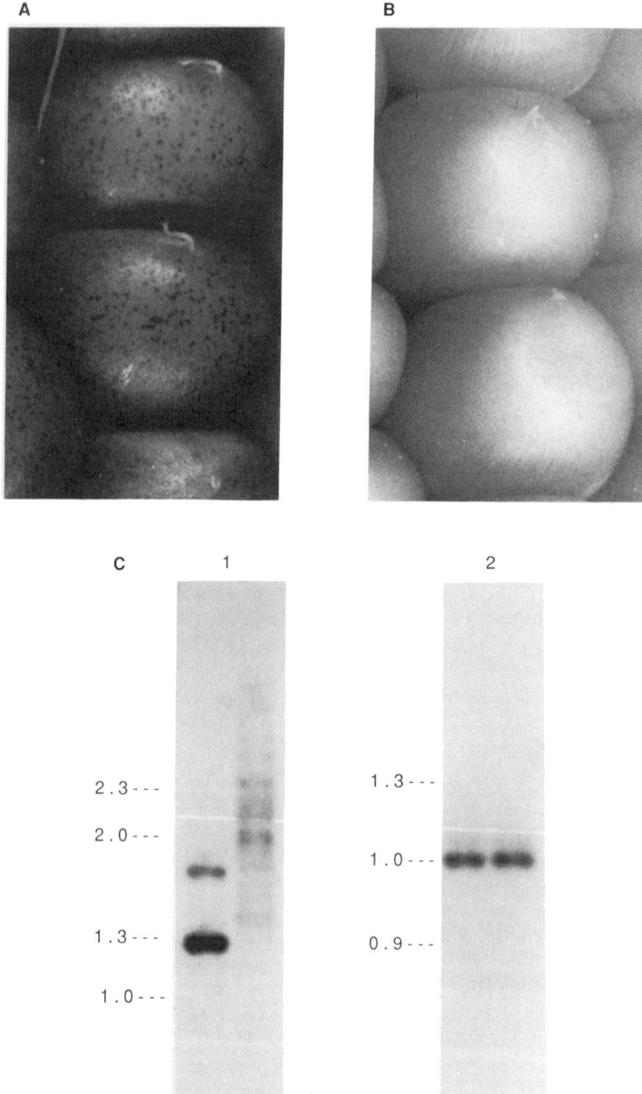

Fig. 1. Somatic instability at bz2-mu1 in active and inactive Mu lines. A. Fine spotting in the aleurone tissue is characteristic of somatic excision from Mu-induced mutable alleles. B. Lack of somatic instability of heterozygous seed (bz2-mu1/bz2) in an inactive Mu line. C. Demonstration of masking of restriction sites in inactive Mu lines. Panel 1 shows HinfI-digests of active (left) and inactive (right) Mu materials probed with plasmid pA/B5; this plasmid contains a 650-bp internal fragment of Mu1 (9). HinfI is sensitive to C methylation, and there is a dramatic difference in the patterns observed in the active and inactive lines. Panel 2 shows that identical hybridization patterns are obtained with TaqI-digested samples probed with pA/B5. This methylation-insensitive enzyme digests near the border of the terminal inverted repeats of both Mu1.4 and Mu1.7 elements.

arise at a high frequency among the progeny. Similarly, L.P. Taylor and V. Walbot (unpubl. data) have demonstrated that parental Mu1.7 elements show simple segregation in outcross progeny, even in cases in which the individual progeny inherited all of the parental fragments and at least three novel fragments. These results suggest that excision of elements from existing sites is not coupled directly to insertion of elements in new sites. Further support for this idea comes from the observation of non-chromosomal copies of Mu1-like elements in maize tissues (32).

Regulation of Mutator Activities

Mutator activities are inherited by most, but not necessarily all, progeny of crosses between Mutator and non-Mutator lines (20). Furthermore, Mutator activity--measured as an increased level of mutant recovery, a reflection of Mu element insertion--can be lost after several generations of self-pollination of Mutator plants (23,25). Using an alternative assay--loss of the spotted kernel phenotype diagnostic for Mu element excision from a reporter gene as illustrated in Fig. 1--we found that some individuals lose Mutator activity during either selfing or outcrossing to non-Mutator lines (37). Loss of somatic instability correlates with an increase in the level of DNA modification of Mu elements, measured as masking of many restriction enzymes sensitive to methylation of C residues in the recognition site (9). Similarly, selfed lines in which the high forward mutation rate characteristic of Mutator lines is lost also show an increased level of DNA modification (4).

Once established, the inactive state of Mutator lines is relatively stable (5,39). When used as the ear parent, inactive lines can suppress the somatic instability of reporter alleles transmitted through the pollen (37); in reciprocal crosses the reporter allele can be unaffected by the modified Mu elements introduced from the pollen. Similarly, the state of DNA modification also shows maternal dominance in crosses between active and inactive lines (4). DNA modification appears to be correlated with loss of activity in Ac elements as well (10,30) and may be a general feature of maize transposable element regulation.

Interestingly, DNA modification has the property of being potentially reversible. Reversible cycling between active and inactive behavior was first noted by McClintock as a property of maize controlling elements and termed "changes of phase" (reviewed in Ref. 12). Such behavior could be readily explained by changes in the level of DNA modification of transposable elements; however, much remains to be learned about the regulation of the modification process.

Approximately 25% of the C residues in maize nuclear DNA (49% G&C content) are methylated (14). In other higher plant species, 5'-C methylation is restricted to residues in 5'-CG-3' and 5'-CNG-3' sequences (13); thus, many of the C residues in these substrate sites in maize DNA are methylated. Once established, methylation patterns are hypothesized to be maintained through mitosis for two reasons. First, each daughter duplex is hemimethylated, containing one modified parental strand; this strand serves as part of the signal or template for modification of the symmetric sites on the other strand. Second, the rate of modification in hemimethylated DNA is much higher than de novo modification of substrate sites in an unmethylated duplex (19).

RESULTS AND DISCUSSION

Cloning of the bz2-mu1 Mutable Allele

The mutable bz2-mu1 allele was recovered in 1982 from a large mutagenesis experiment in which Mutator plants having all of the genes required for a purple aleurone were crossed to various tester stocks lacking required genes (39). This mutable allele was propagated by outcrossing to non-Mutator lines in attempts to dilute the number of Mu elements per genome. To identify a single Mu-containing fragment that segregated with the mutable phenotype, DNA was prepared from individual progeny plants, digested with restriction enzymes that should cleave outside the Mu element sequences, and probed with Mu-specific probes. Using this approach, no fragment could be unambiguously identified as a candidate for bz2-mu1. Consequently, a different strategy was employed (17) and is summarized here.

First, a genomic library was constructed from a partial Sau3A digest of maize DNA containing this allele (Fig. 2). The C230-7 (40) lineage chosen had about 12-15 copies of Mu elements, a relatively low number. The library was screened with an internal fragment of Mu1 and 25 phage recovered. Twenty of these were analyzed and shown to contain either a 1.4-kb or 1.7-kb Mu element. These phage were then nick-translated and used as probes to northern blots containing alternate lanes of RNA isolated from purple (Bz2, B, Pl) or green (bz2, b, pl) husk tissues. Only one phage (#15) was identified that hybridized to RNA from purple but not green tissue.

For confirmation, phage #15 was hybridized to a Southern blot containing DNA from plants with several types of Bronze2 alleles digested with either EcoRI or BamHI, two enzymes that do not digest Mu1. The phage hybridized to a single fragment in a Bz2 stock but not to DNA from the an-bz2-6923 mutant in which a two map unit deletion eliminates Bronze2 function. As expected, phage #15 hybridized to larger fragments in the DNA extracted from the bz2-mu1 sample than from progenitor stock; similarly, the phage hybridized to a larger fragment in the DNA from bz2-m1 (a mutable allele containing a Ds insertion; reviewed in Ref. 12) than from a Bz2 revertant stock (17).

Hybridization to northern blots of single-stranded DNA probes derived from sequences flanking the Mu insert in bz2-mu1 established the direction of transcription and extent of the approximately 850-bp transcript. Fine structure restriction mapping of subclones of phage #15 determined the approximate site of insertion of a 1.4-kb Mu1-like element in the bz2-mu1 allele, and genomic Southern mapping was used to establish the approximate site of insertion and orientation of a 1.4-kb insertion in a second mutant, bz2-mu2. In Fig. 3, a restriction map summarizes these data.

Proposed Structure of bz2-mu1

As shown in Fig. 4, the DNA sequence of the 1.4-kb Mu1-like insertion in bz2-mu1 and approximately 1.1 kb of flanking region has been determined. The Mu element is 1,370 bp in length and is thus 6 bp smaller than the 1,376-bp Mu1 element (2). The two elements are 99% homologous at the sequence level; the position of each difference is enumerated in the legend to Fig. 4. As with all other Mu elements investigated, a

CLONING *bz2-mu1* BY TRANSPOSON TAGGING AND DIFFERENTIAL RNA HYBRIDIZATION

Construct library from *bz2-mu1* plants with low *Mu* copy number

Screen with pA/B5 internal *Mu1* probe

Restrict selected phage with *Hin*fI to determine size of *Mu* element inserts

Hybridize phage to Northern blot of RNA from purple (*Bz2 B Pl)* and green (*bz2 b pl)* husk tissues. One phage showed differential hybridization.

Purple

Green

Hybridize putative *bz2-mu1* phage to stocks with known *Bronze2* alleles. Hybridization pattern is consistent with identification as an allele of this locus.

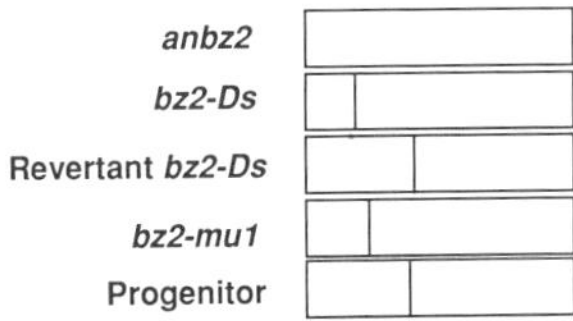

Fig. 2. Strategy used to clone bz2-mu1.

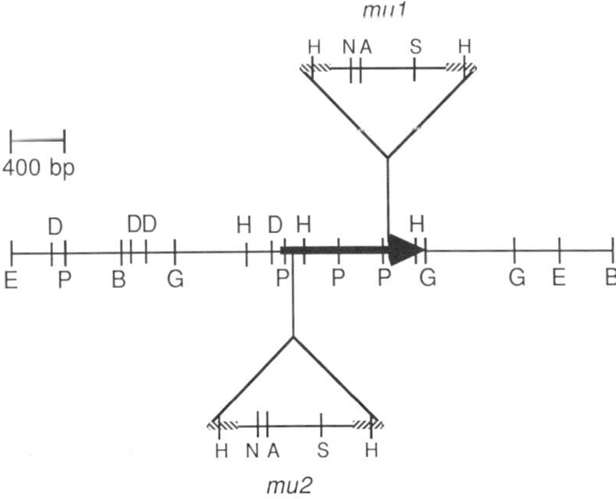

Fig. 3. Restriction map of the cloned bz2-mu1 allele. The Mu1.4 elements in this allele and in bz2-mu2 are shown as triangles inserted into the Bronze2 sequence (thin line). The terminal inverted repeats of the Mu elements are represented by cross-hatched bars. The extent and direction of transcription are indicated by the thick arrow on the gene. Only a subset of the HinfI sites is shown. Restriction enzymes are indicated as A = AvaI, B = BamHI, D = DraI, E = EcoRI, H = HinfI, G = BglII, N = NotI, P = PstI, and S = BstNI.

9-bp duplication (GCCAGACAC) of genomic DNA was created at the point of insertion. Surprisingly, the right terminal inverted repeat in the cloned sequence is truncated and is missing seven of the terminal eight base pairs. If this structure exists in vivo, it suggests that this terminus of Mu is not required for excision from the mutable bz2-mu1 allele.

We hypothesize that the Bronze2 gene has the structure shown in Fig. 5. Hybridization of subcloned regions of the bz2-mu1 allele to northern blots of poly(A)+ RNA from purple tissues identifies an 850-bp transcript as shown by the thick arrow in Fig. 3; probes flanking this region show no hybridization. Unless there are unusually large introns (>800 bp), all of the transcribed region is shown in the sequence presented. No long open reading frame was present in the transcribed region identified. The removal of a 44-bp region surrounded by sequences having close matches with intron donor and acceptor splice sites (as defined in Ref. 7), however, creates an open reading frame coding for a 273-amino acid polypeptide (approximately 30 kD) and an RNA of at least 819 bases. S_1 mapping and primer extension experiments are in progress to confirm the gene model presented in Fig. 5. If the proposed structure is correct, the 1.4-kb Mu insertions of both mutants are within the open reading frame of the Bz2 gene product.

Further Studies of the Modification of Mu Elements

Now that the bz2-mu1 allele has been cloned, gene-specific probes have been used to study the pattern of DNA modification in the Mu1.4 element inserted in this allele (C. Warren and V. Walbot, unpubl. data). A detailed description of these results will be presented elsewhere. The most important finding is that the modification status of this particular Mu element parallels the status of the Mutator system precisely: the element's HinfI sites are digested in active and reactivated plants but are not digested in inactive plants. These results confirm that the somatic mutability assay we employ to gauge the activity state of Mutator lines accurately reflects the molecular status of the Mu elements in general and at the reporter gene specifically.

We are also studying the global modification of Mu elements in inactive lines using a new experimental protocol. The assay of DNA modification by restriction digest, while providing precise data on the degree of modification of specific sites, is necessarily limited to the analysis of a particular subset of restriction sites of previously cloned sequences. We have applied equilibrium density gradient centrifugation in neutral CsCl as an alternative assay of the extent of DNA modification.

The density of any DNA molecule depends upon both its GC content and its degree of modification. Schildkraut et al. (29), in a study of 51 DNAs of known GC content, empirically determined the relationship between mole fraction guanine plus cytosine (GC) and density (ρ) to be ρ = 1.660 + 0.098 (GC).

Because the insertion of a methyl group increases the volume (as well as the mass) of the DNA molecule, Kirk (16) predicted a decrease in density with methylation of 0.6 mg/ml per mole fraction methylated residue. However, the somewhat scant data on DNAs of known methyl content (11, 33) have indicated a decrease of 1 mg/ml per mole fraction methylated residue. Our laboratory has previously demonstrated (9) a correspondence between *Mu* sequence modification status, measured as masking of restriction digestion sites, and excision activity. To determine the global modification status of *Mu* elements, we are now measuring the buoyant density of *Mu* sequences. Because the GC content and DNA sequence of several types of *Mu* elements are known, we can use density determinations in different samples to estimate changes in DNA methylation. This analysis assumes that the only modification difference between active and inactive Mutator lines is the degree of 5'-C methylation.

1 GAGATAATTG CCATTATGGA CGAAGAGGGA AGGGGATTCG ACGAAATGGA
51 GGCGTTGGCG TTGGCTTCTC TGTTTTGGAG ACGCACGCGA CAGCCAAACT
101 CCAAAACGGA TACGAGACAG CTCTTGGGGC TGCGTAAACA GGTATTAGTT
151 TTCTGTCCCC GTTTACCGTT CCCGTGCGCA GACGCCGTCA CGCGTACTCT
201 TCTTGTCTCC GTCGCCGCGC TCTACGGCAG CACGGCGGTG CTGGCGTACG
251 TGCTG**C**ACTC CGCGCCGGCA GAGGCCGCGC GCGCGTCGCC CGCGGACGAC
301 GCCACGCCGC TCCACCTCGC CGCGGCCGCG CACATCCTCC TCGCCGCGAG
351 CGCGTCCGCG GACGCGCGCG CCTTCTCGGG CCTCCACGCC GGGGACCTCC
401 TCCTCCCGCG CGCCAACGAG GCCGCCGCCG CCGCGGACCG GGC**GC**TCCGC
451 GTGCTCCTCA AGTTCCCCGC GGTGTCACTG TCGTCCTCGC CCAAGAAGTC
501 CGCCTCGCCG CCGCCGGCCC CGGAGGCGAG GAAGGAGTAC CCGCCCGACC
551 TGACGCTGCC GGACCTAAGA GCGGGCTGTT AGCACCGACG AGCAACGGAC
601 GACCTTCCAC CTGTCCTAGC ACTAATTACT CCTAAA**C**TTC AT**T**ATACACC
651 AAAGTTTCAA TTAGTAAAAG GTTTTGGTAT TATTTTCTTT ACAAGACTAA
701 AAGCATCCAC TCGTATTTGC CATGGAAATA TTGCCAAAAT GGTTACCGAA
751 GGAACCAATG AAAAATGGCG GTATCCGTCC ACCTGTAGCT TGTCCAAGGC
801 TCTTCCAGCA GAGTAGTGCC GCCATGACGA TTGACAGAGA CACGAGACGA
851 AACAAGCTGA AGGTCCCCGC GGCGTCACTG TCGTCCTCGC CGAAGAAGTC
901 CGCCTCGCCG CCGCCGGCCC CGGAGGCGAG GAAGGAGTAC CCGCCCGACC
951 TGACGCTGCC CACCGTGTCG CCGCCGCCGC CCAACGGCCT CGGCGACATG
1001 CTCAGCCCAG CGGCCTGGCC CTCCTCCCCC GCGAGCAGGC TCAACAAGGC
1051 CGCGCTCGGC GGCGGCCGGG ACCT**G**GACCT GGACCAGTAC CAGCACATGT
1101 TGTT**G**CGACA AGGTGTCGTC GTCGAGGGCC AGTAGGAGAC AAGAAGAGTA
1151 CGCGTGACTG AGATGCGACG GAGAAAAAGG GTACGCGTGA CGGCGTCTGT
1201 ACACGGGAAC GGTAAATGAG GACAGGAAAC TAATAACTGT TTACGCAGCC
1251 CCAAGTGCTG TCTTGTGTCC GTTTTAGAGT TTGGCTGTCG CGTACGTCTC
1301 TAAAACAGAG AAGCCAACGC CAACGCCTCT ATTTCGTCGA ATCCGCTTCT
1351 CTCTTCGTCC ATAATGGCA**C**

Fig. 4. Sequence of the *Mu* element inserted within *bz2-mu1*. Only one strand was sequenced and used for the comparison to the *Mu1.4* element sequenced by Barker et al. (2); differences are shown in bold type. Base pair changes occur at positions 256, 643, 1075, and 1370. Single nucleotide insertions occur at positions 637 and 1105. A two-nucleotide insertion occurs at positions 444-445. Single base pair deletions occur following nucleotides 566, 580, and 601. A 7-bp segment of sequence (following bp 1369) is missing from the right terminal inverted repeat; the TIR region capable of base pairing is underlined.

1 TTTAAATTGC CGGCCGCATC AAGCTGAGGG CTGAGGCAGA GAGGTGCCAA

51 CAGAAGTCAC GTCGAGACCA GCCGGCCGGC CGGACTGCAG CTCGTCGTCG

101 TCG**ATG**ACGG CCGGGACCAT GCGTGTGCTA GGCGGGGAGG TCAGCCCGTT

151 CACGGCGCGG GCGCGTCTGG CGCTGGATCT GCGCGGCGTG GCGTACGAGC

201 TCCTCGACGA GCCGCTGGGG CCCAAGAAAA GCGACAGGCT CCTCGCTGCC

251 AACCCCGTCT ACGGGAAGAT CCCCGTGCTG CTCCTCCCCG ACGGCCGCGC

301 CATATGCGAG TCCGCAGTCA TCGTCCAGTA CATCGAGGAC GTGGCGCGTG

351 AAAGCGGCGG CGCCGAGGCT GGCAGCCTGC TGCTCCCGGA CGACCCCTAC

401 GAGCGCGCCA TGCACCGCTT CTGGACCGCC TTCATCGACG AC*AAGGTGAG*

451 *CACCGAGCAG AGCACGGCAG CAAGTGTCTT TTCTTCAAGG* TAACGTGCAA

501 CAAACTGCCG TGCCTTTCTG CAGTTTTGGC CGGCGCTGGA TGCCGTCTCC

551 CTGGCGCCGA CCCCGGGAGC ACGCGCGCAG GCCGCGGAAG ACACCCGCGC

601 CGCGCTGAGC CTCCTGGAGG AGGCGTTCAA GGACCGCAGC AACGGCAGGG

651 CTTTCTTCTC CGGTGGCGAC GCCGCGCCAG GCCTCCTGGA CCTGGCCCTC

701 GGATGCTTCC TACCGGCGCT CAGGGCCTGC GAGCGGCTCC ACGGCCTCTC

751 ACTCATCGAC GCGTCCGCGA CGCCGCTGCT GGACGGGTGG AGCCAGCGCT

801 TCGCCGCGCA CCCTGCAGCC AAGCGCGTCC TGCCAGACAC GGAGAAGGTG

851 GTGCAGTTCA CGAGGTTCCT CCAGGTCCAG GCACAGTTCA GGGTCCACGT

901 GTCCTAAATG ATTTAGCGTC TCCAAAAGAT TGTCAAAATC TCCGTCTAAA

951 GTCTGTTTTA GAAACT**TAA**A TCACTTTCAG GATTCTCGGA AATTGAGAGA

1001 AA**AATAAA**AT AACTTTTCAC TAATCTTCAG AAATCTAAGA GAATTTCAGT

1051 TTTCAAACTA GCTCTAAATG TGGTGGAGAG AAGGAAAAAA AGATCT

Fig. 5. Sequence analysis of bz2-mu1. Both strands of the transcribed region between the DraI and BglII sites shown in Fig. 3 were sequenced. The Mu element and a single 9-bp duplication sequence (dotted underline) have been removed to recreate the putative wild-type Bz2 allele. The proposed regulatory sequences are in bold type: ATG start codon, TAA termination codon, and poly(A) addition site. A 44-bp intron region (shown in italics) is flanked by consensus 5' and 3' splice junctions (underlined); other close matches with the 3' splice site consensus also occur downstream of the one chosen here. An open reading frame coding for a 273 amino acid polypeptide is created by removal of the 44-bp region. The PstI sites are boxed.

In a typical experiment, DNA is prepared from several pooled seedlings or an immature cob of active and inactive lines. About 50 μg of these DNAs are then digested to completion with the methylation-insensitive enzyme TaqI, which has three recognition sites within the Mu1 transposon (see Fig. 6). The digests are centrifuged to equilibrium in CsCl gradients and collected into 25 fractions. The density of each fraction is then determined from its refractive index. While the fractions carrying the bulk of the DNA can be identified spectrophotometrically, slot blots are used to determine the position in the gradient to which specific sequences have migrated. The blots are hybridized in turn to pA/B5, an internal 650-bp fragment of Mu1, and probes to various standards of known DNA sequence. Standards include fragments of lambda DNA mixed with the maize DNA prior to centrifugation, or internal standards such as maize mitochondrial sequences which lack all DNA modification and are present in high copy number (>1,000 copies/haploid genome). By aligning the slot

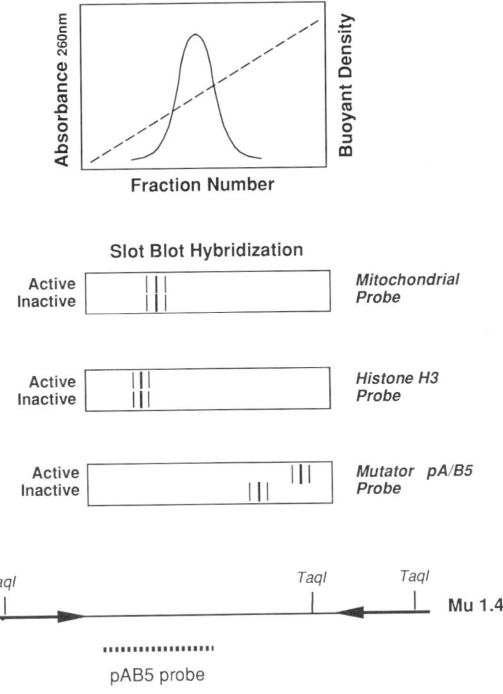

Fig. 6. Strategy for determination of the extent of Mu element DNA modification using CsCl buoyant density centrifugation.

blots based on the position of hybridization of these standards and the buoyant densities determined by refractometry, the relative position of the Mu-hybridizing fractions can be assessed.

Our preliminary results (Tab. 1) indicate no variation (within a margin of error of ± 5%) in the overall level of methylation of the maize genome between active and inactive lines. For example, the average degree of methylation of the histone H3 gene family (∿50 copies/haploid genome) TaqI fragment is invariant; about 65% of all C residues or 100% of the C residues in CG and CNG substrate sequences would have to be modified to account for the buoyant density of these genes. Mu sequences, however, are apparently methylated at only 47% of all C nucleotides or 70% of the CG and CNG sequences in active lines; in contrast, Mu sequences in inactive lines are methylated at 70% of all C residues or essentially 100% of the substrate sites.

Our results suggest that Mu sequences, in comparison to histone genes, are hypomethylated in active Mutator lines. This conclusion extends the prior work with methylation-sensitive restriction enzymes to demonstrate that most substrate sites in Mu elements are subject to DNA modification in inactive lines. Analysis of the distribution of C-methylation in active lines and in lines switching from active to inactive status may provide some clues about the role of specific sequences in Mu elements required for element activities.

Tab. 1. Experimental results.

	Apparent GC content	Known GC content	Percent methylation	Methylation of CG and CNG
Mu1, active line	50.9	63.8	12.9%	66.5%
Mu1, inactive line	43.6	63.8	20.2%	100%
Histone H3, active line	50.9	73.7	23.8%	100%
Histone H3, inactive line	52.0	73.7	21.7%	91%

All values adjusted to mitochondrial standard.

Reactivation of Inactive Mutator Stocks

Our current experimental approach to studying the regulation of Mutator activities is to determine what kinds of agents or treatments will reactivate lines containing an inactive Mutator system and a cryptic mutable allele. In most cases, the bz2-mu1 allele is under observation. Inactive lines rarely reactivate spontaneously on crossing to a bz2 tester stock (W23 and K55 backgrounds); no cases of reactivation have been noted when inactive plants were used as the female parent (37). Furthermore, inactive lines have been shown to suppress somatic instability in mutable alleles transmitted from an active male parent (39). Molecular support for maternal effects on the level of DNA modification in Mu elements has also been reported (4).

In 1985, inactive Mutator lines were treated with γ-irradiation at the dry seed stage; subsequently, the plants were crossed by bz2 tester and scored for reactivation of somatic mutability at bz2-mu1. With 12,000 Roentgen (r) treatment, but not at lower dosage, a few cases of reactivation were found. In 1986, a more extensive experiment was performed, and a clear dose-dependence for reactivation was established: at 15,000r, 0.1% of progeny kernels were reactivated (38).

Molecular analysis of the plants grown from the reactivated kernels established that the Mu elements as a whole, and specifically the Mu1.4 element inserted in bz2-mu1, had regained the HinfI sites typical of elements in active lines. By all criteria the reactivated plants were identical to the original, active Mutator material. We have speculated that the lower level of DNA modification associated with reactivation arises as a consequence of DNA repair following the irradiation treatments (38). However, we have no evidence to support this model.

Experiments in progress include attempts to reactivate with 5-azacytidine, an agent known to block methylation of C residues (19). Preliminary evidence from a small experiment in 1986 suggests that this drug can reactivate somatic instability at bz2-mu1 (S. Otto and V. Walbot, unpubl. data). In addition, we are testing treatments that plants might be expected to experience during a natural lifespan, i.e., ultraviolet radiation, heat, cold, and nutrient deprivation.

The mechanism by which increased C modification (most likely 5'-methylation) limits Mu element excision is not known. Although many studies correlate the extent of C-methylation and gene expression (19), there is no evidence that the Mu1.4 element in bz2-mu1 encodes a product required for element movement. Indeed, this element is probably nonautonomous. It is possible that the lower level of C modification in Mu elements compared to bulk maize DNA (Tab. 1, and V. Walbot, C. Warren, and A.B. Britt, unpubl. data) is maintained by the exclusion of DNA-methylating enzymes by a factor such as transposase that is bound to all Mu elements. Because the heritability of the level of DNA methylation is relatively high through mitosis, such a bound factor could be a transient feature of Mu elements and still be quite effective, resulting in hypomethylation of Mu elements relative to flanking DNA sequences.

A second possibility is that extrachromosomal replication (32) of Mu elements occurs without subsequent DNA methylation. When these newly amplified Mu elements insert into chromosomal DNA, they would be in a completely unmethylated state. Over time, however, de novo methylation might gradually increase their level of modification. This hypothesis offers the possibility of a stochastic "clock" regulating Mutator activities. First, a burst of Mu element replication creates a population of new, unmodified elements, but then, inevitably, the activities of these elements are suppressed by imposition of DNA modification by the host methylation system. Escape from this suppression could involve protection by transposase as well as the production of a new population of unmodified elements.

It is important to consider that the C-modification that appears to be so closely correlated with inactive Mu elements (and now also with inactivity in other maize transposable elements) may be a consequence rather than a cause of element stability. Inactivation of an autonomous element could precede, by many cell generations, the accumulation of methylation in nonautonomous elements of that family. Thus, the modification status of the Mu1.4 elements would merely be a reflection of more important regulatory events. It will be challenging to answer the questions raised about the role of Mu element modification. To do so may require development of assays for Mutator activities in tissue culture or protoplasts in which DNA replication, repair, and modification can be manipulated more easily than in whole plants.

ACKNOWLEDGEMENTS

Research was supported by grants from the United States Department of Agriculture (86-00172) and the National Institutes of Health (GM 32422). A.B.B. is a postdoctoral fellow in the Training Program in Cancer Biology (USPHS CA09302) at Stanford University, K.L. is a postdoctoral fellow of the American Cancer Society, and M.M. was supported by a National Science Foundation postdoctoral fellowship in Plant Biology.

REFERENCES

1. Alleman, M., and M. Freeling (1986) The Mu transposable elements of maize: Evidence for transposition and copy number regulation during development. Genetics 112:107-119.
2. Barker, R.F., D.V. Thompson, D.R. Talbot, J. Swanson, and J.L. Bennetzen (1984) Nucleotide sequence of the maize transposable ele-

ment Mul. Nucl. Acids Res. 12:5955-5967.

3. Bennetzen, J.L. (1984) Transposable element Mul is found in multiple copies only in Robertson's mutator maize lines. J. Molec. Appl. Genet. 2:519-524.
4. Bennetzen, J.L. (1987) Covalent DNA modification and the regulation of Mutator element transposition in maize. Molec. Gen. Genet. 208:45-51.
5. Bennetzen, J.L., R.P. Fracasso, D.W. Morris, D.S. Robertson, and M.J. Skögen-Hagenson (1987) Concomitant regulation of Mul transposition and Mutator activity in maize. Molec. Gen. Genet. 208:57-62.
6. Bennetzen, J.L., J. Swanson, W.C. Taylor, and M. Freeling (1984) DNA insertion in the first intron of maize Adhl affects message levels: Cloning of progenitor and mutant Adhl alleles. Proc. Natl. Acad. Sci., USA 81:4125-4128.
7. Brown, J.W.S., G. Feix, and D. Frendewey (1986) Accurate in vitro splicing of two pre-mRNA plant introns in a HeLa cell nuclear extract. EMBO J. 5:2749-2758.
8. Chandler, V.L., C.J. Rivin, and V. Walbot (1986) Stable non-Mutator lines of maize have elements homologous to the Mul transposable element. Genetics 114:1007-1021.
9. Chandler, V.L., and V. Walbot (1986) DNA modification of a maize transposable element correlates with loss of activity. Proc. Natl. Acad. Sci., USA 83:1767-1771.
10. Chomet, P.S., S. Wessler, and S.L. Dellaporta (1987) Inactivation of the maize transposable element Activator (Ac) is associated with its DNA modification. EMBO J. 6:295-302.
11. Dawid, I.B., D.D. Brown, and R.H. Reeder (1970) Composition and structure of chromosomal and amplified ribosomal DNA's of Xenopus laevis. J. Molec. Biol. 51:341-360.
12. Fedoroff, N.V. (1983) Controlling elements in maize. In Mobile Genetics Elements, J. Shapiro, ed. Academic Press, New York, pp.1-63.
13. Gruenbaum, Y., T. Naveh-Many, H. Cedar, and A. Razin (1981) Sequence specificity of methylation in higher plant DNA. Nature 292: 860-862.
14. Hake, S., and V. Walbot (1980) The genome of Zea mays, its organization and homology to related grasses. Chromosoma (Berl.) 79:251-270.
15. Kim, H.-Y., J.W. Schiefelbein, V. Raboy, D.B. Furtek, and O. Nelson, Jr. (1987) Looking for it. Proc. Natl. Acad. Sci., USA (in press).
16. Kirk, J.T.O. (1967) Effect of methylation of cytosine residues on the buoyant density of DNA in caesium chloride solution. J. Molec. Biol. 28:171-172.
17. McLaughlin, M., and V. Walbot (1987) Cloning of a mutable bz2 allele of maize by transposon tagging and differential hybridization. Genetics (in press).
18. O'Reilly, C., N.S. Shepherd, A. Pereira, Zs. Schwarz-Sommer, I. Bertram, D.S. Robertson, P.A. Peterson, and H. Saedler (1985) Molecular cloning of the al locus of Zea mays using the transposable elements En and Mul. EMBO J. 4:877-882.
19. Razin, A., H. Cedar, and A.D. Riggs (1984) DNA Methylation: Biochemistry and Biological Significance. Springer-Verlag, New York.
20. Robertson, D.S. (1987) Characterization of a mutator system in maize. Mutat. Res. 51:21-28.
21. Robertson, D.S. (1980) The timing of Mu activity in maize. Genetics 94:969-978.

22. Robertson, D.S. (1981) Mutator activity in maize: Timing of its activation in ontogeny. Science 213:1515-1517.
23. Robertson, D.S. (1983) A possible dose-dependent inactivation of mutator (Mu) in maize. Molec. Gen. Genet. 191:86-90.
24. Robertson, D.S. (1985) Differential activity of the maize mutator Mu at different loci and in different cell lineages. Molec. Gen. Genet. 200:9-13.
25. Robertson, D.S. (1986) Genetic studies on the loss of Mu mutator activity in maize. Genetics 113:765-773.
26. Robertson, D.S., and P.N. Mascia (1981) Tests of 4 controlling-element systems of maize for mutator activity and their interaction with Mu mutator. Mutat. Res. 84:283-289.
27. Robertson, D.S., and P.S. Stinard (1987) Genetic evidence of Mutator-induced deletions in the short arm of chromosome 9 of maize. Genetics 115:353-361.
28. Rowland, L.J., and J.N. Strommer (1985) Insertion of an unstable element in an intervening sequence of maize Adh1 affects transcription but not processing. Proc. Natl. Acad. Sci., USA 82:2875-2879.
29. Schildkraut, C., J. Marmur, and P. Doty (1962) Determination of the base composition of deoxyribonucleic acid from its buoyant density in CsCl. J. Molec. Biol. 4:430-443.
30. Schwartz, D., and E.A. Dennis (1986) Transposase activity of the Ac controlling element in maize is regulated by its degree of methylation. Molec. Gen. Genet. 205:476-482.
31. Strommer, J.N., S. Hake, J. Bennetzen, W.C. Taylor, and M. Freeling (1982) Regulatory mutants of the maize Adh1 gene caused by DNA insertions. Nature 300:542-544.
32. Sundaresan, V., and M. Freeling (1987) An extrachromosomal form of the Mu transposons of maize. Proc. Natl. Acad. Sci., USA 84:4924-4928.
33. Takahashi, I., and J. Marmur (1963) Replacement of thymidylic acid by deoxyuridylic acid in the deoxyribonucleic acid of a transducing phage for Bacillus subtilis. Nature 197:794-795.
34. Taylor, L.P., V. Chandler, and V. Walbot (1986) Insertion of 1.4kb and 1.7kb Mu elements into the Bronze1 gene of Zea mays. Maydica 31:31-45.
35. Taylor, L.P., and V. Walbot (1985) A deletion adjacent to the maize transposable element Mu-1 accompanies loss of Adh1 expression. EMBO J. 4:869-876.
36. Taylor, L.P., and V. Walbot (1987) Isolation and characterization of a 1.7kb transposable element from a Mutator line of maize. Genetics (in press).
37. Walbot, V. (1986) Inheritance of Mutator activity in Zea mays as assayed by somatic instability of the bz2-mu1 allele. Genetics 114:1293-1312.
38. Walbot, V. (1987) Activation of a cryptic maize transposable element following gamma irradiation of seeds. Science (submitted for publication).
39. Walbot, V., C.P. Briggs, and V. Chandler (1986) Properties of mutable alleles recovered from mutator stocks of Zea mays L. In Genetics, Development, and Evolution, J.P. Gustafson, ed. Plenum Press, New York, pp. 115-142.
40. Walbot, V., and C. Warren (1987) Regulation of Mu element copy number in maize lines with an active or inactive Mutator transposable element system. Molec. Gen. Genet. (in press).
41. Wessler, S.R., G. Baran, and M. Varagona (1987) The maize transposable element Ds is spliced from RNA. Science 237:916-918.

THE USE OF MUTATOR FOR GENE-TAGGING: CROSS-REFERENCING BETWEEN TRANSPOSABLE ELEMENT SYSTEMS

N.S. Shepherd,[1] W.F. Sheridan,[2]
M.G. Mattes,[1] and G. Deno[1]

[1]Central Research and Development Department
E.I. du Pont de Nemours and Co.
Wilmington, Delaware 19898

[2]Department of Biology
University of North Dakota
Grand Forks, North Dakota 58202

ABSTRACT

A Mutator system, resulting in an approximately 30-fold increase in spontaneous mutation rate in maize, was first described by Donald S. Robertson in 1978 (15). This phenomenon was associated with the movement of a transposon, when a Mutator-derived mutant allele of the Adh1 locus was shown to contain a DNA insertion (24). The transposon, called Mu1, was found to be 1,367 bp in length, to have nearly perfect terminal inverted repeats, and to denerate a 9-bp direct repeat of the Adh1 target site upon insertion (2). Using this initial correlation between a mutant phenotype arising from a Mutator line and the presence of a particular DNA sequence (Mu1) within the mutated gene, it seemed reasonable to assume that other Mutator-induced mutations might also be due to the same DNA sequence. If true, the Mu1 sequence might be used as a hybridization probe to molecularly identify and clone the mutated gene by a process termed gene-tagging. Since the Mu1 sequence is present in multiple copies in the genome of Mutator-derived plants (2,3), the problem would really be one of identifying which Mu1-hybridizing clone also contained DNA of the mutated gene of interest.

Such a gene-tagging approach was used successfully to clone the Mutator-induced a-Mum2 allele of the maize A1 gene (13). Approximately 35 different Mu1-hybridizing recombinant phage clones prepared from plant material homozygous for the recessive a-Mum2 mutation were screened. One clone was identified as containing at least a portion of the sought gene, for this DNA fragment contained an En-transposon insertion when plant material containing an En-induced mutable A1 allele was cloned. Such cross-referencing or identification of the sought gene by using more than one transposon-induced mutant allele is common in gene-tagging experiments (see Ref. 23 for review).

The Mu1-hybridizing element within the a-Mum2 allele was found to be of similar size and structure as Mu1 (13). Further characterization of the A1 gene transcription unit (22) places the Mu1-like element in opposite orientation with respect to the direction of transcription as compared to the original Mu1 element in the Adh1 gene. But, like that element, it generates a 9-bp duplication of target site sequence (Fig. 1).

Therefore the Mutator system seemed to have several characteristics which would lend themselves to a more extensive gene-tagging program: mutants are generated at an exceptionally high rate, with approximately 35% of the new mutants exhibiting a mutable phenotype indicative of a transposon insertion (15); there was a correlation (albeit limited) between a Mutator-derived mutant and the presence of the Mu1 DNA sequence within the mutated gene (13,24); and, finally, that sequence was small in size and of a relatively well-defined structure, thus facilitating cloning.

The following is a preliminary report of a gene-tagging experiment begun in the summer of 1985 using the Mutator system. The basic premise of the experiment is quite simple and a nontargeted approach to gene-tagging (23). It relies upon the assumption that many of the "new" mutants derived from a Mutator line are just as interesting as those already named and mapped to chromosome position (perhaps even allelic), and that, just as in the case of the Adh1 and A1 mutants described above, insertion of a Mu1-like transposon into a gene may be the cause of the mutant phenotype. This latter postulate may not be true for every mutant; however, a Mu1-like element has recently been found in the bz1-mu1 allele of the maize Bz gene (25), two other independent mutations of the Adh1 gene (19), a Mutator-induced mutant of the Sh1 gene (20), and a bz2-mu1 allele of the Bz2 gene (Walbot et al., this Volume).

The problem then becomes one of identifying mutant phenotypes of interest and quickly screening those mutants for the presence of Mu1-hybridizing fragments closely linked with the mutant phenotype. As in any gene-tagging experiment, the final identification of the mutant gene fragment is not an easy task and may rely upon further studies of the mutant plant phenotype, other mutant alleles (cross-referencing), and perhaps complementation of the mutation through transformation.

MUTANT INDUCTION

To induce mutations using the Mutator system, the Standard Mutator Test as originally outlined by Robertson (15) was used. In this test a standard line is crossed by a line thought to contain Mutator activity to form F_1 seeds. Each F_1 kernel is then grown to maturity and self-pollinated to form an F_2 family, which would allow identification of recessive mutations in the progeny. Only mutations present in the F_2 generation, but not present in selfed progeny of either of the original parents, is scored as a new mutation. To facilitate a later gene-tagging experiment, several slight but perhaps significant modifications were performed.

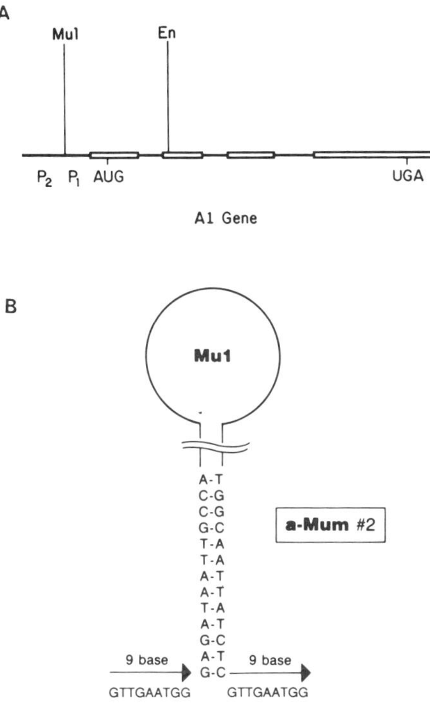

Fig. 1. Molecular characterization of the Mutator element in a-Mum2. A. The position of the Mu1-like element in the clone c10Mu (13) from a-Mum2 was determined by DNA sequence analysis and comparison to that of the wild-type A1 gene (22). B. The target site duplication (9 bp) is identical in size, but different in sequence, to that caused by Mu1 insertion in the Adh1 gene (2). The point of insertion of the Mu1-like element is nearly identical to the insertion point of the I transposon in the a1-m2 allele of maize, where the I element insertion induced a 3-bp duplication of the AAT portion of this target sequence (22). This may indicate a role of chromatin structure in transposon insertion sites. The sequence analysis was performed using synthetic oligonucleotide probes as primers for the dideoxy chain-termination method (21) on a 3.4-kb KpnI fragment of c10Mu subcloned into the mp19 vector, or by direct plasmid sequencing (26) of the Mu1-hybridizing region of the c10Mu clone subcloned into the pUC9 vector.

First, the original cross was performed using a single Mutator plant as pollen donor onto recipient silks of many plants of a standard line. [The standard line used was either Line C (a W22 derivative) or a North Dakota line.] This was done to later simplify the problem of recognizing which positions of the Mu1 element were parental in nature and, thus, not responsible for a new mutation.

Second, leaf material was collected from each of the parent plants. From this material, genomic DNA could be isolated and the distribution and position of parental Mu1 copies could be determined. Furthermore, if a particular Mu1 element was later thought to be responsible for a particular mutant phenotype, it would be important to show that the parental plants lacked Mu1 in this position.

Third, since Mu1 copy number may double in self-pollinated plants (1), and since the extra generation to form the F_2 allows even more transposition and/or rearrangement of the elements, leaf samples were collected from each of the F_1 plants even before a mutation was recognized. Analysis of this F_1 leaf sample may be informative, for the Mu1 copy number may be lower and the newly induced mutant locus may be heterozygous for the Mu1 element.

Finally, since early genetic studies of the Mutator system showed that Mutator activity could be lost in approximately 10% of the Mutator progeny (15,16), and since Mutator activity could not reliably be determined a priori, the experiment was actually performed in triplicate, using three different Mutator stocks as male parent. The intent was to produce F_2 material from each of the three lines, determine which of the three exhibited the highest mutation rate, and to follow only this one into a large grow-out field experiment to look for mutations of a mature plant trait in the F_2 population.

For this initial experiment, approximately 2,000 F_1 kernels were produced using a single Mutator plant as the pollen donor. Each was planted in a winter nursery along with kernels resulting from self-pollination of the parental plants. This was done to screen as quickly as possible for unwanted mutations already present in the parents. The parents were not found to contain any mutation preventing further analysis. However, one of the three Mutator pollen parents used in the original cross was found to be heterozygous for a lethal mutation. In general, the F_1 population varied greatly in plant height, exhibited a few plants containing yellow stripes, and exhibited several plants of small, thin stature, possibly due to a major chromosome aberration. Although the population was thoroughly screened for possible mutant phenotypes arising from a dominant mutation, no heritable dominant mutation was found (approximate F_1 population size of 3 x 2,000 = 6,000 plants). This is in contrast to the finding of four dominant viable plant mutations out of 3,461 M1 plants derived from a pollen mutagenesis experiment using ethyl methanesulfonate (EMS) (11).

ESTIMATE OF MUTATOR ACTIVITY

To determine which of the three parallel experiments was likely to have the highest mutation rate and, thus, be more useful for a grow-out experiment, a seedling screen was conducted on 50 F_2 families of each experiment. Each screen consisted of growing 50 kernels of the F_2 family in the greenhouse. Two of the experiments showed 2/50 families exhibiting a clear-cut new mutation segregating at approximately one-fourth mutant and three-fourths wild-type. This suggested a mutation rate of approximately 4% rather than the 10-11% typical for Mu outcrosses (16,17). The third experiment showed no mutants in this small sampling, and again no mutants when 50 more F_2 families were tested.

Plant DNA isolated from the three parental Mutator plants was digested with a restriction enzyme that does not cut within the element, and hybridized in a Southern blot experiment using a 650-bp AvaI/BstNI internal fragment of Mu1 as a probe. The two plants exhibiting the 4%

mutation rate had approximately 5-15 copies of this fragment, while the third parent, which did not exhibit any mutants, had fewer major hybridizing bands (data not shown). This suggested that the low mutation frequency of the third experiment was due to fewer copies of Mu1. However, since DNA methylation of the elements has been implicated in Mu1 transposition activity (4,5), the DNAs were digested with the methylation-sensitive enzyme HinfI, which cuts twice within the element near the termini (2,9). The two parents exhibiting similar mutation frequencies seemed to have a similar number of elements cut by the enzyme, but differed in that one seemed to have fewer methylated copies (Fig. 2). The Mutator parent for the third experiment, where no mutants were found out of 100 families, had fewer Mu1-size elements cut by HinfI and a majority of the larger Mu1.7-size elements (data not shown).

The conclusion of these limited experiments is that the larger Mu1.7 elements are not sufficient for the induction of mutants at a high frequency, and that the number of Mu1-size elements cut by the enzyme HinfI may be an indication of the Mutator activity of a given plant (see "Discussion" section). The extra methylated copies (i.e., uncut by the enzyme) may not be contributing to the mutation rate and, thus, may pose an extra problem when trying to identify a Mu1 hybridizing band segregating with a mutant phenotype. Therefore, of the three parallel experiments, the one exhibiting 2/50 families segregating for a seedling phenotype (genomic DNA, as in lane 1 of Fig. 2) was chosen for a large-scale grow-out experiment using the F_2 population to search for mutations in mature plant traits.

SURVEY OF THE F_2 POPULATION

To have a relatively high probability of detecting a recessive trait if present in an F_2 family, 12 kernels of each of 1,770 F_2 families were planted. The plants were observed as seedlings, seven weeks after planting, and after tasseling. Due to an initial dry period, many families exhibited fewer than 12 seedlings, such that analysis of seedling lethals in the field population was not possible. However, many families segregating for albino, pale yellow, or other easily scorable seedling traits were identified and confirmed by replanting in the greenhouse. Some of the seedling mutants did exhibit sectoring (e.g., pale yellow plus green sectors), suggesting a transposon insertion.

Plants exhibiting mutant phenotypes at maturity were also found. Such plants were both self-pollinated and outcrossed to a non-Mutator line. Examples of some seedling and heritable mature plant traits are shown in Fig. 3.

DISCUSSION

In general, the types of mutants recovered from the experiment were exciting, but the number of mutants recovered from the field screen of the F_2 population was much lower than expected. As mentioned previously, this may be due in part to the activity of the single Mutator plant used as pollen donor--a factor which can only be reliably determined at present by a large seedling screen of F_2 progeny. To determine whether there was indeed a good correlation between Mutator activity, as measured in a seedling screen, and the number of unmethylated Mu1 elements in the genome, new experiments were begun using five new sources of Mutator activity as the single pollen donor. The results of a Southern blot hybridization experiment on the genomic DNAs of the five mutator plants are shown in Fig. 4. In general, the copy number of the Mu1 elements is higher in these five parents than in the original experiment discussed above.

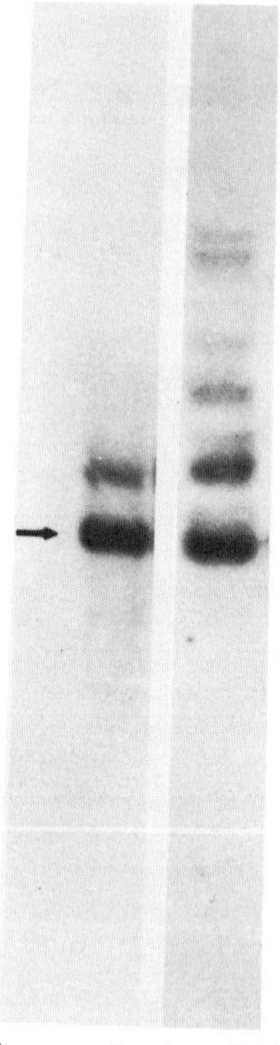

Fig. 2. Gel blot hybridization analysis of Mu element modification pattern of single Mutator plants used as pollen donors in the initial study. Total plant DNA was isolated from each of two Mutator plants, digested with HinfI, and electrophoresed through a 1% agarose gel (lanes 1 and 2). The DNA was transferred to nitrocellulose and hybridized with a nick-translated, ^{32}P-labeled radioactive probe in a Southern hybridization experiment (experimental details as in Ref. 13). The hybridization probe was a 650-bp AvaI/BstNI internal fragment of the Mu1 element (pA/B5) (5). The arrow on the autoradiogram indicates the size of Mu1 elements when both internal HinfI sites of the Mu1 elements are digested with the restriction enzyme. The higher molecular weight, major hybridizing band in each lane is of the size expected for Mu1.7-size elements when digested with HinfI (25). To assure the pattern reflected that of a complete digest with the enzyme, the filter was washed and reprobed with a unique fragment of the maize A1 gene (data not shown). The single kernels of Mutator stock used in this study were either Pl-aleurone/Pl-Mu from a pooled seed stock by Robertson (lane 1) or from Robertson's 83-9239-1/83-9238-9 (lane 2).

Fig. 3. Examples of mutant plant phenotypes arising in the F_2 population from a cross with a Mutator plant as pollen donor. The four mutant phenotypes shown are heritable and recessive. Upper left: les*-1217, a disease lesion mimic mutant similar in phenotype to such mutants described by Neuffer et al. (12). Upper right: wrn*-583, leaves on the mature plant contain longitudinal wrinkles and sometimes ragged edges. Lower left: nec*-1165, chlorotic regions which become necrotic are present in a zebra stripe periodic pattern horizontally across the leaf surface. Lower right: gnv*-86224-5, green near veins on an otherwise white seedling. Allelism tests with known maize mutants (6,10) have not yet been conducted.

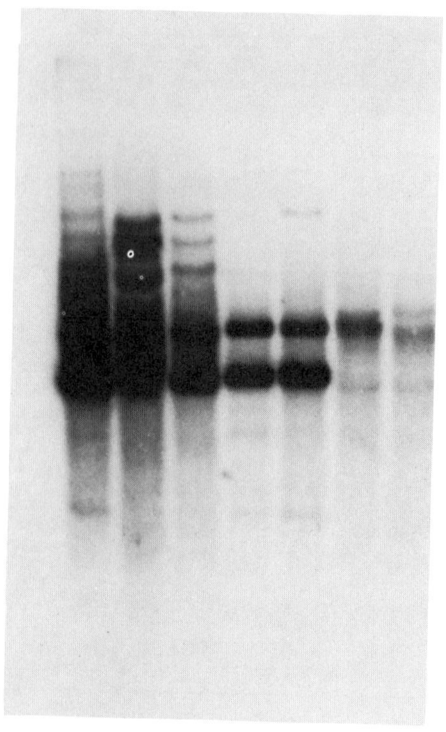

Fig. 4. Mu1 element modification pattern from individual plants. An autoradiogram from a HinfI restriction enzyme digest of individual plant DNAs probed with an internal fragment of Mu1 in a Southern hybridization experiment is shown (see Fig. 2 for details). The individual plant DNAs of lanes 1-5 were from fully colored (Pl) active Mutator stocks designated Pl aleurone/Pl-Mu (Robertson's stock 79-9025-8/79-8028-2 in lane 1; 83-5041-4/83-4041-5 in lane 2; 83-9198-4/83-9197-7 in lane 3; 84-1343-9/84-2340-8 in lane 4; and 85-083-5/85-582-5 in lane 5). For comparison of Mu1 content, lane 6 is from a Pl aleurone, non-Mutator stock (Robertson 71-8283-1/71-9283-6) and lane 7 contains DNA from the inbred B73 (Foundation Research: FRB73). Again, as in Fig. 2, the lower major hybridizing band is of the size expected when both HinfI sites within Mu1 are digested by the enzyme.

The results of the seedling screen of 50 F_2 families from crosses of each of the five parents suggest that the parent shown in lane 1 exhibits the greater number of mutant families segregating in near-Mendelian fashion (at least five different new mutations/50 F_2 families = 10% mutation rate). Although this parent has both methylated and unmethylated copies of Mu1, it does seem to have a large number of Mu1 copies cut by HinfI. In conjunction with the 4% mutation rate identified for plants shown in Fig. 2, this limited survey suggests that the number of unmethylated Mu1 copies may be an important factor in determining mutation rate. However, the Mutator parent shown in lane 2 seems to have a similar number of unmethylated Mu1 elements as that in lane 1, but the seedling mutation rate was not as great.

Clearly, a more extensive seedling screen of larger populations should be performed to be conclusive. In general, our studies suggest that a high copy number of unmethylated Mu1 elements may be necessary but not sufficient to produce a high seedling mutation rate. A large number of unmethylated Mu1 elements may provide a higher mutation rate but may also complicate identification of a Mu1-hybridizing band co-segregating with the mutant plant phenotype.

Another very plausible explanation for the low number of mutants found in the field screen of the F_2 population is that new mutants do not always segregate initially as three normal plants to one recessive mutant. Mutant plants are likely to be present in a seedling screen of approximately 50 kernels from each family but missed when only 12 or fewer plants are examined in the field situation. Such a lack of Mendelian segregation is a common phenomenon when chemical mutagenesis of seeds is used. It is due to the number of cells affected by the mutagen that eventually contribute to the total number of cells in the inflorescence (14). If the transposition of the Mu1 element does not occur until several generations after the formation of the zygote, then less than one-fourth of the progeny from a given family may exhibit the recessive mutant phenotype in the F_2. This was often observed in the seedling screens performed in the greenhouse and has also been mentioned by Robertson (18).

We are hopeful that some of the newly identified mutants will be useful in cloning the mutant gene. Molecular analysis of the Mu1 element distribution within segregating progeny of some of the F_2 families exhibiting newly found mutations is in progress. Recent identification of elements similar to Mu1 at the termini, but containing different internal sequences (Freeling et al. and Chandler et al., this Volume), suggests that the molecular analysis of the F_2 generation should also include these elements as probes. Tests of allelism with previously characterized mutants (cross-referencing) may prove helpful in identifying the correct clone, as will the use of maize monosomics (8) and recombinant inbreds (7).

Regardless of this aspect, the mutants themselves should be valuable in studying plant development, just as mutants derived from more classical methods of mutagenesis have been in the past.

ACKNOWLEDGEMENTS

We are grateful to Dr. M. Pearson (du Pont) for his enthusiasm toward and extra financial support of this nontargeted approach of gene-tagging. We would also like to thank Dr. D.S. Robertson (Iowa State University) for supplying the Mutator stocks used in this study; Dr. V. Chandler (University of Oregon) for the Mu1-specific probe; and Dr. M.G. Neuffer (University of Missouri) for discussions concerning the experiment and resulting mutants. The successful field experiments were the result of efforts by the staffs of Hawaiian Research Ltd. and the du Pont Stine Research Farm Facility.

REFERENCES

1. Alleman, M., and M. Freeling (1986) The Mu transposable elements of maize: Evidence for transposition and copy number regulation during development. Genetics 112:107-119.

2. Barker, R.F., D.V. Thompson, D.R. Talbot, J. Swanson, and J.L. Bennetzen (1984) Nucleotide sequence of the maize transposable element Mu1. Nucl. Acids Res. 12:5355-5967.
3. Bennetzen, J.L. (1984) Transposable element Mu1 is found in multiple copies only in Robertson's Mutator maize lines. J. Molec. Appl. Genet. 2:519-524.
4. Bennetzen, J.L. (1987) Covalent DNA modification and the regulation of Mutator element transposition in maize. Molec. Gen. Genet. (in press).
5. Chandler, V.L., and V. Walbot (1986) DNA modification of a maize transposable element correlates with loss of activity. Proc. Natl. Acad. Sci., USA 83:1767-1771.
6. Coe, Jr., E.H., D.A. Hoisington, and M.G. Neuffer (1987) Linkage map of corn (maize). In Maize Genetics Cooperative Newsletter, E. Coe, ed. Dept. of Agronomy and U.S. Department of Agriculture, University of Missouri, Columbia, Missouri, 61:116-147.
7. Evola, S.V., F.A. Burr, and B. Burr (1986) The suitability of restriction fragment length polymorphisms as genetic markers in maize. Theor. Appl. Genet. 71:765-771.
8. Helentjaris, T., D.F. Weber, and S. Wright (1986) Use of monosomics to map cloned DNA fragments in maize. Proc. Natl. Acad. Sci., USA 83:6035-6039.
9. Kessler, C., and H. Hoeltke (1986) Specificity of restriction endonucleases and methylases: A review (Edition 2). Gene 47:1-110.
10. Neuffer, M.G., L. Jones, and M.S. Zuber (1968) The Mutants of Maize. Crop Sci. Soc. Amer., Madison, Wisconsin, 74 pp.
11. Neuffer, M.G., and W.F. Sheridan (1980) Defective kernel mutants of maize. I. Genetic and lethality studies. Genetics 95:929-944.
12. Neuffer, M.G., D.A. Hoisington, and V. Walbot (1985) The lesion mutants of maize. In Plant Genetics, M.F. Freeling, ed. Alan R. Liss, Inc., New York, pp. 830-833.
13. O'Reilly, C., N.S. Shepherd, A. Pereira, Zs. Schwarz-Sommer, I. Bertram. D.S. Robertson, P.A. Peterson, and H. Saedler (1985) Molecular cloning of the a1 locus of Zea mays using the transposable elements En and Mu1. EMBO J. 4:877-882.
14. Redei, G.P. (1974) Economy in mutation experiments. Z. Pflanzenzuechtg. 73:87-96.
15. Robertson, D.S. (1978) Characterization of a Mutator system in maize. Mutator Research 51:21-28.
16. Robertson, D.S. (1980) The timing of Mu activity in maize. Genetics 94:969-978.
17. Robertson, D.S. (1983) A possible dose-dependent inactivation of Mutator (Mu) in maize. Molec. Gen. Genet. 191:86-90.
18. Robertson, D.S. (1986) New Information on the timing of Mu's mutator activity. In Maize Genet. Coop. Newsletter 60:12-14.
19. Rowland, L.J., and J.N. Strommer (1985) Insertion of an unstable element in an intervening sequence of maize Adh1 affects transcription but not processing. Proc. Natl. Acad. Sci., USA 82:2875-2879.
20. Rowland, J., and J. Strommer (1986) Insertion of Robertson's Mutator in an exon affects transcript stability. Maize Genet. Coop. Newsletter 60:17.
21. Sanger, F., S. Nicklen, and A.R. Coulson (1977) DNA sequencing with chain terminating inhibitors. Proc. Natl. Acad. Sci., USA 74:5463-5467.
22. Schwarz-Sommer, Zs., N. Shepherd, E. Tacke, A. Gierl, W. Rohde, L. Leclercq, M. Mattes, R. Berndtgen, P.A. Peterson, and H.

Saedler (1987) Influence of transposable elements on the structure and function of the A1 gene of Zea mays. EMBO J. 6:287-294.

23. Shepherd, N.S. (1988) Transposable elements and gene-tagging. In Plant Molecular Biology: A Practical Approach, C.H. Shaw, ed. IRL Press, London (in press).

24. Strommer, J.M., S. Hake, J. Bennetzen, W.C. Taylor, and M. Freeling (1982) Regulatory mutants of the maize Adh1 gene caused by DNA insertions. Nature 300:542-544.

25. Taylor, L.P., V.L. Chandler, and V. Walbot (1986) Insertion of 1.4 kb and 1.7 kb Mu elements into the Bronze-1 gene of Zea mays L. Maydica 31:31-45.

26. Zagursky, R.J., K. Baumeister, N. Lomax, and M.L. Berman (1985) Rapid and easy sequencing of large linear double-stranded DNA and supercoiled plasmid DNA. Gene Analyt. Techn. 2:89-94.

ADVANTAGES AND LIMITATIONS OF USING Spm AS A TRANSPOSON TAG

Karen C. Cone, Robert J. Schmidt,
Benjamin Burr, and Frances A. Burr

Biology Department
Brookhaven National Laboratory
Upton, New York 11973

ABSTRACT

Transposon tagging has become the method of choice for isolating genes whose products are in low abundance. We have recently used the transposable element Spm to tag and clone maize regulatory loci. Our choice of Spm was dictated by several factors: The frequency of transposition of Spm is high enough to obtain detectable transposition events, into loci affecting kernel traits, in populations of $<10^6$ seed. Although the copy number of Spm is high in the maize genome, insertions into the gene of interest can be distinguished from other Spm copies by digesting DNAs from segregating populations with methyl-sensitive restriction enzymes, and hybridizing with Spm-specific probes. Since all members of the Spm family thus far examined share DNA homology, hybridization with appropriate probes allows detection of insertions of both autonomous and defective elements. Thus, if a mutable allele can be shown to be under Spm control, one can be reasonably confident of successfully cloning that allele.

INTRODUCTION

A key problem in isolating many genes is that their protein products are generally unidentifiable due to their very low abundance. Therefore, cloning these genes is not possible by traditional routes employing antibodies raised against the purified gene product. Recently, this difficulty has been circumvented by transposon tagging (see Ref. 22 for a current review). If the gene of interest can be tagged by a transposable element for which a molecular probe exists, then it can be cloned even if nothing is known about the molecular or biochemical nature of the gene product.

Molecular probes are now available for three maize transposable elements, including Ac-Ds (Activator-Dissociation), Mu1 and Mu1.7 (associated with Robertson's Mutator), and Spm [Suppressor-Mutator, also known as En (Enhancer)]. Thus, choosing among transposable element systems for the purpose of tagging a specific gene involves determining which element will allow identification and cloning of the tagged gene with relative ease. From our experiences in tagging and cloning two maize regulatory genes, o2 and c1, we feel that Spm is a good choice.

The purpose of this paper is to illustrate that transposon tagging in maize with Spm can be a highly successful venture. In the following discussion we present some of the advantages and limitations of using Spm as a transposon tag by reviewing several factors that should be considered in designing appropriate crosses, analyzing potential mutants, and cloning genes tagged with Spm.

FACTORS THAT INFLUENCE Spm TRANSPOSITION FREQUENCY

In trying to tag a gene with any transposable element, the most efficient cross is designed to maximize the number of transpositions to the gene of interest and thereby minimize the number of progeny screened to detect these transposition events. A number of factors have been shown to influence the frequency of transposition. However, since these factors do not always have parallel effects on different transposable element systems, comparisons between Spm and other systems are made where appropriate in the following discussion.

Direction of the Cross

The sex of the Spm-containing parent appears to play an important role in determining the overall frequency of transposition of Spm to the gene of interest. This was demonstrated by Nelson and Klein (15) in their isolation of Spm insertions at the bz1 locus. In their experiment, the Spm donor--c1-m5(Spm) Bz1 wx-m8(dSpm)--was used either as a male or as a female. No bronze-mutable alleles were recovered from screening 122,200 kernels from the cross in which the Spm donor was used as the female. However, two mutable alleles were recovered from screening 4,582 kernels from the cross in which the Spm was present in the male parent. The frequency of recovery of these mutable alleles was 4.4×10^{-4}, much higher than other reported frequencies of transposition of Spm to other loci (summarized in Tab. 1).

In some cases of mutations induced by Mu, the frequency of mutation can be influenced by the sex of the Mu parent. Robertson (20) demonstrated that the frequency of mutations at the wx locus was enhanced approximately ten-fold when the Mu plants were crossed as males rather than females. On the other hand, the frequency of mutation at the y1 locus did not differ with respect to the sex of the Mu parent. In our own attempts to obtain Mu insertions into o2 (Tab. 2), we performed crosses with a Mu line as the female parent and a non-Mu line homozygous for a recessive o2 allele as the male. (The o2 line was used as the male, rather than as the female, to rule out the possibility of self-contamination that would yield spurious opaque kernels in the F_1.) From 103,200 kernels, no stable opaques or opaque-mutables were obtained.

We successfully tagged o2 (21) using an Spm-containing line as the male parent. In the summer of 1985, crosses were performed using a McClintock line carrying an active Spm at the c1 locus (c1-m5) to pollinate plants that carried a recessive allele of o2 in either Oh43 or W22 background. From 530,000 kernels, three o2-mutable alleles were isolated. All

Tab. 1. Transposition frequencies of Spm.

Locus	Active element present in: Female parent	Male parent	Reference
bz1	0 (0/122,200)	440×10^{-6} (2/4582)[a]	(15)
a2[b]	3.6×10^{-6} (1/275,000)	--	(18)
c1[b]	10.8×10^{-6} (4/368,000)	--	(18)
c1	0 (0/10,000)	46×10^{-6} (3/65,000)[c]	this chapter
o2	--	5.7×10^{-6} (3/530,000)[c]	(21)

[a]Number of confirmed mutable alleles/Number of gametes.

[b]Mutable alleles induced by En.

[c]Only two of the mutable alleles were Spm-controlled.

three came from crosses onto Oh43 females. Upon test crossing, one of these mutable alleles proved to contain an autonomous Spm at o2, one contained a dSpm at o2, and the third o2-mutable allele was not Spm-related.

In a previous effort to detect Spm-induced opaques, a similarly designed cross was set up in which the plants used as males contained not only an active Spm, but dSpms at a1 and wx as well. Female plants were homozygous for a recessive allele of o2. From 125,200 kernels, we did not detect any mutable kernels, but we did obtain two stable opaques. Based on the relatively low frequency with which we obtained Spm insertions at o2 in the cross described above, the sample size in this experiment may have been too small. We should also point out that screening of this cross was difficult because (unknown to us) the female parent carried a recessive r allele, and the progeny kernels were R-mottled. This underscores

Tab. 2. Experiments to tag o2.

Element	Female parent	Male parent	Gametes	Mutable alleles
Ac	O2, P-vv(Ac)	o2, P-vv	489,000	0
Mu	O2, Mu	o2	103,200	0
Spm	o2	O2, a1-m1(dSpm), wx-m8(dSpm), Spm	125,200	0
Spm	o2	O2, c1-m5(Spm)	530,000	3

the importance of selecting appropriate genotypes in constructing the lines for a gene tagging experiment. Small-scale pilot crosses could be useful to ensure that the cross yields phenotypes that will allow easy scoring of the desired mutable allele.

Proximity of Active Element(s) to the Gene of Interest

Several lines of evidence indicate that transposition frequently occurs to sites that are linked to the original site of the element. In examining transpositions of Spm away from a1-m2, McClintock found that 33% (four out of 12) were to sites linked to a1 (13). Peterson found that transpositions of En away from a1-m (pa-pu) were to linked sites 25% of the time, and that there was a preference for transposition to sites 6-20 map units from a1; these transpositions seemed to occur to sites both proximal and distal of the original site of insertion (17). This may be one explanation for the very high rate of recovery of bz1-mutables obtained by Nelson and Klein (Ref. 15; see entry 1 in Tab. 1). In their experiment, the target site was flanked by an Spm at c1 and a dSpm at wx.

Transposition to linked sites has also been observed for movement of Ac from P-vv (67% to linked sites; Ref. 23) and from bz1-m2 (34% to linked sites; Ref. 12). In the case of Ac transposition from P-vv, transposition seems to occur to sites distal on the chromosome (8), whereas transpositions of Ac from bz1-m2 appear to occur to both distal and proximal sites (12).

This trend for Ac to transpose to linked sites may underlie previous failures to detect Ac-induced o2 alleles. In one experiment, we attempted to detect Ac transpositions into o2 from an unlinked location at P-vv (Tab. 2). The Ac-containing line was used as the female, and was pollinated by a tester that was homozygous for a recessive o2 allele. From 489,000 kernels, we obtained no mutable alleles and only one new stable opaque mutant. Salamini and his colleagues (14) also attempted to isolate Ac-induced o2 alleles. From a cross in which the male contained a Ds at the bz2 locus and an unlinked Ac, they detected four mutable o2 alleles out of 200,000 kernels. However, all of these mutables proved to be due to the presence of Ds elements, and therefore were not immediately useful for cloning purposes (discussed below).

Target Site Preference

Several examples of apparent target site preference have been reported. Peterson compared the frequency of occurrence of En-induced alleles at c1 and a2, using various En-containing lines as donors, and demonstrated a preference for En insertion in c1 as opposed to a2 (Ref. 18; refer to entries 2 and 3 of Tab. 1). Robertson's comparison of Mu-induced mutations demonstrated a much higher mutation frequency at y1 than at wx (20).

Other examples of possible insertion site preference are illustrated by failures to detect transpositions of certain elements into defined loci. For instance, no Spm-induced sh1 mutations have been described, and in our own attempts to obtain Spm transpositions into sh1, we were unsuccessful. Salamini and colleagues were unable to detect transposition of the Bg (Bergamo) element into either the sh1 or the wx locus (11); in fact, o2 is the only locus where Bg insertions have been detected.

Another locus which seems to be refractory to some types of insertion is the c1 locus. Although Ds insertions in c1 have been described, no Ac-induced c1-mutable alleles have been isolated. In our own attempt to tag c1 with Ac, we failed to detect Ac insertions, and instead obtained two dSpm insertions. The cross was designed to take advantage of the observation, mentioned above, that the frequency of transposition of Ac is higher to closely linked--as opposed to unlinked--sites, especially if the target is distal, with respect to the centromere, to the original position of the Ac. The Ac-containing parent carried the following markers on chromosome 9S: C1 Sh1 bz1-m2(Ac) Wx; c1 is located about five map units distal to bz1. Crosses with plants carrying recessive markers on 9S were performed using the Ac-containing parent as both male and female. No c1-mutable alleles were recovered in screening 10,000 kernels in which the Ac-containing parent was the female. However, three c1-mutable alleles were isolated in screening 65,000 kernels in which the Ac-containing parent was the male. Surprisingly, none of the mutable alleles proved to be Ac-controlled. Instead, test crosses showed that two of the new mutable alleles contained dSpms at c1; the third c1-mutable allele is not controlled by Spm or Ac. We have since determined that the bz1-m2 parent carried an active Spm at an unknown location, not closely linked to c1.

IDENTIFYING AND CLONING THE TAGGED GENE

Genetic Analysis

In any transposon tagging experiment, many exceptional events representing putative mutants will be selected from the F_1. Invariably, most of these will not breed true. In screening for o2 mutants, we have noticed that trauma during kernel development can frequently cause a "mutable" and/or opaque appearance. From our second attempt to tag o2 with Spm, we selected about 200 kernels as putative mutants; only three were actually mutable, and four bred true as stable opaques. Therefore, one should be prepared to show the heritability of a new phenotype by self-pollinating and obtaining an appropriate segregation ratio of mutable to tester phenotypes.

Other problems in selecting putative mutants from the F_1 may arise. For example, mutable alleles may be characterized by very small sectors of wild-type tissue on a mutant background. This is the case with some states of Spm-induced alleles and especially with Mu-induced insertions [probably due to the late timing of Mu transposition (19)]. If the target gene is one, like o2, that affects the appearance of the endosperm, it may be impossible to detect tiny sectors of wild-type tissue in the background of mutant tissue, and the mutant phenotype will appear stable. In the case of mutable alleles induced by Ds insertion, patterns of variegation can be of two types: either mutant background with sectors of wild-type tissue due to Ac-induced transposition events, or wild-type background with sectors of mutant tissue due to chromosome losses from Ac-induced breaks at some Ds elements. The latter pattern can be misleading, since such losses of wild-type function can be mediated by chromosome breaks at a Ds that is not actually at the locus of interest.

Once a mutant has been obtained and the heritability of its phenotype verified by self-pollination, two steps are required to prepare for molecular analysis. First, a segregating population must be constructed so that the phenotype of the mutant can be correlated later with representative molecular markers. The segregating population can be made either by backcrossing the mutant to the recessive parent used in the initial tagging, or by utilizing the F_2 from the selfed mutant plant. If the phenotype being scored is not mutable, then neither strategy is viable unless appropriate flanking markers are present to allow one to distinguish between plants containing the new mutant allele and those containing the original mutant allele. The second step in the genetic analysis is to identify the element responsible for the instability. If the mutability is due to insertion of an autonomous Spm, then one should be able to demonstrate that Spm activity always segregates with the mutable phenotype. If the mutation is due to insertion of a nonautonomous Spm, then the phenotype should be stable when Spm is not present, but mutable once Spm is reintroduced.

Molecular Analysis

Probe selection. One of the chief advantages of Spm as a transposable element is that thus far all members of the Spm family have proven to be homologous to each other at the DNA level. The dSpm elements that have been examined are internal deletion derivatives of autonomous elements. Therefore, molecular probes derived from the ends of Spm can be used to detect insertions of nonautonomous as well as autonomous elements. This is not always the case with the Ac-Ds system, where Ds elements sometimes bear little homology to Ac (reviewed in Ref. 22). Likewise, one Mutator-associated element, Mu3 (3), has little homology to the internal sequences of Mu1 that are frequently used as hybridization probes (K. Oishi, pers. comm.).

Identification of a restriction fragment associated with the mutable allele. One disadvantage associated with using transposons as tags is that the copy number in the maize genome is generally high [about 50-100 copies for Spm, about four to ten copies for internal sequences of Ac (7), about 10-50 copies for Mu1 [11]]. Therefore, identifying the one copy that is associated with the mutable allele of interest can be problematical. A strategy that is generally used in the case of Mu-induced mutations is to reduced the Mu copy number in the background by outcrossing one or more times to a non-Mu line, while following the phenotype. This method can be time-consuming and burdensome, especially if the phenotype is not mutable.

We have developed a scheme to circumvent the copy number difficulty (4) by taking advantage of the fact that sequences in and around active transposable elements appear to be undermethylated with respect to the bulk of maize DNA (2,5,7). Briefly, DNAs are prepared from individual plants in a segregating backcross or F_2 population. The DNAs are digested with restriction enzymes that are sensitive to cytosine methylation and therefore will not digest the bulk of the DNA. Southern hybridization of these digests is performed using probes generated from the ends of Spm, with the aim of identifying a restriction fragment that segregates with the mutable phenotype of interest.

Figure 1 shows an example of this type of segregation analysis that was performed on one of the mutable alleles (o2-m21) from our attempt to tag o2 with Spm. DNAs were prepared from individual plants grown from kernels of the F_2 population that were previously classified as opaque or opaque-mutable. In this population, some of the opaque kernels are homozygous for the recessive tester allele used in the tagging; but others,

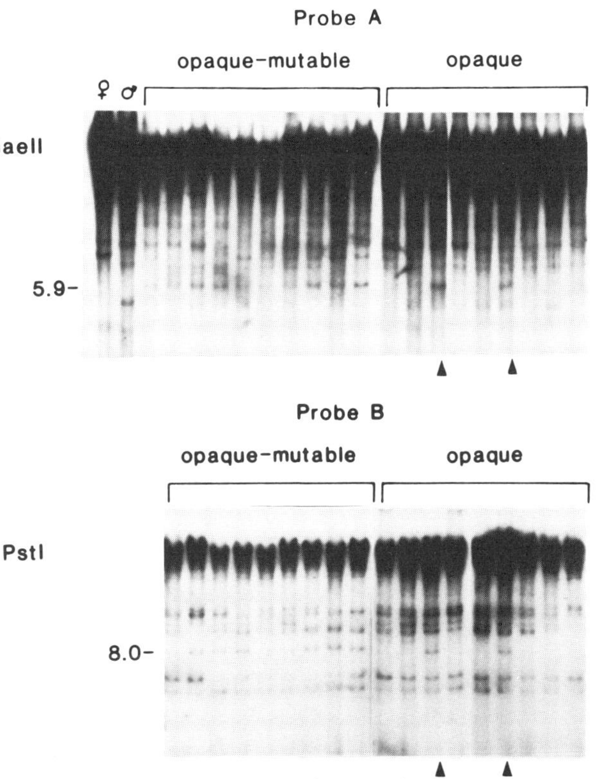

Fig. 1. Segregation analysis of o2-m21. An F_2 population, segregating opaque and opaque-mutable kernels, was generated by self-pollinating a plant derived from one of the newly created o2-mutable kernels selected from the tagging experiment. This kernel was heterozygous for a new mutable allele, o2-m21, and a standard recessive o2 allele. Plants were grown from kernels that were classified as either opaque-mutable or opaque. DNAs were prepared from individual plants, digested with methyl-sensitive restriction enzymes, and hybridized with probes generated from the ends of Spm. Upper panel: DNAs were digested with HaeII and hybridized with Probe A (shown in Fig. 2). This panel also contains DNAs prepared from the female and male parents (left lanes) from the tagging experiment that were digested with the same enzyme. Lower panel: DNAs were digested with PstI and hybridized with Probe B (shown in Fig. 2). The sizes, in kb, of fragments that segregated with the mutable phenotype are indicated to left of each panel. Two individuals (see lanes marked by ▲ in each panel) that were classified as opaque actually proved to contain the restriction fragments that segregated with the opaque-mutable phenotype. These plants, upon self-pollination, bred true for a stable opaque phenotype. The most likely explanation for this phenotype is that in these kernels, the autonomous Spm element had segregated independently from the dSpm at o2, yielding a kernel with a stable opaque phenotype.

although phenotypically opaque, might be carrying the mutable allele in which very late transposition events generate tiny revertant sectors that are difficult to detect. Also, if mutability is due to the presence of a dSpm, some opaque kernels could contain the mutable allele, but appear to be stable due to independent segregation of an unlinked autonomous element. Thus, by hybridization with an Spm-specific probe, we hope to detect a restriction fragment present in all the lanes corresponding to DNAs from opaque-mutable kernels, but missing in some of the lanes corresponding to DNAs from opaque kernels.

When probes from either end of Spm are used to hybridize to methyl-sensitive enzyme digests, most of the hybridization on the blots is at the top of the gel and represents Spm sequences associated with DNA that is poorly digested. However, on both HaeII and PstI digests, restriction fragments of 5.9 kb and 8.0 kb, respectively, were identified that segregated with the opaque-mutable phenotype. Since these fragments were not present in either of the parents used for the tagging experiment, they were presumed to represent a new transposition event. We have subsequently shown by cloning and by genetic analysis that the insertion associated with the o2-m21 allele is a 6.8-kb dSpm. As expected from the segregation analysis, this dSpm contains sequences homologous to probes from both ends of Spm.

We have used a number of methyl-sensitive restriction enzymes for such analyses. They include: AvaI, ClaI, EcoRII, HaeII, NruI, PstI, PvuI, PvuII, SalI, SmaI, SstII, and XhoI. We have found that, although Spm contains restriction sites for most of these enzymes (Fig. 2), some of the methyl-sensitive sites are cleaved in digests of genomic DNA, while others are not. The undermethylation of Spm elements does not appear to be uniform throughout the element. For example, one SalI site located about 2 kb from the 5' end of Spm is not cleaved in digests of genomic DNA (4).

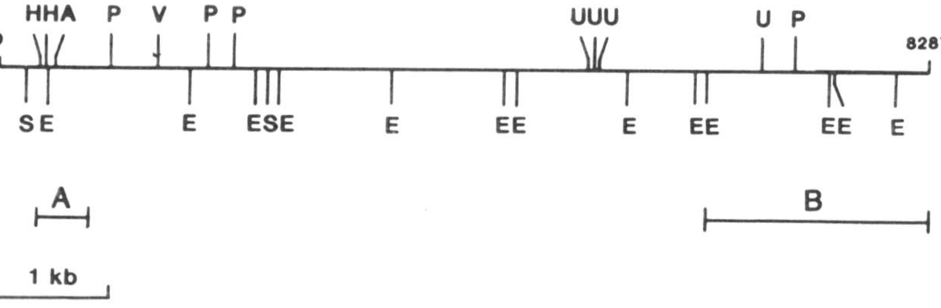

Fig. 2. Restriction map of Spm. The sites of some methyl-sensitive restriction enzyme recognition sequences are indicated. The positions of these sites were taken from the published nucleotide sequence of En-1 (16). A, AvaI; E, EcoRII; H, HaeII; P, PstI; V, PvuI; U, PvuII. There are no sites for ClaI, NruI, SmaI, SstII, or XhoI in Spm. The bars beneath the map represent fragments that we have used as molecular probes. Probe A: 457 bp BanII (355) - XbaI (812) fragment derived from the dSpm at wx-m8. Probe B: 2.3 kb XbaI (6304) - SalI derived from the Spm at c1-m5 (the SalI site is in c1 sequences). Numbers on the map and numbers in parentheses refer to positions, in base pairs, in the sequence of En-1.

Theoretically, if a mutable allele is due to the insertion of a member of the Spm family, it may be possible to detect a restriction fragment associated with the element by digestion with at least one of the methyl-sensitive restriction enzymes. However, it is conceivable that methyl-sensitive sites will not be suitably located flanking the insertions. Figure 3 shows a hypothetical Spm insertion. A single ClaI site is indicated to the right of the insertion. Since there are no ClaI sites within Spm, digestion of this DNA with ClaI would not produce a restriction fragment small enough to be resolved on Southern blots from the other copies of Spm present in the bulk of poorly digested DNA. However, digestion with both SalI and ClaI would generate a restriction fragment of about 9 kb which could be easily detected. Therefore, double digests should be tried in cases where single digests do not initially yield a restriction fragment that correlates with the mutable phenotype.

Cloning the identified restriction fragment. Some of the methyl-sensitive restriction enzymes generate ends that are compatible with conventional cloning vectors; for example, SalI and XhoI fragments can be cloned into the SalI site of λgtWES•λB (10). However, other fragments can be cloned only after addition of appropriate linkers; we have successfully cloned ClaI fragments into the EcoRI site of λgtWES•λB using EcoRI linkers (21).

PROVING THE IDENTITY OF THE CLONED GENE

To determine that the cloned Spm element is associated with the gene of interest, unique sequences flanking the insertion should be used as hybridization probes to show patterns of hybridization consistent with the gene of interest. The DNA of the mutable allele can be compared to DNA from the progenitors, from other mutable alleles, from germinal revertants, and/or from genetically defined deletions of the gene of interest. Also, the probes can be used to analyze RNA blots; for example, RNAs from induced vs uninduced tissues or RNAs from mutant and wild-type tissues can be compared by northern hybridization.

In the event that other mutable alleles or germinal revertants are not available, linkage of the cloned gene to appropriate markers can be established by mapping restriction fragment length polymorphisms (RFLPs). RFLPs associated with the gene of interest often prove useful in analyzing populations that are segregating for wild-type and mutant phenotypes. In addition, RFLPs can serve as the basis for showing that the cloned sequence is located on either the appropriate chromosome, through analysis of monosomic lines (9), or the appropriate chromosome arm, by molecular analysis of B-A translocations (6).

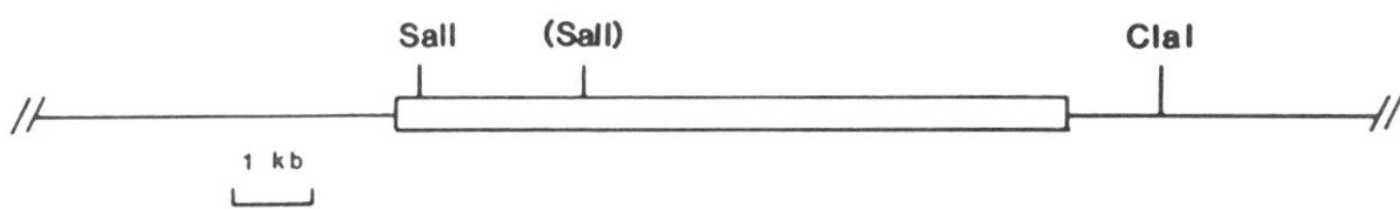

Fig. 3. Restriction map of a hypothetical Spm insertion. Spm sequences are represented by the box. The SalI site shown in parentheses is not cleaved in digests of genomic DNA.

SUMMARY

Transposon tagging can work. Even though most of our understanding about the factors that contribute to a successful tagging experiment has been accumulated from a limited number of experiments using different transposable elements in different genetic backgrounds, it is still possible to draw some conclusions regarding the best experimental strategies for gene tagging. In our experience, Spm has proved to be a good element for transposon tagging. The frequency of recovering mutable alleles induced by Spm is not significantly different from that for Ac-Ds or for Mu (summarized in Ref. 22) and varies from about 10^{-6} to 10^{-4}. Spm has the unique advantage, however, in that all of the members of this family that have been examined thus far are homologous to each other at the DNA level. Therefore, by combining molecular analysis with genetic segregation, it is possible to identify and isolate alleles that are due to insertions of either autonomous or nonautonomous Spm elements.

There are definite steps one can take to increase the chances of detecting a transposition into the gene of interest. The most important step is to select a genetic background in which the desired phenotype will be easy to screen. If the phenotype is not likely to be mutable, then tester lines should be constructed so as to contain flanking markers that can aid in subsequent segregation analyses. Including other unlinked markers in constructing genetic stocks can also prove useful in ruling out possible contamination. Finally, everything possible should be done to maximize the frequency of transposition to the gene of interest. In the case of Spm, using a male parent that contains Spm, preferably linked to the gene of interest, appears to be the best approach.

ACKNOWLEDGEMENTS

We thank Karen Archer for excellent technical assistance and Keith Thompson for computer analysis of the Spm nucleotide sequence. KCC was supported, in part, by a postdoctoral grant from Pioneer Hi-Bred International, Inc. RJS was supported by a National Science Foundation Postdoctoral Fellowship in Plant Molecular Biology. Research support was provided by Grant GM31093 from the National Institutes of Health and by the Office of Basic Energy Science of the U.S. Department of Energy.

REFERENCES

1. Alleman, M., and M. Freeling (1986) The Mu transposable elements of maize: Evidence for transposition and copy number regulation during development. Genetics 112:107-119.
2. Chandler, V.L., and V. Walbot (1986) DNA modification of a maize transposable element correlates with loss of activity. Proc. Natl. Acad. Sci., USA 83:1767-1771.
3. Chen, C.-H., K. Oishi, B. Kloeckner-Gruissem, and M. Freeling (1987) Organ-specific expression of maize Adh1 is altered after a Mu transposon insertion. Genetics 116:469-477.
4. Cone, K.C., F.A. Burr, and B. Burr (1986) Molecular analysis of the maize anthocyanin regulatory locus C1. Proc. Natl. Acad. Sci., USA 83:9631-9635.

5. Dellaporta, S.L., and P.S. Chomet (1985) The activation of maize controlling elements. In Plant Gene Research: Genetic Flux in Plants, B. Hohn and E. Dennis, eds. Springer, New York, pp. 170-217.
6. Evola, S.V., F.A. Burr, and B. Burr (1986) The suitability of restriction fragment length polymorphisms as genetic markers in maize. Theor. Appl. Genet. 71:765-771.
7. Fedoroff, N., S. Wessler, and M. Shure (1983) Isolation of the transposable maize controlling elements Ac and Ds. Cell 35:235-242.
8. Greenblatt, I.M. (1984) A chromosome replication pattern deduced from pericarp phenotypes resulting from movements of the transposable element, Modulator, in maize. Genetics 108:471-485.
9. Helentjaris, T., D.F. Weber, and S. Wright (1986) Use of monosomics to map cloned DNA fragments in maize. Proc. Natl. Acad. Sci., USA 83:6035-6039.
10. Leder, P., D. Tiemeier, and L. Enquist (1977) EK2 derivatives of bacteriophage lambda useful in the cloning of DNA from higher organisms: The λgtWES system. Science 196:175-177.
11. Marotta, R., G. Ponziani, M. Motto, H. Hartings, A. Gierl, N. Di Fonzo, C. Soave, A. Bianchi, and F. Salamini (1986) Genetic instability at the shrunken and waxy loci in the o2-m(r)-Bg strain of maize. Maydica 31:131-151.
12. McClintock, B. (1956) Controlling elements and the gene. Cold Spring Harbor Symposia on Quantitative Biology 21:197-216.
13. McClintock, B. (1962) Topographical relations between elements of control systems in maize. Carnegie Institution of Washington Yearbook 61:448-461.
14. Motto, M., R. Marotta, N. Di Fonzo, C. Soave, and F. Salamini (1986) Ds-induced alleles at the opaque-2 locus of maize. Genetics 112:121-133.
15. Nelson, O.E., and A.S. Klein (1984) Characterization of an Spm-controlled bronze-mutable allele in maize. Genetics 106:769-779.
16. Pereira, A., H. Cuypers, A. Gierl, Zs. Schwarz-Sommer, and H. Saedler (1986) Molecular analysis of the En/Spm transposable element system of Zea mays. EMBO J. 5:835-841.
17. Peterson, P.A. (1970) The En mutable system in maize. III. Transposition associated with mutational events. Theor. Appl. Genet. 40:367-377.
18. Peterson, P.A. (1978) Controlling elements: The induction of mutability at the A2 and C loci in maize. In Maize Breeding and Genetics, D.B. Walden, ed. John Wiley and Sons, New York, pp. 601-631.
19. Robertson, D.S. (1980) The timing of Mu activity in maize. Genetics 94:969-978.
20. Robertson, D.S. (1985) Differential activity of the maize mutator Mu at different loci and in different cell lineages. Molec. Gen. Genet. 200:9-13.
21. Schmidt, R.J., F.A. Burr, and B. Burr (1987) Transposon tagging and molecular analysis of the maize regulatory locus opaque-2. Science 238:960-963.
22. Shepherd, N.S. (1987) Transposable elements and gene-tagging. In Plant Molecular Biology: A Practical Approach, C.H. Shaw, ed. IRL Press, London (in press).
23. van Schaik, N.W., and R.A. Brink (1959) Transpositions of Modulator, A component of the variegated pericarp allele in maize. Genetics 44:725-738.

TRANSPOSITION OF _Ac_ IN TOBACCO

Barbara Baker,[1] George Coupland,[2] Reinhard Hehl,[1] Nina Fedoroff,[3] Horst Lörz,[4] Peter Czernilofsky,[5] Peter Starlinger,[2] and Jeff Schell[4]

[1]U.S. Department of Agriculture
Plant Gene Expression Center
Albany, California 94710
[2]Institut für Genetik
Universitat zu Köln
D-5000 Köln 41, Federal Republic of Germany
[3]Department of Embryology
Carnegie Institution of Washington
Baltimore, Maryland 21210
[4]Max-Planck-Institut für Züchtungsforschung
D-5000 Köln 30, Federal Republic of Germany
[5]California Biotechnology, Inc.
2450 Bayshore Parkway
Mountain View, California 94043

INTRODUCTION

The maize transposable element _Activator_ (_Ac_) was first identified and studied genetically by Barbara McClintock (15; for review, see Ref. 9). The _Ac_ element is capable of transposing autonomously, and it can also _trans_-activate the transposition of a group of elements collectively designated _Dissociation_ (_Ds_) elements. _Ac_ and _Ds_ elements comprise a maize transposon family. Many elements of this family have been cloned and subjected to structural analysis (3,6,10). The _Ac_ element is a small, 4.6-kilobase (kb) transposon that has an 11-base-pair (bp) terminal inverted repetition and generates an 8-bp duplication upon insertion. Sequence analysis of the element has revealed the presence of three major open reading frames (ORFs) (16,19). Recently, an RNA transcript of 3.5 kb was identified and found exclusively in maize lines that carried an active _Ac_ (13). Overlapping cDNA clones spanning most of the mRNA were sequenced. The transcript contains a 600-700 nucleotide long untranslated leader, an open reading frame encoding 807 amino acids, and an untranslated 3' sequence of 239 nucleotides. Four introns with a combined length of 654 bases are removed from the primary transcript.

Insight into element-encoded functions has also been gained from the study of Ds elements. Genetically, Ds elements are defined by their ability to transpose only in the presence of an Ac element. The results of molecular analyses have shown that Ds elements comprise a structurally heterogeneous group of elements, all of which have similar or identical 11-bp terminal inverted repetitions, but only some of which are closely related to Ac in structure (for review, see Ref. 7 and 8). Several Ds elements that arose directly from an Ac element by spontaneous mutations have been analyzed and found to have sustained internal deletions, the smallest of which removes 194 nucleotides from the third exon of the 3.5-kb transcript (10,13,19).

To facilitate the further genetic analysis of Ac-encoded functions, as well as to explore the possibility of using Ac as a mutagen and gene tag in plants other than maize, we introduced a cloned Ac element into tobacco cells on an Agrobacterium tumefaciens Ti-plasmid and showed that the element can excise from its original location in the T-DNA and integrate elsewhere in the tobacco genome (1). The occurrence of typical Ac footprints (19-22) suggested that excision of Ac in tobacco occurred by a mechanism similar to that in maize (1). The ability to introduce foreign DNA into tobacco makes it an attractive system in which to study the sequences required for the activity of Ac.

However, a disadvantage of this system is the lack of a phenotypic assay for Ac excisions. In maize, the genetic studies of Ac have been greatly facilitated by the use of easily visualized endosperm markers, which allowed Ac insertions and excisions to be scored. We therefore designed a phenotypic assay which could monitor Ac excisions in tobacco. An NPT II gene whose expression is prevented by the insertion of Ac was constructed in vitro. After its introduction into tobacco, Ac can excise, resulting in NPT II gene expression and, consequently, in kanamycin-resistant (kan^r) tobacco cells. The frequency of Ac excision can, therefore, be monitored by the frequency with which kan^r calli appear after transformation.

This phenotypic assay has also been used in conjunction with tobacco lines which harbored either Ac or Ds to perform functional analysis of the Ac excision process (Starlinger, this Volume). Several features of the phenotypic assay have been employed in construction of vectors that will be used in attempts to tag tobacco mosaic virus-resistance genes present in certain tobacco and tomato cultivars.

Plasmid Constructions

In order to clone the Ac element within the region coding for the untranslated leader of the NPT II gene, it was first necessary to construct an appropriate gene fusion containing a restriction endonuclease cleavage site within the untranslated leader. This was achieved by replacing the nopaline synthase promoter, which expresses the NPT II gene in pLGV1103neo (5), with the 1' promoter of octopine TR-DNA (23). The resulting plasmid, pKU2, contains a unique BamHI site within the leader sequence (Fig. 1A).

In the wx-m7 allele of maize, Ac is flanked by nearby BssHII restriction endonuclease cleavage sites (3,12,16). Cleavage with this enzyme yields a 4.6-kb fragment which contains Ac plus 60 bp of the wx locus.

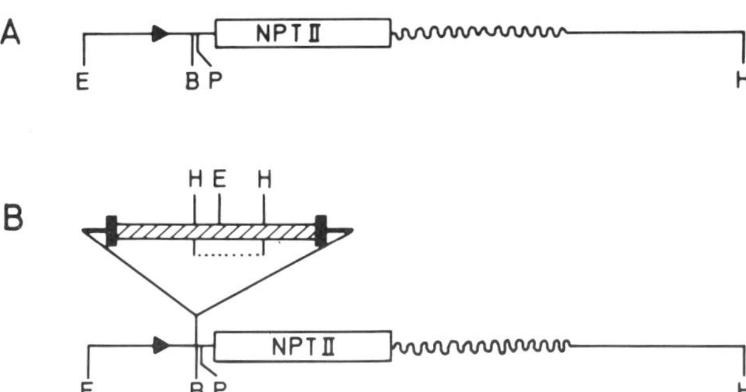

Fig. 1. Partial restriction enzyme map of the region encoding the NPT II gene of plasmid pKU2 and derivatives.

A. pKU2: The 1' promoter of octopine Ti-plasmid TR-DNA is located at least 100 bp downstream from the EcoRI restriction enzyme site. The transcription initiation site, indicated by the solid arrow, is located approximately 50 bp upstream from the BamHI restriction enzyme site. The translation initiation codon of the NPT II gene is located 30 bp downstream from the BamHI restriction enzyme site. The NPT II gene coding sequences are represented by the open box. The waved line denotes 3' untranslated sequences and includes the polyadenylation signal from the octopine synthase gene.

B. pKU3 and pKU27 are Ac insertion derivatives cloned at the BamHI site of pKU2. The orientation of the 4.6-kb Ac insert (hatched boxed area) in pKU3 is such that the long open reading frame of Ac reads from left to right (13). The element was cloned in the opposite orientation in pKU27. The Ac element is drawn to one-fifth scale of the NPT II gene plasmid sequences. The element contains 11-bp terminal inverted repeats indicated by the solid rectangles (not to scale). The 60 bp of maize wx sequence that flank the element are indicated by the thicker solid lines adjacent to the inverted repeats (not to scale).

The pKU4 and pKU11 were constructed by deletion of the internal 1.6-kb HindIII fragment of the Ac element of pKU3 and pKU27, respectively. The sequences deleted in pKU4 and pKU11 are indicated by the dotted line drawn below the Ac element. Restriction enzymes were abbreviated as follows: EcoRI, E; BamHI, B; PstI, P; and HindIII, H.

This fragment was inserted in both orientations into the BamHI site of pKU2 to produce the plasmids pKU3 and pKU27 (Fig. 1B). It was reasoned that the insertion of this 4.6-kb fragment would inactivate the NPT II gene, and that the excision of Ac, leaving only the 60-bp fragment of wx DNA in the untranslated leader, would restore NPT II activity and resistance to kanamycin.

Plasmids pKU3 and pKU27 contain the entire NPT II coding sequence separated from the 1' promoter by the insertion of Ac, and were expected to produce kan^r calli by specific excision of Ac. In order to determine the frequency with which kan^r calli could be produced by a mechanism other than Ac-controlled excision, the internal HindIII fragment of Ac was deleted from pKU3 and pKU27, resulting in plasmids pKU4 and pKU11 (Fig. 1B). This HindIII fragment contains the 194 bp which were deleted in the inactive Ds-9 element (10,19); therefore, the Ac deletion derivatives present in pKU4 and pKU11 should not be capable of transposon-encoded excision from the NPT II gene.

Ac Insertion Within the Untranslated Leader of the NPT II Gene Prevents Expression of the Gene

To determine whether Ac insertion did indeed inactivate the NPT II gene, the ability of pKU3 and pKU4 to promote NPT II expression was determined in a transient expression assay.

As expected, pKU2 and pLGV1103neo, which contain fusions of plant promoters to the NPT II coding sequence, were able to promote NPT II expression in tobacco protoplasts (Fig. 2, lanes 1 and 4). However, neither pKU3 nor pKU4, which contain, respectively, a complete Ac and AcΔ element in the untranslated leader, was able to promote NPT II expression in this transient assay (Fig. 2, lanes 2 and 3).

Ac Excision in Stably Transformed Tobacco Cells Restores NPT II Gene Expression

The plasmids described above (pKU2, pKU3, pKU4, pKU27, pKU11; Fig. 1) were transferred to Agrobacterium tumefaciens containing Ti-plasmid vectors. These strains were used in eight independent co-cultivation experiments, and the results of all these experiments are summarized in Tab. 1.

In experiments numbered 1-4 (Tab. 1), A. tumefaciens strains harboring pGV3850 recombinants of the above plasmids were used for transformation. After co-cultivation with A. tumefaciens containing pGV3850::pKU2, approximately 1% of the total number of protoplasts originally exposed to the bacteria produced kan^r calli. This confirmed that the 1' promoter-NPT II gene fusion present in pKU2 allows selection of stably transformed calli.

In the co-cultivation experiments numbered 5-8 (Tab. 1), the A. tumefaciens containing recombinants of the above plasmids and pGV3850-HPT, a Ti-plasmid vector carrying a hygromycin phosphotransferase gene (HPT) active in plant cells, were used for transformation. The frequency of transformation could therefore be determined by selecting a portion of the protoplasts with hygromycin. The frequency of transformation by each A. tumefaciens strain was similar in these co-cultivation experiments regardless of the expression of the NPT II gene.

The NPT II gene disrupted by Ac was transferred to tobacco via A. tumefaciens (pGV3850::pKU3 or pGV3850HPT::pKU3) and produced kan^r calli at approximately 25% (average of all experiments in Tab. 1) of the frequency found with pGV3850::pKU2 or pGV3850HPT::pKU2. These data suggested that 25% of all transformed cells gave rise, after only 10-12 days of growth, to microcalli which had sustained an Ac excision from at least one of the several Ac-carrying NPT II genes present in most transformed cells, leading to reactivation of the NPT II gene. Co-cultivation

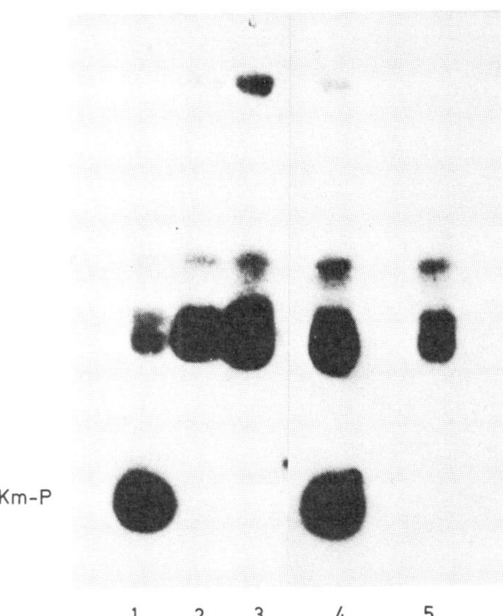

Fig. 2. Transient NPT II enzyme activity in tobacco protoplasts. NPT II enzyme activity expressed in tobacco protoplasts two days after transformation with E. coli plasmids. The position of the kanamycin phosphate is indicated. Each lane represents the measurement of NPT II activity in 1 x 10^5 protoplasts transformed with 10 μg of plasmid DNA: pKU2, lane 1; pKU3, lane 2; pKU4, lane 3; pLGV1103neo, lane 4; and control with no DNA, lane 5.

experiments with pGV3850::pKU27 and pGV3850::pKU11 gave results similar to those obtained with pGV3850::pKU3 and pGV3850::pKU4, respectively. These data indicated that the frequency of excision of Ac was independent of the element's orientation with respect to the NPT II gene promoter sequences (data not shown). The levels of NPT II activity in kan^r calli obtained after transformation with either pGV3850::pKU2 or pGV3850::pKU3 were approximately equivalent (Fig. 3, lanes 1-4).

NPT II Gene Structure Is Restored After Ac Excision

Analysis of DNA from transformed callus tissue was performed in order to confirm that kanamycin resistance of calli transformed with pGV3850::pKU3 was a consequence of Ac excision. We analyzed DNA from three pGV3850::pKU2, four pGV3850::pKU3, and five pGV3850::pKU4 transformed kan^r callus lines which were initially selected with, and grown in the presence of, 100 μg/ml kanamycin. DNA was also analyzed from one pGV3850::pKU2 and three pGV3850::pKU3 kan^r plants regenerated in the presence of 100 or 200 μg/ml kanamycin.

DNA isolated from transformed tobacco tissue was digested with restriction enzymes which produce DNA fragments characteristic of originally constructed, Ac-interrupted or reconstituted NPT II gene structures. The DNA fragments used as hybridization probes (p1' probe, Tn5 probe, and Ac probe) were comprised of sequences which would distinguish the three

Tab. 1. Number of kanamycin-resistant colonies derived from tobacco protoplasts infected with Agrobacterium tumefaciens strains.

Experiment	Ti-plasmid vector	Number of infected protoplasts	Km µg/ml	pKU2		pKU3		pKU4	
				A	B	A	B	A	B
1	pGV3850	2.1×10^4	100	185	(100%)	50	(27%)	17	(9%)
2	pGV3850	2.1×10^4	100	78	(100%)	16	(21%)	0	(0)
3	pGV3850	2.1×10^4	100	140	(100%)	27	(19%)	4	(2.5%)
4	pGV3850	3.15×10^4	200	274	(100%)	70	(25.5%)	0	(0)
5	pGV3850HPT	5.6×10^4	200	224	(100%)	91	(40%)	0	(0)
6	pGV3850HPT	1.3×10^5	200	140	(100%)	51	(36.4%)	0	(0)
7	pGV3850HPT	1.3×10^5	200	120	(100%)	42	(35%)	0	(0)
8	pGV3850HPT	1.3×10^5	200	197	(100%)	26	(13%)	0	(0)

A. Total number of colonies that grew after selection with kanamycin. For experiments 1 and 3, colonies were counted that continued to grow after transfer from bead culture to solidified MS agar medium containing 100 µg/ml kanamycin. The others were counted directly in bead cultures.

B. The number of kanamycin-resistant tranformants expressed as a percentage of the pGV3850::pKU2 number in each experiment.

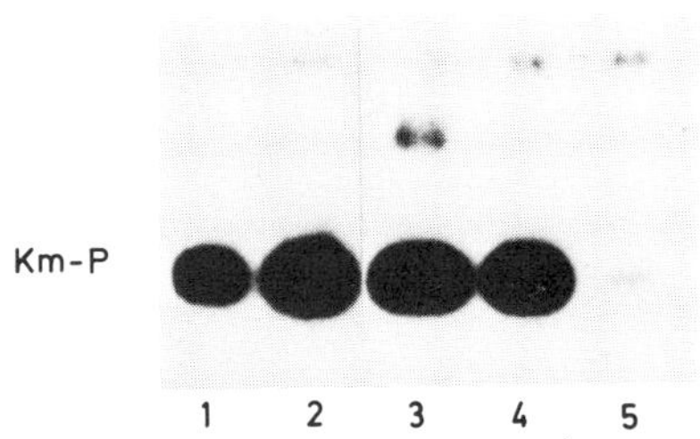

Fig. 3. Neomycin phosphotransferase II assay of kanamycin-resistant tobacco calli. Lane 1, pGV3850::pKU2-transformed tissue (100 μg/ml kanamycin); lane 2, pGV3850::pKU2-transformed tissue (200 μg/ml kanamycin); lanes 3 and 4, pGV3850::pKU3-transformed tissue (200 μg/ml); lane 5, pGV3850::pKU4-transformed tissue (100 μg/ml kanamycin). The approximate concentration of protein (mg/ml), determined by comparison to BSA standards, was the following, respectively, for lanes 1-5: 0.25, 0.25, 1.0, 0.25, and 0.5.

different NPT II gene fragments (see Fig. 4D). DNA isolated from the *A. tumefaciens* strains was analyzed in parallel with transformed tobacco DNAs.

The NPT II coding region and 5' and 3' flanking sequences are present on an *EcoRI-HindIII* T-DNA fragment (2.9 kb) in pGV3850::pKU2 Ti-plasmid DNA and in all four tobacco DNAs transformed with this *Agrobacterium* strain. The 2.9-kb fragment hybridizes to the p1' probe (Fig. 4, Panel A, lanes 1 and 4) and Tn5 probe (data not shown). The 5' flanking leader region of the wild-type NPT II gene is present on an *EcoRI-PstI* restriction enzyme fragment (0.5 kb) in both *A. tumefaciens* and transformed tobacco DNA, and is detected by hybridization to p1' probe (Fig. 4B, lanes 1 and 4, and Fig. 4D, lane 1).

After insertion of *Ac* or *AcΔ* into the leader region of the NPT II gene, the *EcoRI-HindIII* and *EcoRI-PstI* restriction fragments are interrupted by *Ac* sequences, so that *EcoRI-HindIII* and *EcoRI-PstI* digestion of pGV3850::pKU3 and pGV3850::pKU4 *Agrobacterium* DNA yield fragments comprised of leader sequences (Fig. 4A and B, lanes 5 and 6; Fig. 4D, lanes 2 and 4) and *Ac* sequences (Fig. 4C, lane 3; Fig. 4D, lanes 2 and 4). Excision of *Ac* in transformed cells is expected to yield DNA fragments of size and composition similar to that found for pGV3850::pKU2 *Agrobacterium* and transformed tobacco DNA, but the fragments containing the untranslated leader are expected to be slightly larger due to the 60 bp of *wx* sequence that were cloned with the element, and are expected to remain in the leader region after excision (Fig. 4D, lane 3). The reconstituted NPT II gene fragments were indeed observed in the DNA of all seven kan^r lines obtained after transformation with pGV3850::pKU3. An example of the results of p1' probe hybridization to the DNA of one callus line is shown in Fig. 4A and B, lane 2.

The results of hybridization with p1' probe to *EcoRI-HindIII* digestions of all callus and plant DNAs transformed with pGV3850::pKU3 revealed that the reconstituted NPT II gene fragment (3.0 kb) was of greater intensity than the *Ac*-interrupted fragment (2.3 kb). The ratio of

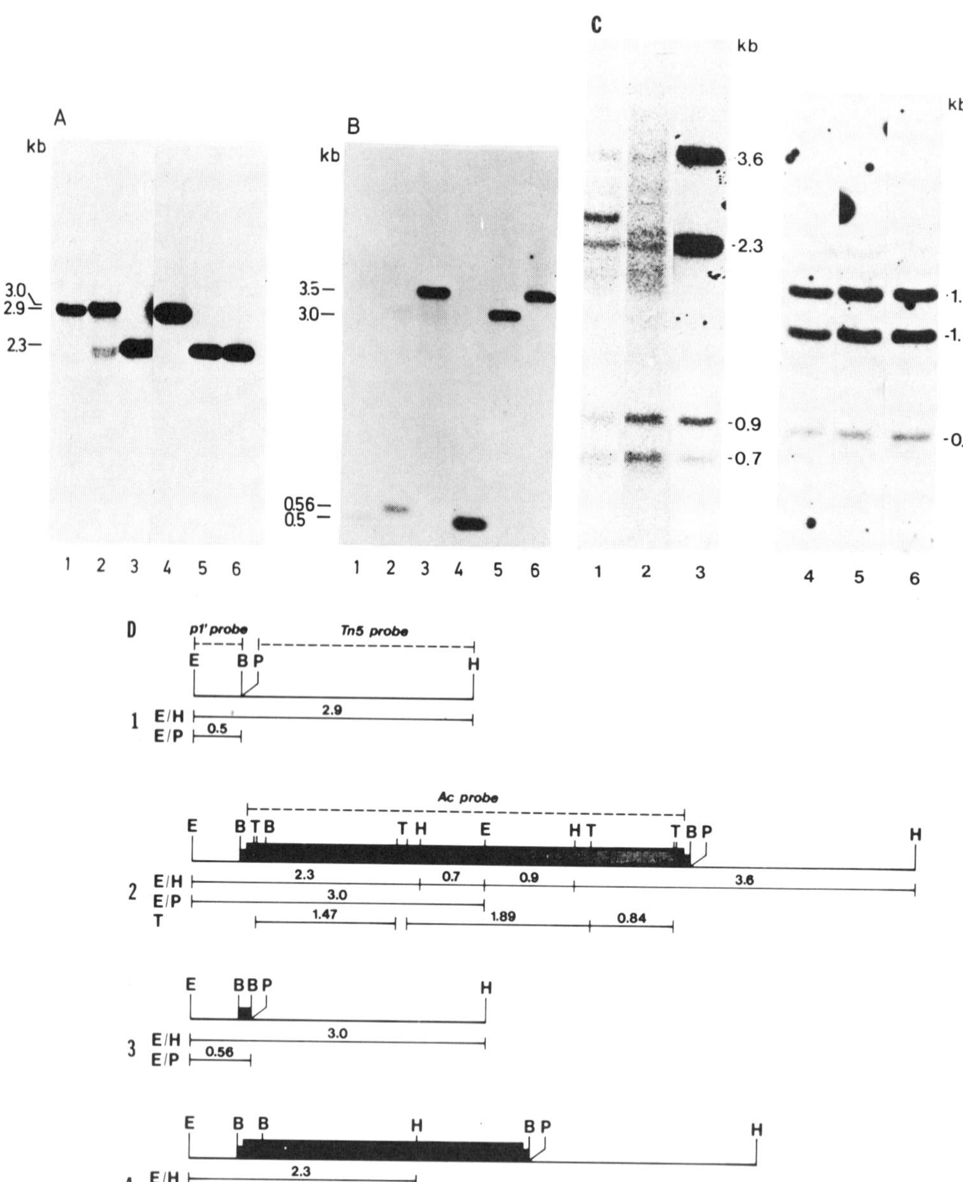
A
kb
3.0
2.9
2.3
1 2 3 4 5 6
B
kb
3.5
3.0
0.56
0.5
1 2 3 4 5 6
C
kb
3.6
2.3
0.9
0.7
1 2 3
kb
1.89
1.47
0.84
4 5 6
D
p1' probe
Tn5 probe
E B P H
1
E/H 2.9
E/P 0.5
Ac probe
E B T B T H E H T T B P H
2
E/H 2.3 0.7 0.9 3.6
E/P 3.0
T 1.47 1.89 0.84
E B B P H
3
E/H 3.0
E/P 0.56
E B B H B P H
4
E/H 2.3
E/P 3.5

the 3.0-kb fragment to the 2.3-kb fragment was approximately 5 to 1, as determined by densitometer scanning, for all callus DNAs. This suggests that *Ac* excised from a majority of the T-DNAs in the transformed kan^{r} *tissues* (Fig. 4A, lane 2).

None of the pGV3850::pKU4-transformed tobacco DNAs contained the fragments expected if the *Ac* had excised from the leader of the NPT II gene. In three callus lines, *Ac*Δ and surrounding T-DNA sequences were arranged as originally constructed (Fig. 4D, lane 4); an example of one of these is shown in Fig. 4A and B, lane 3. In two lines the T-DNA and *Ac* sequences appeared rearranged. However, as all NPT II gene fragments were linked to *Ac* sequences, neither line showed evidence for excision of the element (data not shown).

Similar analyses were performed using DNA isolated from two pGV3850::pKU2, three pGV3850::pKU3, and two pGV3850::pKU4 transformed nopaline-positive callus lines (Tab. 1, experiment 2) that had not

Fig. 4. Southern blot analysis of DNA isolated from transformed lines of tobacco callus and total DNA isolated from *Agrobacterium tumefaciens*. Tobacco DNA was isolated from individually propagated callus lines derived from transformed protoplasts initially selected with 100 μg/ml kanamycin. Total *Agrobacterium* DNA was isolated from overnight cultures. Then 10 μg of tobacco DNA or 5 μg of *Agrobacterium* DNA were digested with restriction enzymes.

Panel A: Tobacco DNA, lanes 1-3, and *Agrobacterium*, lanes 4-6, were digested with *Eco*RI and *Hin*dIII and hybridized to p1' probe; lanes 1, 2, and 3 contain tobacco DNA transformed with pGV3850::pKU2, pGV3850::pKU3, and pGV3850::pKU4, respectively, and lanes 4, 5, and 6 contain *A. tumefaciens* DNA harboring pGV3850::pKU2, pGV3850::pKU3, and pGV3850::pKU4, respectively.

Panel B: DNA samples and hybridization probe as in panel A, but DNAs digested with *Eco*RI and *Pst*I.

Panel C: Lanes 1 and 4 contain tobacco DNA isolated from one callus line; lanes 2 and 5 contain tobacco DNA isolated from another, independently derived callus line; lanes 3 and 6 contain *A. tumefaciens* DNA harboring pGV3850::pKU3; lanes 1-3 were digested with *Eco*RI and *Hin*dIII and lanes 4-6 with *Taq*I. The size of the restriction enzyme fragments is given in kilobase pairs (kb).

Panel D: Schematic summary of the results of Southern blot hybridization with p1', Tn5, and *Ac* probes to tobacco transformed with *Agrobacterium tumefaciens* harboring the following Ti-plasmids: 1) pGV3850::pKU2; 2) pGV3850::pKU3; 3) pGV3850::pKU3 with *Ac* excised; and 4) pGV3850::pKU3. Hybridization probes are indicated by the dashed lines. Abbreviations are as in Fig. 1, and T indicates *Taq*I. Numbers indicate the size of restriction enzyme fragments in kb.

been selected on kanamycin, in order to determine if selection with kanamycin had indeed "enriched" for a population of cells in which Ac had excised from the NPT II gene. The results of analysis of DNA isolated from the pGV3850::pKU2 and pGV3850::pKU4 unselected transformants were similar to those described above for the kan^r tissues (data not shown).

The three pGV3850::pKU3 transformants contained the EcoRI-HindIII NPT II gene fragment characteristic of Ac excision. However, the ratio of the intensities, as determined by densitometer scanning, of hybridization of pl' probe to the reconstituted (3.0 kb) and Ac-interrupted (2.3 kb) NPT II gene fragments was different in each of the callus lines examined. In one line, the ratio of the 3.0-kb fragment to the 2.3-kb fragment was approximately 10 to 1 (data not shown), similar to that described above for the kanamycin-selected calli (Fig. 4A, lane 2). In another line, the reconstituted and interrupted NPT II gene fragments were of similar intensity, and the ratio was determined to be 1 to 1. In the third line, the ratio of the 3.0-kb fragment to the 2.3-kb fragment was approximately 0.05 to 1 (data not shown).

These results differ from those described above for kan^r callus lines, where in all cases the intensity of the reconstituted 3.0-kb fragment exceeded that of the 2.3-kb fragment. These data suggest that kanamycin selection had indeed "enriched" for a population of cells in which Ac had excised from the NPT II gene.

Ac Excision is Accompanied by Integration of Elements Elsewhere in the Tobacco Genome

To determine if excised Ac elements had integrated elsewhere in the genome of pGV3850::pKU3 kan^r lines, DNA isolated from kan^r callus and plant tissue was digested with restriction enzymes and hybridized with Ac probe. The hybridization results of EcoRI-HindIII digestions showed that internal Ac restriction enzyme fragments (0.9 kb and 0.7 kb) produced by these enzymes were present in all four calli and two of the three plants, but that flanking Ac-NPT II gene fragments (3.6 kb and 2.3 kb) were of far lower abundance than expected if Ac remained in the NPT II gene (Fig. 4D, lane 2). New Ac-homologous fragments were visualized in six transformants as several bands of various sizes and intensities or as a smear of less intense hybridization. The overall pattern of Ac probe hybridization to these new fragments, presumably new Ac-tobacco DNA flanking fragments, was different in each transformant. An example of these results is shown for two different callus lines and corresponding A. tumefaciens control in Fig. 4C, lanes 1, 2, and 3. As noted above, one of the regenerated plants did not contain any Ac-homologous sequences, although it did contain the T-DNA fragments corresponding to an excision product (data not shown).

We investigated the overall constitution and integrity of the transposed Ac elements in the kan^r callus lines by digestion of callus DNAs with TaqI. There are seven TaqI sites within the Ac element (16,19). These are distributed such that three major internal fragments are produced (1.89 kb, 1.47 kb, and 0.84 kb), and comprise 92% of the Ac element (Fig. 4D, lane 2). The relative intensities and sizes of these three fragments were similar after Ac hybridization in all callus DNAs to those of the corresponding Agrobacterium DNA. An example of the results obtained with two callus DNAs and the corresponding A. tumefaciens DNA is shown in Fig. 4C, lanes 4, 5, and 6. These data indicate that the structure of Ac elements in these lines (the majority of which are transposed) is, for the most part, unperturbed.

Transgenic Tobacco With a Single Copy of Ac

To determine the fate and activity of transposed Ac in tobacco, we analyzed the DNA of several tobacco plants that arose from selfing of a progenitor plant known to have multiple copies of Ac. We performed Southern blots using BglII, an enzyme which does not cut within Ac, and hybridized the genomic DNA to a probe made from the 1.6-kb internal HindIII fragment of Ac. The number of hybridizing bands should therefore reflect the number of Ac copies in the plant analyzed. Figure 5A, lane 1, shows the result of hybridization of the probe to BglII-digested DNA of the progenitor plant. Due to the high number of hybridizing bands, complete resolution was not achieved, and hence an Ac copy number determination could not be established for the progenitor plant. Lanes 2 to 5 of Fig. 5A show the result of hybridization of the probe to BglII-digested DNA of four different plants isolated by selfing from the multicopy progenitor. A reduction of Ac-homologous bands is evident and indicates the segregation of Ac elements.

Because of the high number of Ac-homologous bands in the progenitor (Fig. 5A, lane 1), we could not determine whether every Ac copy in each of the four progeny plants was present in the progenitor or whether new bands arose due to possible transposition or recombination events. To obtain better resolution, we cut the same genomic DNAs with BglII and EcoRI. EcoRI cuts within Ac and therefore generates two bands per element. Lane 1 in Fig. 5B shows the results of hybridization of the 1.6-kb HindIII Ac probe to the DNA of the progenitor plant and four individual progeny plants (lanes 2 and 5). Almost every band present in each of the progeny plants is present in the progenitor, and only two new bands (indicated by arrows) are detected. This result confirms that segregation of Ac elements has taken place in the selfed progeny. The newly visible bands may represent new transpositions but do not provide direct evidence that secondary transposition has taken place.

In one of the four progeny plants (Fig. 5A and B, lane 3), one copy of Ac is maintained. This copy of Ac is represented by the 13-kb BglII DNA fragment and the 9.2- and 3.8-kb BglII-EcoRI DNA fragments. This Ac copy is present in the progenitor as well as in the other three sibling plants (Fig. 5A and B, lanes 1, 2, 4, and 5). The single-copy Ac was also detected in four progeny R2 plants generated by selfing of the single-copy R1 plant (Fig. 5c and D, lanes 3-1 to 3-4). Preliminary data suggest that this single-copy Ac plant still exhibits transposase activity. This was analyzed by leaf disk transformation of four single-copy Ac plants with Ti-plasmid pGV3850HPT::pKU4 (Fig. 1B) containing an AcΔ construct in the NPT II gene. Kanamycin-resistant shoots were detected after transformation of the single-copy Ac plants, whereas transformation of tobacco plants void of Ac yielded no kan^r shoots.

To confirm the integrity of the single-copy Ac at its new integration site, we have cloned the 13-kb BglII fragment into EMBL4 and are currently analyzing the nature of the genomic sequences flanking the element. This line provides a unique opportunity to study a single transposed Ac element in molecular detail.

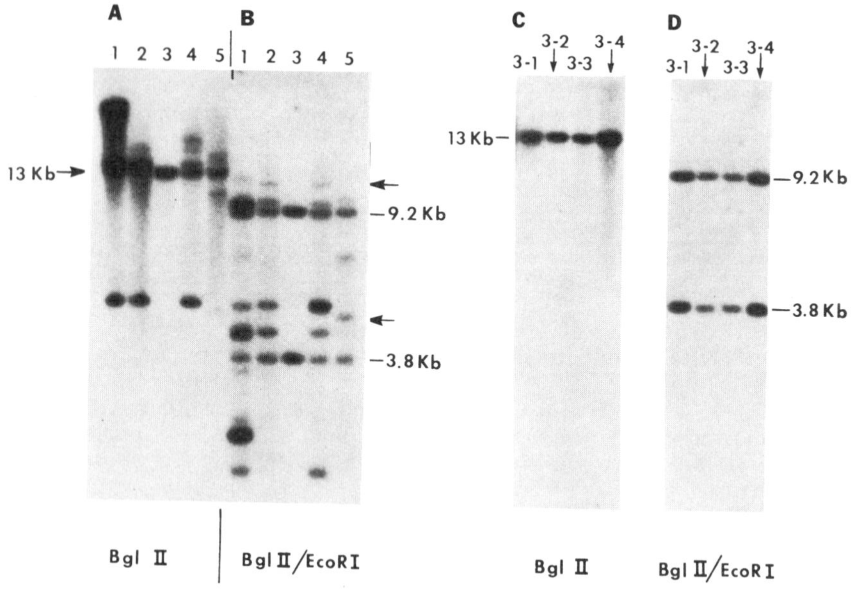

Fig. 5. Southern blot analysis of DNA isolated from transgenic tobacco (R0) carrying transposed Ac and DNA isolated from selfed progeny plants (R1 and R2). DNA was isolated from leaves of individually propagated plants initially transformed with pTiAc (1). DNA (5 μg) was digested with BglII or BglII-EcoRI as indicated, and hybridized with a probe generated from the internal 1.6-kb HindIII DNA fragment of Ac (see Fig. 1). Panels A and B: DNA of the regenerated progenitor plant (R0), lane 1; and its selfed progeny (R1), lanes 2-5. Panels C and D: DNA of selfed progeny (R2) of the R1 plant represented in lane 3 of Panels A and B. Lanes 3-1, 3-2, 3-3, and 3-4 contain DNA isolated from individual R2 plants.

A Strategy for Tagging the TMV Resistance Gene Using Ac in Tobacco

Many genes of known and unknown function have been successfully isolated by transposon-tagging in maize and Antirrhinum majus (4,11,14, 17,18,24). Due to the abundant sequences homologous to the element in the genetic background of maize and Antirrhinum majus, it was difficult to distinguish between element-homologous sequences and the element located at the gene of interest. The relatively low copy number of transposed Ac elements in tobacco and the absence of Ac-homologous sequences will aid in the tagging of genes in this plant.

Our first attempts at gene-tagging in tobacco have been directed toward a dominant viral resistance gene, N, which confers a hypersensitive reaction after tobacco mosaic virus infection and restricts systemic spread of the virus. The strategy is as follows: A cross between the tobacco line homozygous for N and a line homozygous for n and carrying Ac is made. The F_1 plants should show the hypersensitive reaction except in those cases where Ac has been transposed to the N gene and inactivated it. Initial studies have indicated a high degree of N gene instability irrespective of mutagenesis by Ac. Such instability requires further understanding prior to major efforts directed toward the tagging of the N gene.

ACKNOWLEDGEMENTS

We thank F. Kreuzaler for providing the pGV3850HPT strain, J. Velten and B. Gronenborn for valuable discussions, and V. Fantes and B. Hoffman for excellent technical assistance. G.C. was supported by a European Science Exchange Fellowship from the Royal Society, and by an EMBO Long Term Fellowship.

REFERENCES

1. Baker, B., J. Schell, H. Lörz, and N. Fedoroff (1986) Transposition of the maize controlling element "Activator" in tobacco. Proc. Natl. Acad. Sci., USA 83:4844-4848.
2. Baker, B., G. Coupland, N. Fedoroff, P. Starlinger, and J. Schell (1987) Phenotypic assay for excision of the maize controlling element Ac in tobacco. EMBO J. 6:1547-1554.
3. Behrens, U., N. Fedoroff, A. Laird, M. Müller-Neumann, P. Starlinger, and J. Yoder (1984) Cloning of the Zea mays controlling element Ac from the wx-m7 allele. Molec. Gen. Genet. 194:346-347.
4. Cone, K.C., F.A. Burr, and B. Burr (1986) Molecular analysis of the maize anthocyanin regulatory locus C1. Proc. Natl. Acad. Sci., USA 83:9631-9635.
5. Czernilofsky, A.P., R. Hain, B. Baker, and U. Wirtz (1986) Studies of the structure and functional organization of foreign DNA integrated into the genome of Nicotiana tabacum. DNA 5:473-482.
6. Döring, H.P., E. Tillmann, and P. Starlinger (1984) DNA sequence of the maize transposable element Dissociation. Nature 307:127-130.
7. Döring, H.P., and P. Starlinger (1984) Barbara McClintock's controlling elements: Now at the DNA level. Cell 35:253-259.
8. Döring, H.P., and P. Starlinger (1986) Molecular genetics of transposable elements in plants. An. Rev. Genet. 20:175-200.
9. Fedoroff, N.V. (1983) Controlling elements in maize. In Mobile Genetic Elements, J.A. Shapiro, ed. Academic Press, New York, pp. 1-63.
10. Fedoroff, N.V., S. Wessler, and M. Shure (1983) Isolation of the transposable maize controlling elements Ac and Ds. Cell 35:235-242.
11. Fedoroff, N., D.B. Furtek, and O.E. Nelson, Jr. (1984) Cloning of the bronze locus in maize by a simple and generalizable procedure using the transposable controlling element Activator (Ac). Proc. Natl. Acad. Sci., USA 81:3825-3829.
12. Klösgen, R.B., A. Gierl, Zs. Schwarz-Sommer, and H. Saedler (1986) Molecular analysis of the waxy locus of Zea mays. Molec. Gen. Genet. 203:237-244.
13. Kunze, R., U. Stochaj, J. Laufs, and P. Starlinger (1987) Transcription of transposable element Activator (Ac) of Zea mays L. EMBO J. 6:1555-1563.
14. Martin, C., R. Carpenter, H. Sommer, H. Saedler, and E.S. Coen (1985) Molecular analysis of instability in flower pigmentation of Antirrhinum majus, following isolation of the pallida locus by transposon tagging. EMBO J. 7:1625-1630.

15. McClintock, B. (1951) Chromosome organization and genic expression. Cold Spring Harbor Symposia on Quantitative Biology 16:13-47.
16. Müller-Neumann, M., Y.I. Yoder, and P. Starlinger (1984) The DNA sequence of the transposable element Ac of Zea mays L. Molec. Gen. Genet. 198:19-24.
17. O'Reilly, C., N.S. Shepherd, A. Pereira, Zs. Schwarz-Sommer, I. Bertram, D.S. Robertson, P.A. Peterson, and H. Saedler (1985) Molecular cloning of the al locus of Zea mays using the transposable elements En and Mu1. EMBO J. 4:877-882.
18. Paz-Ares, J., U. Wienand, P.A. Peterson, and H. Saedler (1986) Molecular cloning of the C locus of Zea mays: A locus regulating the anthocyanin pathway. EMBO J. 5:829-833.
19. Pohlman, R.F., N.V. Fedoroff, and J. Messing (1984) The nucleotide sequence of the maize controlling element Activator. Cell 37:635-643.
20. Sachs, M.M., W.J. Peacock, E.S. Dennis, and W.L. Gerlach (1983) Maize Ac/Ds controlling elements. A molecular viewpoint. Maydica 28:289-302.
21. Saedler, H., and P. Nevers (1985) Transposition in plants: A molecular model. EMBO J. 4:585-590.
22. Schwarz-Sommer, Zs., A. Gierl, H. Cuypers, P.A. Peterson, and H. Saedler (1985) Plant transposable elements generate the DNA sequence diversity needed in evolution. EMBO J. 4:591-597.
23. Velten, J., L. Velten, R. Hain, and J. Schell (1984) Isolation of a dual plant promoter fragment from the Ti plasmid of Agrobacterium tumefaciens. EMBO J. 3:2723-2730.
24. Wienand, U., U. Weydemann, U. Niesbach-Klösgen, P.A. Peterson, and H. Saedler (1986) Molecular cloning of the C2 locus of Zea mays, the gene coding for chalcone synthase. Molec. Gen. Genet. 203:202-207.

TRANSPOSITION AND RETROTRANSPOSITION IN PLANTS

Zsuzsanna Schwarz-Sommer and Heinz Saedler

Max-Planck-Institut für Züchtungsforschung
D-5000 Koln, Federal Republic of Germany

SOME ASPECTS OF TRANSPOSITION IN PLANTS

The mobile elements thus far identified in bacteria, yeast, animals, and plants possess common features such as the generation of somatic or germinal instability in genes affected by their insertion, the creation of target site duplication upon their integration, and the structural organization of their termini (for review, see Ref. 40).

In two aspects, however, plant transposons reveal unique properties, not exhibited by elements in other organisms. First, the excision of plant transposons only rarely restores the wild-type gene sequence, although the gene function may be phenotypically restored (36,38). Second, functionally defective elements generating a stable mutation in adjacent genes can still interact with "signals" emitted in trans by functionally intact elements of the same family (so called two component systems; for review, see Ref. 26). The consequence of this interaction resembles generation of a novel regulatory unit which replaces the genuine control of genes (41).

In this sense, plant transposable elements deserve the designation "controlling element" (22,24). Both processes result in rapid diversification of expressed genes and, as we shall outline below, may be indicative for the utility of plant transposons during evolution. In fact, molecular analysis of genetically defined plant tansposons revealed the presence of multiple copies of silent elements in the plant genome. Most of these transposons were mobilized in the past by "stress" or "genomic shock" (for reviews, see Ref. 5, 11, 26). The occurrence of silent elements with the potential to become active may indicate that functional transposable elements confer, under certain circumstances, selective advantage to the organism. This advantage could be the mutagenic effect of transposons by insertion--and in plants also by excision--which generate diversity for subsequent selection during evolution. The mechanism of the reversible activation/inactivation of plant transposons is not clear. There is some evidence for the involvement of hypo- and hypermethylation for the induction and suppression of transposition (for review, see Ref. 6).

In contrast to the broadly documented transposition via a DNA intermediate in virtually all living organisms, RNA-mediated transposition seems to be restricted to higher eukaryotes (32). In discussing the evolutionary relevance of plant transposons, we also shall provide evidence for retrotransposition in plants that may also be involved in the evolution of plant genes. In addition, we present a model of the mechanism of integration of a nonviral maize retrotransposon.

PLANT TRANSPOSABLE ELEMENTS MAY GENERATE NOVEL REGULATORY UNITS

Insertion of active plant transposable elements, if it occurs into an expressed gene, generates an unstable mutation. If the element loses its capacity to excise from this location, the mutation becomes stable. Such stable mutations may arise either by internal deletions affecting the coding function of the element (27,30) or by deletions affecting its termini serving as the substrate for proteins involved in excision (18). Both types of deletions require proteins encoded by the element; thus, many types of deletion derivatives may be produced within a single class of elements. Allelic series, so-called "states," can be generated in this way, where every allele represents an independent deletion event affecting the same progenitor element at its original location. Although such series have been obtained with several transposon families, we shall restrict our discussion to the En(Spm) family (21,28) because genetic and molecular analyses were most intensive here. Interest has mainly focused on mutable alleles of the A1 gene of maize encoding an NADPH-dependent reductase and involved in anthocyanin biosynthesis of the plant (2,33).

Genetic analysis of states of the a1-m1 and a1-m2 alleles revealed that the phenotypic expression of color in mutants varies within the allelic series (23,31). Through cloning and sequence analysis of various states of each allele, the relationship between the internal structure of the elements residing at identical locations within the same gene and their effect on gene expression could be established (39,42,44). Clearly, the extent of interference by the insert with transcription, splicing, and translation of the affected gene depends on signals carried by the element. Deletion of such signals may result in expression of an altered but still functional gene product (the a1-m1 5719A-1 allele), whereas their presence prohibits phenotypic gene expression [the a1-m16078 allele, Tacke et al. (44) and see references cited above]. The immobile "receptor" elements (termed Inhibitor or I) in these mutants per se exhibit properties of cis-acting modules on gene expression.

The presence of immobilized elements as modules within a gene can be detected in two ways. Either allelic variants of the gene exist which do or do not contain the insert (and may or may not be equally expressed), or the insert can be mobilized by a trans-acting "signal." In the first case, one deals with two wild-type genes in which an insert became an integral part of a transcription unit. In the second case, the stability of the insert is transient and is due, for example, to lack of a functional transposase. Two-component systems in maize reveal this second property. Their detection is due to the observation that crosses between plants with a virtually stable mutation and plants carrying the functional ("autonomous" or "regulator") element uncover instability at the given locus.

The mobilization of defective elements by gene products of a functional element is highly specific. By this means, several families of plant transposons can be defined (29). The interaction between the Spm (regulator) element and the I (receptor) element, however, is not restricted to mobilization of the transposition-deficient receptor alone. The regulator in trans may also alter the receptor's influence on gene expression. This includes suppression or induction of gene expression as displayed by the mutable alleles a1-m1 5719A-1 and a1-m2 8004, respectively (23). Based on the analysis of transcripts of mutant alleles in the absence and in the presence of a regulator, we assume that suppression is the consequence of blocking transcriptional read-through after binding of a regulator-encoded protein to the termini of the receptor (14,39). This assumption is substantiated by the finding that the En-encoded polypeptide may bind to the termini of the element (see Gierl et al., this Volume). Such an interaction will result in negative control of the gene if the structure (and position) of the particular receptor evokes little or no interference with the expression of the affected gene.

Protein signals emitted by Spm may also induce expression of an adjacent gene, thereby conferring positive control on that gene. Molecular analysis of the a1-m2 8004 allele, which exhibits induction of gene expression by Spm activity, indicates that in this case the particular location of the receptor element rather than its structure is crucial for the phenomenon. In this allele the receptor is located upstream of the CAAT and TATA boxes of the A1 gene promoter, thereby displacing upstream regulatory sequences (42). A1 gene expression in this mutant is abolished.

From a formal point of view, the induction of A1 gene activity by Spm replaces the wild-type induction mechanism. Which Spm-encoded protein governs this induction, and which DNA region within the receptor (or within the A1 gene promoter) is involved in protein binding and initiation of A1 gene transcription, is not known. It is likely that the basic mechanism of suppression and induction by Spm of expression of adjacent genes affected by I inserts is similar (42). Hence, whether negative or positive control is exhibited by the regulator element depends on the composition of the receptor insert and on its position with respect to the affected transcription unit.

If the gene control established by a transposable element should become of value for evolution, both the regulator and the receptor components have to be stabilized at their locations. As mentioned already, small deletions affecting the termini of elements perfectly immobilize them without affecting their coding function or their response to trans-acting signals. In theory, transposable elements possess the potential to become integral parts of genes as cis-acting modules which respond to trans-acting signals. Inserts of this kind may replace the genuine control of genes.

De facto, none of the genetically identified transposable elements were found to be an integral part of a gene. McClintock's emphasis on the involvement of transposons as "controlling elements" in gene regulation and plant development has therefore not yet received support at the molecular level. However, after gene control is established, the transposon-derived elements of control (receptor and regulator) may undergo mutational changes, masking their original structure but not affecting basic properties needed for their regulatory interaction. Inserts altered in this way would no longer resemble transposons and may escape detection as such.

Genetically unidentified inserts altering the structure of transcription units have recently been found by comparing alleles of wild-type plant genes (7,35,50). In these cases it is not known which consequences these inserts have on the expression of the affected genes.

PLANT TRANSPOSABLE ELEMENTS AS TOOLS FOR PROTEIN EVOLUTION

DNA sequence analysis of several revertant genes arising from excision of a plant transposable element revealed that the excision process is rather imprecise but still follows certain rules (3,36). As a consequence, at the target of insertion, altered sequences ("footprints") are left behind which only occasionally are identical with the wild-type sequence. In most of the cases sequenced, these alterations are not simple base substitutions, but rather small insertions or deletions (30,34,38). If the insertion of the transposable element occurs within a gene encoding a protein, its footprint generated upon excision will alter the structure of that protein in the resulting revertant (see citations above). In the cases mentioned above, the function of the wild-type gene was restored such that no difference between the phenotypic expression of wild-type and revertant could be detected. But the alterations may also affect the function of the protein, as documented by analysis of several excision products (1,48). These observations indicate that transposable elements can serve as generators of protein sequence diversity.

There is some indication that alterations which can be attributed to visitation of genes by transposable elements may have occurred during evolution. By comparing the sequence of alleles, one can detect "footprints" as perfect or imperfect duplications present in one allele but absent in the other. Footprints most frequently occur within intron sequences, which apparently tolerate various kinds of DNA alterations (for review, see Ref. 40). But footprints can also be tolerated within exons (33,40), resulting in allelic variants of wild-type proteins. One can therefore speculate that visitation of genes by transposable elements generates frequent mutations, not only by their integration but also if the element leaves that location. The consequence of such a visit is sequence diversity of the affected gene(s) within a population on which, subsequently, selection can operate.

RETROTRANSPOSITION IN PLANTS?

The Cin4 Element of Maize is a Nonviral Retrotransposon

We argued in the previous sections that the DNA-mediated transposition of insertion elements in plants can become of value for the species during evolution. By generating novel regulatory units upon insertion, or by generating sequence diversity within a gene upon excision, plant transposons create a playground for subsequent selection. In the following section we wish to draw attention to mutagenesis by insertion of DNA via an RNA intermediate.

Until recently, nonviral retrotransposons--in particular, processed pseudogenes and similar dispersed repetitive, intronless sequences terminating in a poly(A) track--have only been found in mammals (32). There is, however, circumstantial evidence for the occurrence of pseudogenes in yeast (12), and a single-copy processed actin pseudogene in potato was also reported (8). Until now there has been no evidence for the occurrence of reverse transcriptase encoded by the plant genome, except that some plant insertion elements resemble in structure viral retroposons possessing long terminal repeats (15,16; and see compilation in Ref. 40). Recently we have discovered a retrotransposon-like element in maize which encodes for a putative reverse transcriptase protein (42,43).

The Cin4 element in maize was detected as an insert altering the structure of the wild-type A1 transcription unit (42,43; and unpubl. results). Cin4-1 at the A1 locus is 1.1 kb long. It has no terminal structure, but ends in 12 adenosine residues. One of the A residues can be part of a 7-bp long duplication flanking Cin4-1 and occurring only once in the other wild-type A1 allele. The copy number of Cin4-1-related sequences is 50-100 per diploid maize genome, and the Cin4 family is composed of 5'-truncated family members. Cloning and sequencing of five independent truncated Cin4 elements revealed that the length of the poly(A) track (defining the 3' end of the element) varies, and also the length of the direct duplication flanking the elements differs (see below). The elements contain a long open reading frame (ORF). The length of the ORF in the longest (but still truncated) copy comprises 3,198 amino acids (43). In all these features, Cin4 resembles processed pseudogenes (46).

Recently a class of elements, termed nonviral retrotransposons, has been detected in several organisms. The L1 family of mammals (9), the I element of *Drosophila* (10), and the Ingi element of *Trypanosoma* (19) share no extended sequence homology at the nucleic acid level. Yet all these elements show striking similarities in their overall organization (Fig. 1) and in the conservation of two regions at the amino acid level within their long ORFs. One of these regions is homologous to the conserved amino acid sequence within the pol region of retroviral reverse transcriptases (Fig. 2). The other conserved region is similar in structure to retroviral DNA-binding "fingers," although nonviral retrotransposons differ with respect to position within the longer ORF of the putative "finger" from that of viral retrotransposons (9).

Sequence analysis revealed that the Cin4 element of maize is homologous to nonviral retrotransposons, on the basis of those criteria mentioned above (Fig. 1). This homology includes the spacing between homology blocks within the conserved "reverse transcriptase" region (Fig. 2) and also the position of the two conserved domains within the long ORF compiled in Fig. 1.

A Model of the Mechanism of Cin4 Integration

Except for the idea that processed pseudogenes and nonviral retroposons integrate in a random manner into preformed chromosomal staggered nicks, details of the mechanism of integration are a subject of speculations (32,47,49). The nature of the intermediate that integrates and how and where reverse transcription occurs remain open questions.

The experimental approach to answering these questions is limited to analysis of the elements that are already integrated. The data obtained after sequence analysis of five independent Cin4 elements indicate that sequence homology between the target and the 5' region of the element could play a role in the integration process (Fig. 3). The sequence of

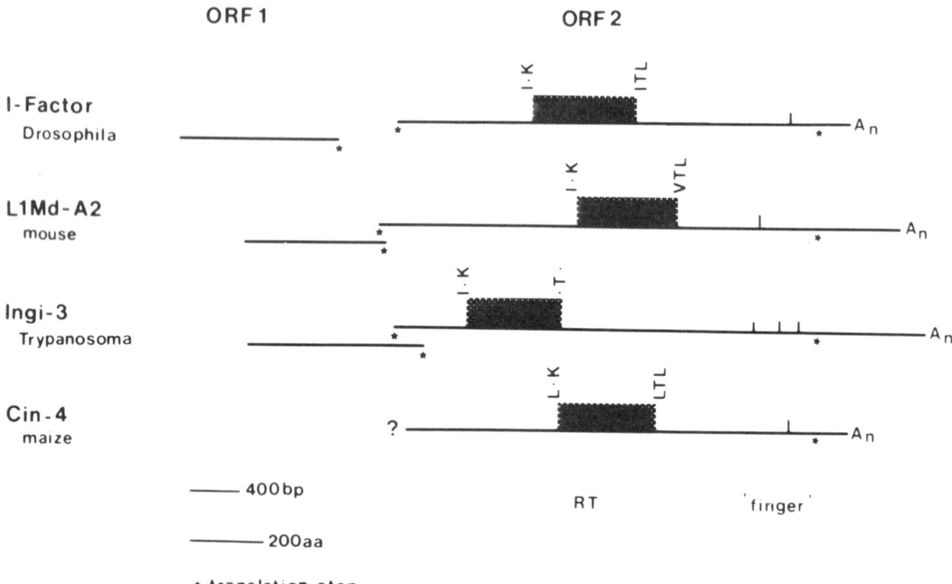

Fig. 1. Conserved structural features of nonviral retrotransposons. The published sequences of I (10), L1Md-A2 (20), Ingi-3 (19), and Cin4 elements (Schwarz-Sommer et al., ms. in prep.) are compiled in a linear form reflecting true size and distance relations. For comparison, the translation stop (*) within the longer ORF of the elements is aligned to an identical position. The conserved domain with homology to retroviral reverse transcriptase (RT, also see Fig. 2) is represented by boxes. Capital letters indicate the first and last amino acid of this domain. Horizontal lines show the position of putative DNA-binding "fingers." The question mark indicates that the structure of the nontruncated Cin4 element is not yet known. In this figure, the longest analyzed truncated Cin4-198 element is depicted.

each target shows homology to sequences present in the longer Cin4 elements upstream of the truncation endpoint of each integrated element. The varying sizes of the flanking direct duplications, on the other hand, confirm that these molecules integrate into preformed staggered nicks. In addition, the mechanism of Cin4 integration must be compatible with the following observations on integrated Cin4 elements:

- The duplications flanking an individual element are always identical, indicating that chromosomal staggered nicks did not undergo changes during integration;

- The oligo(A) track at the 3' end of the element is shorter than the poly(A) tail of a eukaryotic message;

- There is no evidence for the truncation of Cin4 at its 3' end; and

- The element encodes a protein containing a putative reverse transcriptase domain.

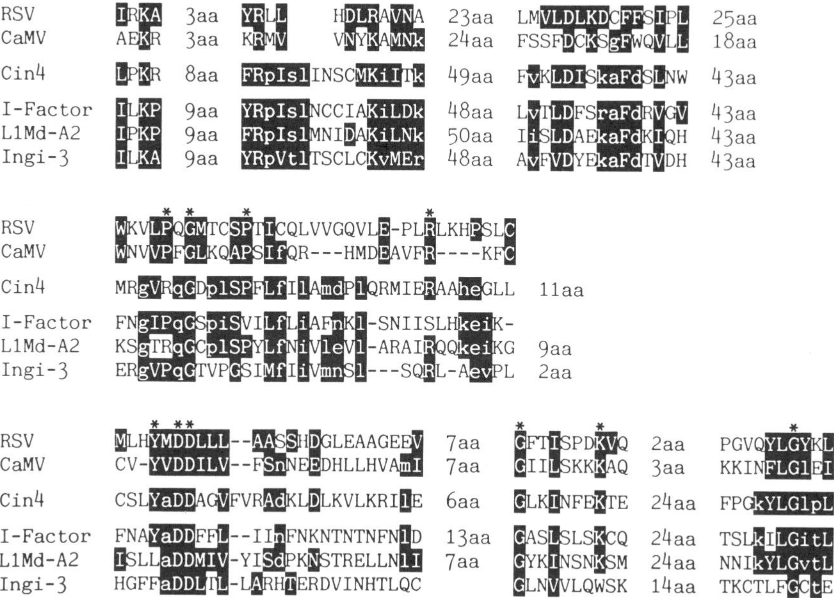

Fig. 2. Amino acid sequence homology within the conserved region of known and putative reserve transcriptases. The amino acid sequence within the conserved domain of Cin4 is aligned, together with the sequences of the conserved domains from other nonviral retrotransposons (17; for sequence references see Fig. 1), with the polymerase gene product of Rous sarcoma virus (RSV, Ref. 37) and with the putative polymerase gene product of cauliflower mosaic virus (CaMV, Ref. 13). Conservative positions are indicated by dark boxes and correspond to the conservation first detected by Toh, Hayashida, and Miyata (45). The ten invariant amino acids found by these authors, comparing the polymerase gene products of retroviruses and CaMV, are indicated by asterisks. Positions conserved in at least three of the four nonviral retrotransposons are shown in dark boxes by small letters. For comparing sequences, the same groups of residues have been used as by Fawcett et al. (10): P, A, G, S (neutral or weakly hydrophobic); Q, N, E, D (hydrophylic, acid amine); H, K, R (hydrophylic, basic); L, I, V, M (hydrophobic); F, Y, W (hydrophobic, aromatic); C (cross-link forming). Gaps were introduced to increase similarity, and numbers give the distance between homology blocks.

The model we propose for the mechanism of Cin4 integration combines several aspects of retroposon integration suggested by others (see references quoted above) and the data obtained with Cin4. In the model depicted in Fig. 4, insertion is initiated by attachment of the poly(A) tail of the putative Cin4 message to the protruding 5' end of a randomly selected, preformed staggered nick of variable size (step I, right side of the molecule in Fig. 4). After endonucleotic restriction or after exonucleotic cleavage of the poly(A) tail, the remaining short oligo(A) track can either be ligated to the chromosomal DNA or fixed without covalent linkage.

	5' flank	element	3' flank	size of the element (bp)
Cin4-1	CT**ATCTCTT** *CTAcaaCTT*	*GTTTG-------------------A(11)*	**ATCTCTT**	1081
Cin4-162	TTTC**AACAAT** *TTggttaAAT*	*CTGTC-------------------A(8)*	**AACAAT**	1239
Cin4-232	**GAGATATTTTAGATGC** *ATTcTAGAaGC*	*TTTCG-------------------A(6)*	**GAGATATTTTAGATGC**	1506
Cin4-151	CATG**GCATCG** *CATGaCtTCG*	*GACCA-------------------A(8)*	**GCATCG**	2947
Cin4-198	**AGGTTAG** AttggAG	CATCG-------------------A(7)	**AGGTTAG**	3423

Fig. 3. Target site duplications flanking individual truncated Cin4 elements. The 5' ends of individual Cin4 elements were determined by sequence comparison with the truncated element Cin4-198, the 5' end of which was defined by sequence comparison with the further upstream extending Cin4-15 element. The total lengths of the truncated elements in base pairs are indicated at the right of the figure. The flanking short duplications are shown in black boxes. Shaded boxes represent sequences contained by Cin4. Capital letters in these boxes indicate identical nucleotides present within the chromosomal target and within the contiguous sequence in longer Cin4 elements. The sequence at the 5' end of individual elements and the number of A residues (in brackets) at the 3' end of the elements are also indicated by capital letters.

Although the details of this linkage remain open, one can assume that the Cin4-encoded protein might exert all the necessary specificities. The fact that we could not detect 3' truncated Cin4 elements in the maize genome may suggest that sequences within the 3' region are required for some processes during integration. These may be related to template specificity of the enzyme and/or to the stabilization of the protein-RNA-DNA complex for the initiation of cDNA synthesis. The free 3' end of the chromosomal DNA at this side of the staggered nick can serve as the primer for reverse transcription toward the 5' end of the message (step II, lower strand in Fig. 4).

We propose that at the still unoccupied side of the staggered nick, hybridization occurs between the mRNA and sequences within the 5' protruding end of the staggered nick (steps I and II, left side of the molecule in Fig. 4). As indicated in Fig. 4, the short, imperfect RNA-DNA hybrid includes several bases upstream of the staggered nick. This invasion of the DNA duplex by the Cin4 mRNA may stabilize the duplex, but it is not obligatory for integration (see the structures of Cin4-198 and 232 in Fig. 3). The RNA-DNA hybrid then prevents reverse transcription from proceeding across RNA sequences located farther upstream. The 3' OH group at the end of the incoming cDNA strand can be ligated to the free 5' end of the chromosomal staggered nick (step II in Fig. 4; see arrow), and integration is completed by repair processes.

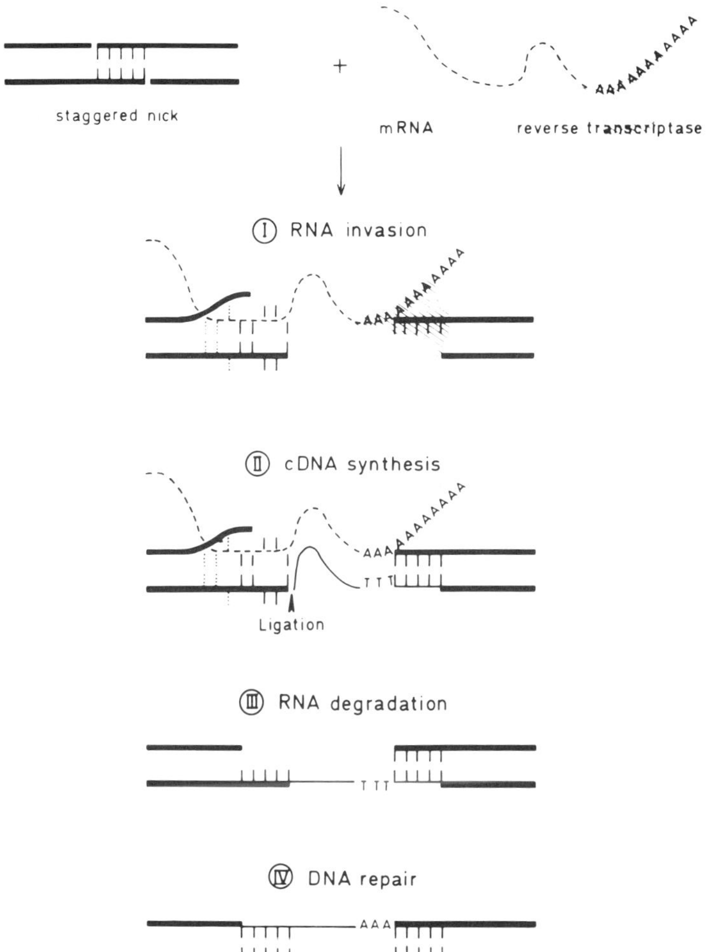

Fig. 4. A molecular model of the mechanism of Cin4 integration. The model is based on the analysis of the structure of integrated Cin4 elements (Fig. 3), and it is explained in detail in the text. The Cin4 mRNA is depicted by staggered lines. Bold horizontal lines represent chromosomal DNA, and thin horizontal lines represent new DNA strands synthesized either during reverse transcription (step II, lower strand) or during DNA repair (step IV, upper strand). The 5' to 3' orientation of the upper DNA strand is left to right. The reverse transcriptase protein is symbolized by a shaded circle. Complementary bases between the two strands of the staggered chromosomal nick are depicted as vertical short lines pointing toward each other. Note that the intermediate hybrid between the 5' region of the mRNA and the protruding end of the chromosomal DNA is not perfect (vertical short lines do not point to each other) and that the hybrid may include bases upstream of the staggered nick (dotted vertical lines).

Repair includes degradation of RNA within the RNA-DNA hybrid by RNase H activity (step III in Fig. 4) which is encoded in the pol region of retroviruses, and perhaps also in the long ORF of nonviral retrotransposons. All other enzymes, like DNA polymerase, ligase, and exonuclease, can be recruited from the repair machinery of the plant cell. Due to the repair process, the staggered nick is filled and the flanking duplication perfectly matches the original chromosomal sequence (step IV in Fig. 4).

In summary, the model of Cin4 integration described above favors in situ cDNA synthesis initiated by random attachment of mRNA into staggered chromosomal nicks with protruding 5' ends. A similar mechanism can be envisaged for the integration of a cDNA into preformed staggered nicks with protruding 3' ends. We favor integration initiated by the mRNA only because this eliminates the problem of a primer needed for cDNA synthesis. Truncation of Cin4 elements occurs upon integration and not necessarily during cDNA synthesis. The basis for the selection of the site of truncation is the homology between the mRNA and the target sequence.

How Recent Are Retrotransposition Events?

There are several aspects to be discussed concerning the presence of a nonviral retrotransposon in plants. For example, the copy number of Cin4 maize is several orders of magnitude lower than the copy number of L1 or Ingi elements. This could mean that the mobility of the element is suppressed and that Cin4 has to be supplied with an "activator" during crosses, in analogy to the activation of I elements in Drosophila during hybrid dysgenesis (10). Alternatively, the low copy number can also reflect that Cin4 was only active for a short period of time and that the element is a remnant of an "infection."

In fact, we did not find evidence for Cin4 activity in maize, except the variability in the chromosomal location of the element, by comparing different maize lines in Southern blot experiments (43). Northern blot analysis with mRNAs from different plant tissues (including leaves, roots, tassels, fertilized and unfertilized ears, and somatic embryos) reveals a smear if the 3' region of Cin4 was used as a probe (data not shown). Also, cDNA cloning gave no indication for the presence of long Cin4 transcripts synthesized from the authentic Cin4 copy.

Although many cDNA clones were found by hybridization to Cin4, all contained additional sequences not related to Cin4, and therefore all seem to be transcribed from external promoter (Schwarz-Sommer et al., ms. in prep.). Further, none of the cDNAs contain sequences homologous to upstream regions of the longest truncated Cin4 copy. These observations argue rather in favor of an early dysgenic cross which mobilized Cin4 or in favor of a past infection by Cin4, leaving the question open as to which organism was the infectious agent. In this light, the differences in the chromosomal locations of Cin4 copies could be in part the consequence of large-scale rearrangements and might not exclusively be due to transposition of elements.

Our failure to detect the full-size Cin4 transcript as an indication for Cin4 activity may reside in not having looked at the proper tissue or not having looked at the appropriate time. The functional relevance of Cin4 for maize (and perhaps the relevance of similar elements in other plants) may be revealed by its homology to nonviral retrotransposons in other organisms. L1, I, Ingi, and Cin4 elements differ in nucleotide sequence. However, certain regions are conserved in all elements at the amino acid level, suggesting that functional constraint conserves the integrity of these regions. Whether this homology indicates the common functional relevance of all nonviral retrotransposons for the organisms or whether it indicates their common origin during evolution remain open questions. Nevertheless, detection of Cin4 as an integral part of some wild-type A1 genes indicates that mobility of retrotransposons adds to the variability of genes and does not only serve to reorganize genomes by large-scale rearrangements.

REFERENCES

1. Chen, C.H., M. Freeling, and A. Merckelbach (1986) Enzymatic and morphological consequences of Ds excisions from maize Adh1. Maydica 31:93-108.
2. Coe, E.H., and M.G. Neuffer (1977) The genetics of corn. In Corn and Corn Improvement, G.F. Sprague, ed. American Society of Agronomy, Inc., Madison, Wisconsin, pp. 11-223.
3. Coen, E.S., R. Carpenter, and C. Martin (1986) Transposable elements generate novel spatial patterns of gene expression in Antirrhinum majus. Cell 47:285-296.
4. Covey, S. (1986) Amino acid sequence homology in gag region of reverse transcribing elements and the coat protein gene of cauliflower mosaic virus. Nucl. Acids Res. 14:623-633.
5. Dellaporta, S.L., and P.S. Chomet (1985) The activation of maize controlling elements. In Genetic Flux in Plants, B. Hohn and E.S. Dennis, eds. Springer Verlag, Wien, New York, pp. 169-216.
6. Döring, H.-P., and P. Starlinger (1986) Molecular genetics of transposable elements in plants. An. Rev. Genet. 120:175-200.
7. Doyle, J.J., M.A. Schuler, W.D. Godette, V. Zenger, and R.N. Beachy (1986) The glycosylated seed storage proteins of Glycine max and Phaseolus vulgaris: Structural homologies of genes and proteins. J. Biol. Chem. 261:9228-9238.
8. Drouin, G., and G.A. Dover (1987) A plant processed pseudogene. Nature 328:557-558.
9. Fanning, T., and M. Singer (1987) The LINE-1 DNA sequences in four mammalian orders predict proteins that conserve homologies to retrovirus proteins. Nucl. Acids Res. 15:2251-2260.
10. Fawcett, D.H., C.K. Lister, E. Kellett, and D.J. Finnegan (1986) Transposable elements controlling I-R hybrid dysgenesis in D. melanogaster are similar to mammalian LINEs. Cell 47:1007-1015.
11. Fedoroff, N.V. (1983) Controlling elements in maize. In Mobile Genetic Elements, J.A. Shapiro, ed. Academic Press, London, New York, pp. 1-63.
12. Fink, G.R. (1987) Pseudogenes in yeast? Cell 49:5-6.
13. Gardner, R.C., A.J. Howarth, P. Hahn, M. Brown-Luedi, R.J. Shepherd, and J. Messing (1981) The complete nucleotide sequence of an infectious clone of cauliflower mosaic virus by M13mp7 shotgun sequencing. Nucl. Acids Res. 9:2871-2887.
14. Gierl, A., Zs. Schwarz-Sommer, and H. Saedler (1985) Molecular interactions between the components of the En-I transposable elements in Zea mays. EMBO J. 4:579-583.
15. Harberd, N.P., R.B. Flavell, and R.D. Thompson (1987) Identification of a transposon-like insertion in a Glu-1 allele of wheat. Molec. Gen. Genet. (in press).

16. Harris, N., and R.B. Flavell (1986) Transposition events in the evolution of interspersed repeated DNA in plants. Heredity 57:276-277.
17. Hattori, M., S. Kuhara, O. Takenaka, and Y. Sakaki (1986) L1 family of repetitive DNA sequences in primates may be derived from a sequence encoding a reverse transcriptase-related protein. Nature 321:625-628.
18. Hehl, R., H. Sommer, and H. Saedler (1987) Interaction between the Tam1 and Tam2 transposable elements of Antirrhinum majus. Molec. Gen. Genet. (in press).
19. Kimmel, B.E., O.K. Ole-Moiyoi, and J.R. Young (1987) Ingi, a 5.2 kb dispersed sequence element from Trypanosoma that carries half of a smaller mobile element at either end and has homology with mammalian LINEs. Molec. Cell. Biol. 7:1465-1475.
20. Loeb, D.D., R.W. Padgett, S.C. Hardies, W.R. Shehee, M.B. Comer, M.B. Edgell, and C.A. Hutchison (1986) The sequence of a large L1Md element reveals tandemly repeated 5' end and several features found in retrotransposons. Molec. Cell. Biol. 6:168-182.
21. McClintock, B. (1954) Mutations in maize and chromosomal aberrations in Neurospora. Carnegie Institution of Washington Yearbook 53:254-260.
22. McClintock, B. (1956) Controlling elements and the gene. Cold Spring Harbor Symposia on Quantitative Biology 21:197-216.
23. McClintock, B. (1965) The control of gene action in maize. In Genetic Control of Differentiation, Brookhaven Symposia in Biology, Vol. 18, pp. 162-184.
24. McClintock, B. (1984) The significance of responses of the genome to challenge. Science 226:792-801.
25. Miller, J., A.D. McLachlan, and A. Klug (1985) Repetitive zinc-binding domains in the protein transcription factor IIIA from Xenopus oocytes. EMBO J. 6:1609-1614.
26. Nevers, P., N.S. Shepherd, and H. Saedler (1986) Plant transposable elements. Advances in Botanical Research 12:103-203.
27. Pereira, A., Zs. Schwarz-Sommer, A. Gierl, I. Bertram, P.A. Peterson, and H. Saedler (1985) Genetic and molecular analysis of the enhancer (En) transposable element system of Zea mays. EMBO J. 4:17-23.
28. Peterson, P.A. (1953) A mutable pale green locus in maize. Genetics 45:682-683.
29. Peterson, P.A. (1981) Instability among the components of a regulatory element transposon in maize. Cold Spring Harbor Symposia on Quantitative Biology, 45:447-454.
30. Pohlman, R.F., N. Fedoroff, and J. Messing (1984) The nucleotide sequence of the maize controlling element activator. Cell 37:635-643.
31. Reddy, L.V., and P.A. Peterson (1984) Enhancer transposable element induced changes at the A locus in maize: The a-m-1 6078 allele. Molec. Gen. Genet. 194:124-137.
32. Rogers, J.H. (1985) The origin and evolution of retrotransposons. International Review of Cytology 93:187-279.
33. Rohde, W., E. Barzen, A. Marocco, Zs. Schwarz-Sommer, H. Saedler, and F. Salamini (1985) Isolation of genes that could serve as traps for transposable elements in Hordeum vulgare. Barley Genetics, Vol. 5 (in press).
34. Sachs, M.M., W.J. Peacock, E.S. Dennis, and W.L. Gerlach (1983) Maize Ac/Ds controlling elements: A molecular viewpoint. Maydica 28:289-302.

35. Sachs, M.M., E.S. Dennis, W.L. Gerlach, and W.J. Peacock (1986) Two alleles of maize alcohol dehydrogenase 1 have 3' structural and poly(A) addition polymorphisms. Genetics 113:449-467.
36. Saedler, H., and P. Nevers (1985) Transposition in plants: A molecular model. EMBO J. 4:585-590.
37. Schwartz, D.E., R. Tizard, and W. Gilbert (1983) Nucleotide sequence of Rous sarcoma virus. Cell 32:853-869.
38. Schwarz-Sommer, Zs., A. Gierl, H. Cuypers, P.A. Peterson, and H. Saedler (1985) Plant transposable elements generate the DNA sequence diversity needed in evolution. EMBO J. 4:591-597.
39. Schwarz-Sommer, Zs., A. Gierl, R. Berndtgen, and H. Saedler (1985) Sequence comparison of "states" of a-m1 suggests a model of Spm(En) action. EMBO J. 4:2439-2443.
40. Schwarz-Sommer, Zs. (1987) The significance of plant transposable elements in biological processes. In Results and Problems in Cell Differentiation 14, Structure and Function of Eukaryotic Chromosomes, W. Hennig, ed. Springer Verlag, Berlin-Heidelberg, pp. 213-221.
41. Schwarz-Sommer, Zs., and H. Saedler (1987) Can plant transposable elements generate novel regulatory units? Molec. Gen. Genet. 209:207-209.
42. Schwarz-Sommer, Zs., N. Shepherd, E. Tacke, A. Gierl, W. Rohde, L. Leclercq, M. Mattes, R. Berndtgen, P.A. Peterson, and H. Saedler (1987) Influence of transposable elements on the structure and function of the A1 gene of Zea mays. EMBO J. 6:287-294.
43. Schwarz-Sommer, Zs., L. Leclercq, and H. Saedler (1987) Cin4, a retrotransposon-like element in Zea mays. In Plant Molecular Biology, NATO Advanced Study Institute, D. von Wettstein and N.-H. Chua, eds. Plenum Publishing Corp., New York, pp. 191-198.
44. Tacke, E., Zs. Schwarz-Sommer, P.A. Peterson, and H. Saedler (1986) Molecular analysis of "states" of the A1 locus of Zea mays. Maydica 31:83-92.
45. Toh, H., H. Hayashida, and T. Miyata (1983) Sequence homology between retroviral reverse transcriptase and putative polymerases of hepatitis B virus and cauliflower mosaic virus. Nature 305:827-829.
46. Vanin, E.F. (1985) Processed pseudogenes: Characteristics and evolution. An. Rev. Genet. 19:253-272.
47. Weiner, A.M., P.L. Deininger, and A. Efstradiatis (1986) Nonviral retroposons: Genes, pseudogenes and transposable elements generated by the reverse flow of genetic information. An. Rev. Biochem. 55:631-661.
48. Wessler, S.R., G. Baran, M. Varagona, and S.L. Dellaporta (1986) Excision of Ds produces waxy proteins with a range of enzymatic activities. EMBO J. 5:2427-2432.
49. Wilde, C.D. (1986) Pseudogenes. CRC Critical Reviews in Biochemistry 19:323-352.
50. Zack, C.D., R.J. Ferl, and L.C. Hannah (1986) DNA sequence of a shrunken allele of maize: Evidence for visitation by insertional sequences. Maydica 31:5-16.

THE MAINTENANCE OF TRANSPOSABLE ELEMENTS IN NATURAL POPULATIONS

Brian Charlesworth

Department of Biology
The University of Chicago
Chicago, Illinois 60637

ABSTRACT

Models of the maintenance of transposable elements in randomly mating host populations are reviewed. It is shown that the data on the distribution of copy numbers between individuals are largely concordant with what is expected on the basis of the Mendelian transmission of elements. The role of regulation of rates of transposition, and of various modes of natural selection, in maintaining an equilibrium in copy numbers in the face of transpositional increase in copy number is discussed. Tests for the role of selection against insertional mutations and against chromosome rearrangements induced by exchange between homologous elements located at nonhomologous chromosome locations are discussed. Reasons for expecting elements to accumulate in chromosome regions where crossing over is restricted are discussed, and data suggesting the existence of such an effect are described. Theory and data on the probability distribution of element frequencies at individual chromosomal sites are described. It is concluded that the available population data are consistent with the notion that element abundances are largely controlled by the interaction of transpositional increase in copy number with opposing forces.

INTRODUCTION

The discovery that the genomes of most species contain families of repeated DNA that have the property of self-replication and movement to novel locations within the genome has triggered a considerable amount of debate concerning their significance in evolution, and the means by which they are maintained within their host populations (9,16a,67a,78a). While much of this debate was initially conducted in the absence of data on the distribution of transposable elements between host individuals in natural populations, and of clearly formulated models of their population biology, these deficiencies have to some extent been remedied in recent years [see reviews by Charlesworth (12), Brookfield (4), and Hartl et al. (37)]. We are presently far from being able to give a conclusive answer to these questions. Nonetheless, the theoretical and empirical framework for pursuing the answers has become considerably clearer, and at least we now have a better idea of what we do not know. This paper reviews both evidence and theories concerning the mechanism of maintenance of transposable elements in their host populations. Many of the data come from studies of Drosophila, partly because of the power and suitability for population studies of the technique of in situ hybridization of element probes to the polytene chromosomes in this group.

Models of the population biology of transposable elements may be divided into the following three categories: (i) Models that consider essentially only a single nuclear gene with non-Mendelian transmission biased in its own favor (25,34), or with an interaction between a nuclear gene and the cytoplasm (79). (ii) Models of bacterial transposable elements, with elements dispersed over multiple genomic sites. These assume only limited possibilities of sexual transmission between individuals, and a correspondingly low frequency of recombination between elements at different genomic locations (72). (iii) Models of randomly mating, diploid populations with elements dispersed over multiple genomic sites, and with relatively high frequencies of recombination between different sites (13,44).

Models of class (i) have been helpful in defining certain basic properties of the population behavior of transposable elements. In particular, Hickey (34) pointed out that the spread of elements by a selfish DNA mechanism is crucially dependent on the existence of sexual or parasexual processes, such that there is a nonzero frequency of exchange of genetic material between individuals. With strictly clonal reproduction, any transpositional increase in copy number of a family of elements would be confined to lineages descended from individuals that possessed at least one element belonging to the family in question. Transpositional increase in copy number thus could not in itself cause an element family that is initially present in only a few members of the population to spread to the majority of the population. Hickey suggested that this explains why transposable elements are absent from genomes with uniparental transmission, such as mitochondria and chloroplasts. Nonetheless, this type of model is inherently unrealistic, in view of the fact that transposable elements are normally present as multiple, dispersed copies, and so cannot be expected to provide detailed predictions that can be tested against population data.

Models of type (ii) escape this objection, but are specifically designed with bacterial populations in mind. Here, genetic exchange between individuals is mediated by plasmids; the frequency of such exchange is extremely low in most cases (of the order of 10^{-12} per cell per generation), and so the frequency of genetic recombination between elements located at different genomic sites is very low in most species (30,33). This has necessitated the development of mathematical models that take into account the possibility of extensive nonrandom associations between the frequencies of elements at different sites (31,33,72). Although transposable elements such as the IS sequences are found on the bacterial chromosome as well as on plasmids, they can probably be transmitted between individuals only in association with the transmission of plasmid genomes. This lead to correlations between the numbers of copies per genome of members of different families of IS sequences, due to the occasional rare simultaneous transmission of members of several IS families on the same plasmid (31,32).

The analysis of the dynamics of bacterial elements thus encounters several formidable problems that are peculiar to their biology. The question of the dynamics of elements in genomes with low frequencies of genetic exchange is, however, highly relevant to what may happen in plants with high frequencies of self-fertilization, where extensive nonrandom association between genotypes at different conventional loci are observed (5). At present, no theoretical work on partially self-fertilizing populations has been carried out.

These difficulties are largely absent from the analysis of randomly mating populations with enough recombination that linkage disequilibrium can be ignored. This is appropriate for outcrossing species of plants and most animal species. Models of class (iii) apply here, and I shall now proceed to describe such models in more detail, omitting mathematical derivations that can be found in the original papers cited. I shall first discuss some static aspects of the distribution of numbers of elements between individuals in a population in the light of model (iii), and then go on to consider dynamic aspects. Data relevant to the testing of the models are reviewed in parallel with the presentation of the models.

THE POPULATION STATISTICS OF TRANSPOSABLE ELEMENTS

Theoretical Considerations

Consider a particular family of elements in a given generation. It is assumed that there is a number m of chromosomal sites in a haploid genome, into which members of the family are capable of inserting. At present, we have no reliable information concerning the value of m for any system, but it seems clear that it is usually a large number, probably of the order of several thousand in a higher eukaryote such as maize or *Drosophila*. The number of elements belonging to a given family in the diploid genome of a given individual is denoted by n.

Studies of populations of yeast, *Drosophila*, and bacteria have shown wide variation among individuals in the sites at which elements are located (see below), indicating that n is very much less that $2m$, the total number of sites available for occupation in a diploid genome. Such studies also indicate some variation between individuals of a population with respect to the value of n; $\bar{n}$ and V_n here denote the mean and variance of n over individuals. The state of a population of size N at a given chromosomal site i with respect to the given family of elements can be described by the frequency x_i with which the $2N$ copies of that site are occupied by an element. One way of describing the state of the population with respect to the element family is thus by a listing of the values of the element frequencies x_i for the entire array of sites $(i=1,2,m)$. If the element frequencies at different sites are independent (no linkage disequilibrium), this is a complete description of the state of the population. The mean copy number $\bar{n}$ is equal to $2\ \Sigma_i x_i$, regardless of the presence of linkage disequilibrium.

Simple quantitative genetics considerations enable some predictions to be made concerning the properties of the distribution of the number of elements between host individuals (12,13,44). The number of elements per individual can be regarded as being a strictly additive, polygenic trait, since it is merely the sum of the number of elements at each site. If

Tab. 1. Tests of the Poisson distribution for three families of elements of *Drosophila melanogaster*.

Chromosome	X		2L		2R		3L		3R	
	$\bar{n}$	V_n	$\bar{n}$	V_n	$\bar{n}$	V_n	$\bar{n}$	V_n	$\bar{n}$	V_n
297	4.45	4.68	3.95	4.58	3.95	5.10	4.65	2.87	5.70	2.22
412	2.65	3.50	4.55	4.15	3.95	5.00	4.10	3.15	5.70	4.22
roo	11.4	3.73	13.2	24.8	11.6	7.50	11.2	7.08	13.8	14.9

Data from Montgomery et al. (60).

there is no linkage disequilibrium, it follows from standard theory (Ref. 7, p. 158) that the variance of copy number is given by:

$$V_n=\bar{n}(1-\bar{x}) - 2m\ \sigma_x^{\ 2} \qquad \text{(a)}$$

where $\bar{x}=\bar{n}(2m)^{-1}$ is the mean element frequency, and $\sigma_x^{\ 2}$ is the variance in element frequency between sites ($\sigma_x^{\ 2}=m^{-1}\Sigma_i x_i^{\ 2}-\bar{x}^2$). If $\sigma_x^{\ 2}$ is small, then only the first term need be considered; this corresponds to a binomial distribution of n across individuals. If $\bar{x}<1$, as frequently seems to be the case (see below), then there is a Poisson distribution of n, with $V_n \approx \bar{n}$. Variation between sites in element frequencies reduces the variance below binomial expectation; if there is a tendency for the frequency of occupation of one site to be higher if a neighboring site is occupied (positive linkage disequilibrium), then there will be a greater-than-binomial variance (13).

Population Data on Copy Number Distribution

Population data to test these predictions seem to be available only for *Drosophila melanogaster* and *Escherichia coli*. The former are mostly from the technique of *in situ* hybridization of labeled element probes to the polytene salivary chromosomes, enabling counts to be made of the numbers of elements in the euchromatin of each chromosome arm. (Elements present in the centromeric heterochromatin cannot be detected by this technique, owing to the under-replication of the heterochromatin during polytenization.) The most reliable results are from experiments in which sets of isogenic chromosomes have been extracted from natural populations, using the balancer chromosomes uniquely available in *Drosophila* (see Tab. 1). Use of stocks that have been made homozygous for a single X, second or third chromosome enables one to determine the number of elements on what corresponds to a single haploid chromosome of the natural population, barring any transpositions subsequent to the construction of the stock. The latter may be tested for comparing different individuals from the same isogenic stock. Owing to the labor involved, relatively few studies of this kind have been made (47,59,60). It has the great advantage of eliminating genetic changes in the stocks due to genetic drift or selection. There are also no ambiguities due to uncertainty as to whether a given chromosomal site is homozygous or heterozygous for the presence of an element. Both of these problems are encountered in surveys of stocks that have not been made isogenic (e.g., Ref. 71).

Equation (a) applies to the distribution across haploid sets, with *n* now representing the number of elements in a haploid genome. It will be seen from Tab. 1 that agreement of the data with the Poisson expectation is generally fairly good, with the exception of the distribution of the element *roo* on the X chromosome. Early studies of the distribution of *Drosophila* elements between laboratory strains impressed investigators with the relative uniformity of copy number, compared with the diversity of chromosomal sites occupied by elements, and led to the suggestion that stabilizing selection on copy number was maintaining this uniformity (78,82). However, reanalysis of these data shows that the variance between lines is approximately equal to the mean, as would be expected from the above considerations (Charlesworth, unpubl. results). Young (82) also commented on the high correlation in the copy number of a given family between two laboratory strains of *D. melanogaster*, using the pairs of copy numbers for the two strains over 17 different families as the *x* and *y* variates in the correlation coefficient. M. Turelli (unpubl. results) has shown that this is expected if each family follows a Poisson distribution, but there are wide differences in mean copy number between families.

Data on bacterial *IS* sequences, obtained by using Southern blot estimates of copy numbers for individual isolates of *E. coli*, show wide deviations from the binomial distribution, as would be expected from the linkage disequilibrium generated by the sporadic nature of genetic transmission in this species (72a). In contrast to the results for *Drosophila*, a considerable fraction of *E. coli* strains lack copies of one or more families of elements (31-33).

At present, there seems little reason to doubt that the nature of the distribution of elements between individuals in natural populations simply reflects the consequences of Mendelian transmission, although more population data for a wider variety of species are needed.

Frequencies of Elements at Individual Chromosomal Sites

Identification of the sites of labeling of probes to sets of isolated salivary chromosomes also permits analysis of the frequencies of elements at individual chromosomal sites. The resolution of this technique is somewhat coarse, since the chromosomal locations of elements can only be determined down to the level of a salivary chromosome band. Nonetheless, a useful picture of the general features of the distribution of element frequencies can be obtained (47,59,60). A convenient summary of the results of such an experiment is provided by the occupancy profile, which gives the numbers of sites at which hybridization is detected at that site in 1,2,3... separate chromosomes of the sample. Some results of this kind of study for elements of *D. melanogaster* are shown in Fig. 1.

It is obvious that most of the time a site is occupied only once in the sample for each element in the family. Because of the lack of resolution of the method, it is possible that at least some instances of apparent multiple

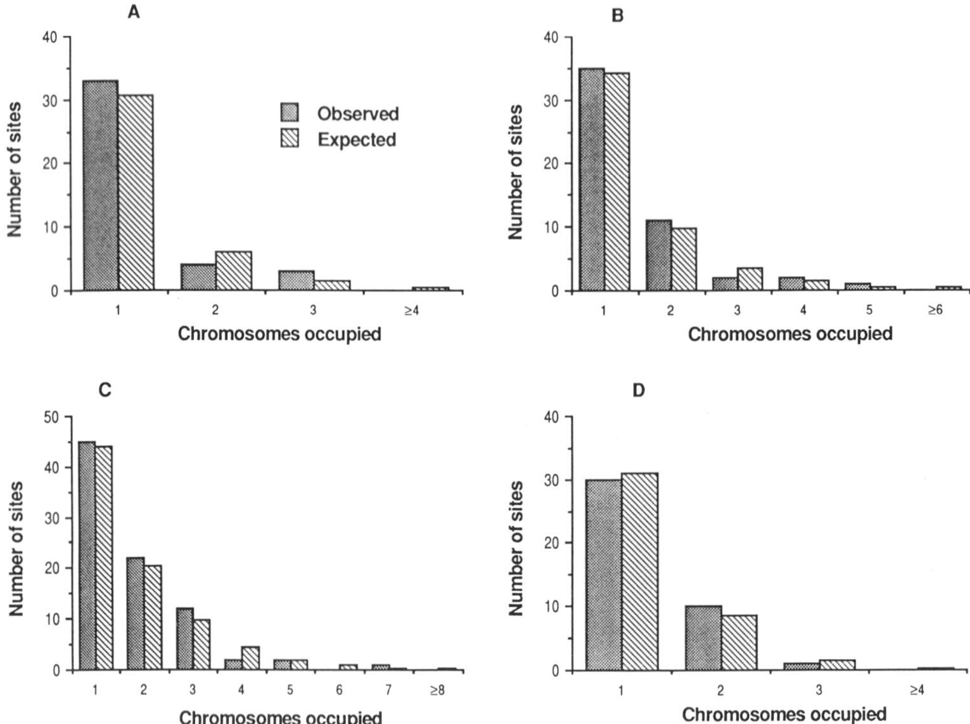

Fig. 1. Observed (shaded bars) and expected (striped bars) occupancy profiles for four families of D. melanogaster elements, for sets of X chromosomes extracted from natural populations. A and B are for 412 and 297 from a North Carolina population (59). C is for roo from a Maryland population (original data). D is for I from a Spanish population (47). Divisions 19 and 20 of the X chromosome were not included in the analysis. The expected profiles were calculated using the parameters estimated by the methods described in the section on small populations, and given in Tab. 2.

occupancy of the same site are the result of insertions into different sites within the same salivary chromosome band (37). An alternative method is to employ restriction fragment mapping of a defined region of the genome to determine with much greater accuracy the locations of insertions of elements in a sample of genomes from a population. In D. melanogaster, this method has been applied to the Adh region (1,43a), to the heat-shock region (46), and to the white locus (42). Each of these studies indicates that element insertions appear to occur at unique locations within a sample of a few tens of genomes, indicating low frequencies at these locations in the population from which the samples were drawn. Southern restriction mapping studies of complete yeast (8) and E. coli (72a) genomes similarly indicate low element frequencies at individual sites.

Further analysis and interpretation of these data require consideration of possible dynamic processes affecting element frequencies and abundances, which will now be discussed.

DYNAMICS OF ELEMENTS IN LARGE POPULATIONS

Basic Population Models

A model of the transmission of a family of elements from generation to generation needs to incorporate the following features.

Change in copy number due to transposition. The probability per generation that a given element in the germline of an individual with n elements of the family in question produces a new copy that is inserted elsewhere in the genome is denoted by u_n. The subscript n is used to take into account the possibility of functional dependence of the probability of transposition of an element on the number of elements in the same cell; there is evidence for self-regulation of transposition probability (such that u_n is a decreasing function of n) in a variety of species (14). If transposition is not accompanied by excision of the parental elements, there is clearly an expected gain of copy number of nu_n in the germline of an individual with n elements. The value of u for most eukaryote elements seems to be of the order of 10^{-4} or less per generation (13); the high values recorded for I and P elements of D. melanogaster in crosses experiencing hybrid dysgenesis are not representative of the frequencies in natural populations (18).

Excision of elements. The probability per generation that an element is excised from its location in the genome, and hence not transmitted to the next generation, is denoted by v (v could obviously also be written as a function of n, but there is little evidence that this is often so). There is experimental evidence for such excision events in bacterial transposable elements (41) and Drosophila P elements (18). There is no firm evidence for excision of the abundant retroviral-like elements of Drosophila, on the other hand (23). In some bacterial systems, it seems that transposition is frequently accompanied by excision (28), but it would generally seem that transposition and excision are sufficiently loosely coupled that there is a net increase in copy number per generation in individuals with sufficiently low copy numbers that $n(u_n-v)$ is positive. In maize, transposition of at least some elements is not strictly replicative, but is correlated with chromosome replication in such a way that there is a net increase in copy number (21).

On the assumption of sufficiently high frequencies of recombination between different sites that linkage disequilibrium can be neglected, the following approximate equation can be written for the change in mean copy number per generation under transposition and excision:

$$\Delta\bar{n} \approx \bar{n}(u_{\bar{n}}-v) \qquad \text{(b)}$$

This equation is only approximate when there is regulation of the rate of transposition, since it neglects the effect of variation in copy number around the mean, but the error is usually small (13). If transposition is regulated, and if the rate of transposition at low copy number (u_0) exceeds v, it will lead to an increase in copy number until $\bar{n} \approx \hat{n}$, where $u_{\hat{n}}=v$, or all available sites have filled up with elements. In the former case, there will be a stable equilibrium at $\hat{n}$, with the frequency of elements at each site equal to $\hat{n}/(2m)$. If there is no regulation, transposition leads to fixation of elements at each site if $u_0>v$. Unless regulation of transposition is so strong that there is no transposition at all above a certain threshold copy number, maintenance of an equilibrium with element frequency polymorphism requires a nonzero excision rate.

Natural selection. There are several ways in which natural selection can affect the abundance of elements within populations. These will be discussed in detail below. All of them have the common feature that the fitness of a host individual with copy number n for a given family is assumed to be a decreasing function of n, w_n. If this is the case, and if similar assumptions to those used for equation (b) are used, the following equation is obtained for the change in copy number per generation (12,13):

$$\Delta\bar{n} \approx \bar{n}(1-\bar{n}/[2m])(\partial \ln \bar{w}/\partial \bar{n}) + \bar{n}(u_{\bar{n}}-v) \qquad \text{(c)}$$

where w is the mean fitness of the population ($\bar{w} \approx w_{\bar{n}}$).

The approximate equilibrium value of $\bar{n}$, $\hat{n}$ is given by the solution of the equation:

$$(1-\hat{n}/[2m])(\partial \ln \bar{w}/\partial \hat{n}) \approx v-u_{\hat{n}} \qquad \text{(d)}$$

It is relatively easy to find functional forms that result in an equilibrium with low element frequencies at each site; for example, with $w_n = \exp -\frac{1}{2}tn^2$, we have $\hat{n} \approx (u_{\hat{n}}-v)/t$. An equilibrium is possible even in the absence of excision ($v=0$). In the absence of regulated transposition, the stability of this equilibrium requires that the logarithm of fitness declines more steeply than linearly with increasing n (12,13), Given that low frequencies of transposition seem to be the norm, this shows that even a weak pressure of selection (as measured by t, which is the slope of the relation between the logarithm of fitness and copy number at a copy number of 1) is capable of maintaining a balance with transpositional increase in copy number. For example, a mean copy number of 50 elements per diploid genome would be maintained if t is one-fiftieth of the excess of rate of transposition over excision. With transposition rates of the order of 10^{-4} or less, this implies that extremely small selection coefficients would be needed to maintain a balance.

Possible Modes of Natural Selection on Copy Number

This section will consider modes of selection that could lead to the stabilization of element copy numbers in the face of transpositional increase in copy numbers.

Mutations associated with insertions of elements. First, it is well-established that the insertion of elements into or near genes alters their expression, usually in a deleterious fashion. Most studies of the mutational consequences of element insertions have concentrated on visible mutations, for reasons of technical convenience (74). However, studies of *Drosophila* mutations affecting viability have shown that the most frequent class of spontaneous mutations with detectable effects are those with relatively small, detrimental effects (75). Recent studies of *P* element mutagenesis in *D. melanogaster* have shown that transposition is frequently accompanied by such detrimental mutations (24,52,83). As argued by Charlesworth (12), a reasonably good description of this mode of selection is obtained by writing the fitness of a host individual carrying $\underline{n}$ elements as a decreasing function $\underline{w}_{\underline{n}}$ of the number of elements.

The finding that a very weak intensity of selection is sufficient to check the spread of elements casts doubt on the possibility that selection against insertional mutations is the main factor in maintaining element frequencies, in the absence of self-regulated transposition or other forces, since the studies of Mukai and colleagues on viability mutations indicate a mean selection coefficient of the order of 2% against homozygous detrimental mutations, and 0.7% against heterozygotes (61). If all insertions resulted in a mutation with this magnitude of effect on fitness, it is clear that elements would never be able to spread in the face of selection, unless transposition rates are higher than seems realistic for most elements, or there is regulation of transposition such that transposition rates are very high for low copy numbers. This condition might well have prevailed during the initial spread of P elements through populations of D. melanogaster (18).

Two classes of sites: selected and neutral. At first sight, a solution to this dilemma would be to take into account the possibility that a large fraction of insertions have no effect on fitness, because they involve sites that are sufficiently remote from active genes. The average impact on fitness of an insertion is thus the product of the probability that it involves a selectively significant site, and the selection coefficient against a mutation induced in such a site. The model described above can be extended to include a set of neutral sites in addition to a set of selected sites (Ref. 12 and unpub. results). Numerical analysis of this model shows, however, that it is difficult to find conditions under which mean copy numbers of the kind characteristic of many transposable elements can be maintained. Unless $\underline{v}$ is rather close to $\underline{u}$ or the number of neutral sites is low, the mean number of elements at neutral sites tends to increase to very high levels, if selection is of the magnitude suggested by the Drosophila data. With a small number of neutral sites or high excision rates, the equilibrium copy numbers tend to be very small.

The majority of elements in an equilibrium population are found at the neutral sites. Many parameter sets generate element frequencies of 50% or more at the neutral sites, particularly if the rate of excision is low compared with the rate of transposition. Mean copy number may become very large in this case, if there is a large number of occupable neutral sites. If there is no excision, then all the neutral sites become fixed. This great dependence on the values of the parameters may be relevant to species and family differences in population behavior, such as the great abundance and high level of fixation at individual sites of the mammalian Alu sequences (4,73), in contrast to the low abundance and low frequencies of elements in Drosophila. But it is hard to believe that the population behavior of, say, most Drosophila retroviral-like elements can be explained in these terms (see below), especially if excision is rare or absent.

Two classes of sites: strongly and weakly selected. A modification to the two-class model that could explain the data would be to assume that there are two classes of selected sites; one class being the sites subject to relatively strong selection of the sort considered above, and the other subject to much weaker selection against insertions. These might correspond to insertions into transcribed sequences and nontranscribed flanking regions, for example. On this kind of model, if weakly selected sites are sufficiently common, the dynamics of elements are essentially controlled by them, and copy numbers will stabilize at values close to those predicted by equation (d), with $\hat{n}$ and $\partial \ln \bar{w}/\partial \hat{n}$ corresponding to values for the weakly selected sites, the abundance of elements in the strongly selected sites being effectively negligible. Comparisons of DNA sequences among species suggests that the rate of nucleotide substitution in 5' and 3' flanking regions of genes is higher than that for nonsynonymous changes in coding regions, but considerably slower than the rate for pseudogenes (48). This suggests that there may indeed be selection against insertions of elements into nontranscribed regions, but it is unclear whether neutral sites are so rare that the problems discussed above can be ignored. It seems more likely that the flanking regions contain a mixture of sites, some being effectively neutral and others being subject to more or less weak selection. High element frequencies at the neutral sites would still be expected.

These considerations lead to the conclusion that, although selection against insertional mutations may be a factor in stabilizing element frequencies in natural populations, it may not be the only force involved. One possibility is that regulation of transposition rates may act in addition to selection. This makes it much easier to achieve stabilization of element abundances at reasonable levels in the mixed-site class of model.

Unequal exchange between elements. Another possibility is that crossing over between homologous elements located at different chromosomal sites will lead to the production of deleterious chromosome rearrangements (45,60). There is experimental evidence for such crossing over and the concomitant production of rearrangements in yeast and *Drosophila* (16,26,58,69). A simple model of this process of unequal exchange is to write the probability per generation of the production of at least one rearrangement in an individual with n elements as $\exp -krn^2$, where k is a constant of proportionality and r is the rate of crossing over. If rearrangements behave as dominant lethals, the log fitness of individuals is thus a quadratic function of copy number, and hence satisfies the condition for stability derived above. The equilibrium copy number is given by $(u-v)/(2kr)$.

More realistic models that take into account heterogeneity in rates of recombination over different parts of the genome have been developed (45). If unequal exchange events between pairs occur with a frequency that is dependent only on the rates of crossing over in their neighborhood, the equilibrium density of elements in a genomic region is inversely proportional to the frequency of crossing over in that region. If unequal exchange is confined to elements at nearby locations, as is suggested by the *Drosophila* data, abundances are inversely proportional to the square root of the frequency of exchange. In principle, this mechanism is capable of accounting for the main facts concerning element frequencies reviewed above.

Induction of advantageous mutations. Finally, it has been suggested that element insertions may occasionally induce favorable mutations or rearrangements that would cause the associated elements to spread through the population, and hence help to maintain the presence of elements without transpositional increase in copy number (78a). Experimental evidence for the induction of such favorable mutations in an *E. coli* chemostat population as a result of the transposition of the element *Tn10* has been presented by Chao et al. (11). If a strain containing this element is present in competition with one that lacks it, it can spread if its initial frequency is sufficiently high. If present initially at a low frequency, however, there is a negligible chance that an advantageous mutation will arise and confer a selective advantage that outweighs the loss in fitness caused by the far more numerous deleterious mutations induced by transposition.

Similarly, the use of P element mutagenesis in hybrid dysgenic strains of D. melanogaster has led to the production of variability in quantitative characters that can be utilized in artificial selection experiments (51). This is, however, accompanied by a large decline in fitness of the strains concerned. As in the bacterial system, the fact that the bulk of insertional mutations are necessarily deleterious means that it is hard to see how an initially rare element could rise to a high frequency purely as a result of the occasional induction of a favorable mutation. As pointed out by Engels (18), surveys of the distribution of P elements in natural populations by the in situ hybridization method indicate low element frequencies at individual chromosome sites, which is inconsistent with a selective advantage to element insertions having been involved in the recent spread of these elements through populations of D. melanogaster. The same obviously applies to the other cases of low element frequencies reviewed above. It therefore seems unlikely that the induction of favorable mutations is an important factor in determining element abundances in natural populations. Similarly, the frequency data are unfavorable to the idea that elements often play a positive functional role in physiology or development, as has sometimes been suggested (9,54).

Longer-term, population-level advantages to the presence of elements within a population, due to increased adaptability in changing environments conferred by their mutagenic effects, have also been invoked as an explanation of their persistence (63,78a). This type of model similarly fails to explain the frequency data, unless the subsidiary hypothesis that favorable mutations result exclusively from imprecise excision events is adopted. It also suffers the usual drawback of models of group selection (55) in that it fails to explain the initial spread of elements within the population, or their maintenance in the face of counter-selection against deleterious consequences of insertional mutations. Random sampling of element frequencies as a result of finite population size effects (genetic drift) could be invoked to overcome this difficulty, as is usually done in models of group selection (80), but the frequency data suggest that drift is relatively ineffective compared to forces opposing the spread of elements by transposition (see below).

Tests of the Models

Given the small probable magnitudes of the forces involved in determining element frequencies, apart from exceptional situations such as hybrid dysgenesis (18), it seems that attempts to discriminate between the various possibilities for the maintenance of stable element frequencies in natural populations by direct measurement of the quantities concerned are doomed to failure. As already pointed out, there is experimental evidence for each of the above possibilities; what is unclear is their relative significance in controlling element abundances. This is a common situation in population biology; the usual response to the dilemma is to devise tests of the models based on their consequences for the statistical properties of populations. There are obvious dangers in this procedure, notably that goodness of fit to a model does not guarantee its truth. All conclusions based on such tests must therefore be regarded as tentative in the absence of supporting evidence.

Tests of the role of insertional mutations. One such test has been devised and applied by Montgomery et al. (60). They noted that the hypothesis that frequencies are stabilized by the deleterious effects of insertional mutations predicts that element frequencies at equilibrium will be lower for X chromosomal sites than for autosomal sites, due to the expression of X chromosomal mutational effects in the hemizygous state in males, compared with the predominantly heterozygous state of rare autosomal genes. The partial recessivity of mutational effects on fitness (75) thus means that there will be a greater selective impact of X-linked transposable elements relative to autosomal elements. A quantitative model of the relative equilibrium mean copy numbers for the X chromosomes and autosomes of *D. melanogaster* was developed by Montgomery et al. (60) for the hypothesis of insertional mutation effects, and for the null hypothesis that X-linked and autosomal elements are eliminated at the same rate, as would be expected with regulation of transposition or induction of dominant lethal rearrangements by unequal exchange. The model assumes that the size of the euchromatic regions of the two autosomes as targets for insertion are equal, and are twice the target size presented by the X. This assumption seems reasonable in view of the genetic evidence for such equivalence, and would only be violated if there was a nonrandom distribution of insertion sites between the chromosomes.

For a haploid set of X, second and third chromosomes, the first hypothesis predicts that approximately 17% of elements in the set should be located on the X, whereas the second hypothesis predicts that about 11% will be on the X. The results of scoring the numbers of copies of the retroviral elements *copia*, *roo*, and *412* on sets of 20 X, second and third chromosomes from a natural population showed no significant difference between the observed abundance of elements on the X and the expectation on the null hypothesis for *copia* and *roo*, whereas *412* showed a significant ($p<0.05$) deviation from the null expectation, but agreed with the expectation on the insertional mutation hypothesis. It is unclear why these differences between families should exist; further data on a wider variety of families are clearly desirable. On the whole, these results suggest that, while there may be some effect of insertional mutations on copy number, it is not necessarily the chief mechanism involved.

Further evidence for a role of selection against insertional mutations is provided by the data from the restriction site studies of limited regions referred to above. For instance, in the study of the *Adh* region by Aquadro et al. (1), the data show clearly that elements *tend* to be confined to noncoding regions, suggesting that insertions into structural genes have such drastic effects on fitness that they are rapidly eliminated by selection. This provides an empirical basis for the two-class model of sites described above; in view of the properties of this model, this evidence is in fact unfavorable to the view that the fitness effects on insertions provide the sole means of stabilizing copy numbers. A further interesting feature of the *Adh* data is discussed by Golding et al. (27). Reconstructions of the phylogeny of the haplotypes detected by restriction site polymorphisms shows that element insertions are essentially confined to the tips of the branches of the phylogeny. This suggests that some deterministic force is actively opposing the spread of elements; analysis of a quantitative model shows that this pattern cannot be explained by constant transposition and excision rates (27). It is not, however, clear whether regulated transposition, such that the rate of insertion of elements into a particular site declines with overall element abundance (and hence with the frequency of the element at the site in question), or a form of selection is involved.

Tests for effects of unequal exchange. A test for the role of selection against chromosome rearrangements produced by unequal exchange between elements is provided by asking whether or not the abundance of elements tends to be higher in regions where exchange is reduced in frequency. *Drosophila melanogaster* again provides useful material for carrying out such a test, since there is pronounced suppression of meiotic crossing over in the telomeric and centromeric regions of the euchromatin of the chromosome arms (49). It is straightforward to use in situ hybridization to determine the distribution of the numbers of elements over the salivary chromosome maps in chromosomes sampled from natural populations, and to compare this with the distribution expected on a null hypothesis of no variation in element abundance with respect to the rate of crossing over. The results of a study of *roo* by Langley et al. (45) showed clear evidence for an excess of elements at the base of the X chromosome, of the order of three-fold over the number of elements that would be expected if elements were inserted in proportion to the physical size of the region. The picture is less clear for the other chromosomes. The data suggest that there is, indeed, a tendency for elements to accumulate at the base of the euchromatin but that the effect is usually not very large. There is no evidence for an accumulation of elements at the tips; rather, there is a slight tendency for elements to be less frequent there than expected. The study of Miklos et al. (57) of DNA cloned from band 19E8, which is located close to the base of the X, also demonstrates an unusual abundance of elements in this region.

These observations should not be taken as providing watertight evidence in favor of the unequal exchange hypothesis, since an alternative mechanism exists for creating a build-up of elements in regions where crossing over is suppressed (see discussion of Muller's ratchet below). Furthermore, the failure to detect an excess of *roo* near the chromosome tips, particularly that of the X where crossing over is strongly reduced, presents problems of interpretation for either model. One possible solution, that crossing over near the tips is induced by the interchromosomal effect on crossing over of inversion heterozygosity (50), is discussed by Langley et al. (45).

An extreme example of the differential concentration of elements in genomic regions where crossing over is suppressed is provided by the numerous *Drosophila* elements that are found to be disproportionately abundant in the centric heterochromatin (23). It is possible that the mechanisms just discussed are responsible for this concentration, since crossing over is virtually absent in this part of the genome (Carpenter et al., 1982). Another possibility is that the rarity of functional genes in this region (35) means that insertional mutations do not oppose the spread of elements within the heterochromatin. The two-class models discussed above suggest, however, that a very large concentration of elements should be found in the heterochromatin if that were the case. In fact, for at least some element families, it seems that such disproportionality is not observed between the euchromatin and heterochromatin (68,78). It is possible that there is a relative dearth of sites for insertion within the heterochromatin, which outweighs the lack of effect of insertions on fitness. Furthermore, it is likely that elements may have difficulty in experiencing new transpositions once inserted into the heterochromatin, owing to the lack of transcription in this region (56). This will tend to retard the build-up of elements in this region, as is found in computer simulations of a model similar to the two-site model discussed above (B. Charlesworth, unpubl. results).

DYNAMICS OF ELEMENTS IN SMALL POPULATIONS

Since the deterministic forces acting on transposable elements seem to be so weak, the full interpretation of statistics on element frequencies in natural populations requires that effects of genetic drift be taken into account. Stochastic models of element frequencies have been constructed by Ohta and Kimura (67), Ohta (65), Ohta (66), Charlesworth and Charlesworth (13), Langley et al. (44), Kaplan and Brookfield (36), Charlesworth (12), and Kaplan et al. (38).

Probability Distribution of Element Frequencies: Theory

One problem which has been satisfactorily solved is the form of the probability distribution of element frequencies for a given family that is attained when the forces of transposition, excision, and selection considered above come into statistical equilibrium with random changes in element frequencies, particularly the stationary distribution of element frequencies [Charlesworth and Charlesworth (13); Langley et al. (44); Kaplan and Brookfield (36); Charlesworth (12)]. Strictly speaking, no such stationary distribution exists, because the state of loss of all elements is an absorbing boundary of the process, to which all populations tend with the probability of one (17,65,67); once all copies of a family have been accidentally lost from a species, there is no way in which they can be restored in the absence of reintroduction from another species. As will be discussed below, the rate of approach to this state of loss is very slow, and the form of the probability distribution of element frequencies conditioned on their presence in the population can be regarded as essentially equivalent to a stationary distribution.

The following parameters provide a complete approximate description of the form of $\phi(x)$, the stationary probability density for element frequency x: N_e is the effective size of a local population; $\tilde{n}$ is the expected value of the mean copy number, which can be approximated by the equilibrium value $\hat{n}$ for a large population; $\alpha = 4N_e u \tilde{n}/(2m-\tilde{n})$ measures the effect of drift, and the effect of transposition in causing insertions into a given site; $\beta = 4N_e(s+v)$, where s is the value of $-\partial \ln \bar{w}/\partial \bar{n}$ at $\bar{n}=\tilde{n}$, measures the joint effects of drift, excision, and selection. It turns out that ϕ is approximated by a β distribution:

$$\phi(x) \approx C\, x^{\alpha-1}(1-x)^{\beta-1} \qquad \text{(e)}$$

where C is a constant of integration, equal to $\Gamma(\alpha+\beta)/\Gamma(\alpha)\Gamma(\beta)$ (13).

This is the formula for a closed population; if $\tilde{n} \ll 2m$, the effect of migration can be included by adding $4N_e M$ to β, where M is the frequency of immigrants into a local population. If m is sufficiently large compared with $\tilde{n}$, α can be neglected in equation (e). This yields the formula derived by Langley et al. (44) for the expected number of sites per haploid genome with element frequency $\underline{x}$, $\Phi(\underline{x})=\underline{m}\phi(\underline{x})$.

Probability Distribution of Element Frequencies: Data

These formulae can be applied to the analysis of in situ hybridization data on element frequencies, using the sampling theory developed by Charlesworth and Charlesworth (13), Langley et al. (44), and Kaplan and Brookfield (37). The model of Langley et al. assumes that $\alpha=0$, and uses an extension of Ewens' (20) sampling theory to estimate β (their θ). As pointed out by Kaplan and Brookfield, this procedure as applied to salivary chromosome in situ data probably underestimates β, because of the overestimation of the frequency of multiple occupancy of the same chromosomal site in a sample. They examined an alternative, in which element frequencies are assumed to be the same at each site that is occupied, and equal to $1/\underline{S}$ where $\underline{S}$ is the number of sampled chromosomes. Multiple occupancy is then due purely to the chance occupation at different sites of the same salivary chromosome band in different chromosomes of the sample.

An essentially equivalent model, with equal frequencies of elements at each site, was suggested by Charlesworth and Charlesworth (13). The most general model, in which α and β are jointly estimated by a minimum χ^2 method of fitting the observed occupancy profile to that predicted from the sampling formula based on equation (e), was proposed by Charlesworth and Charlesworth (13). This method has the advantage that the estimated value of β is not affected by the frequency estimation bias introduced by the in situ technique; the confounding of separate sites within the same salivary chromosome band merely has the effect of reducing the value of $\underline{m}$ in the expression for α, and hence inflating the value of α as far as the true insertion sites are concerned.

Table 2 displays the estimates of α and β obtained with this method for a number of elements scored on chromosomes sampled from natural populations of D. melanogaster. It should be stressed that the fits obtained with this are usually not significantly better than those obtained with the assumption that $\alpha=0$, or that element frequencies are equal at each site. Nevertheless, the generally large values of β obtained are highly significant when compared with small values, reflecting the low frequencies of multiple occupation. This effectively rules out the possibility that element frequencies are controlled by regulated transposition in the absence of excision, such that transposition rates are zero in an equilibrium population (see above). An excellent fit to the observed occupancy profiles is obtained (Fig. 1) with the minimum χ^2 estimates of the parameters.

For natural population data such as these, the effects of immigration must be included in β. Estimates of the value of $4\underline{N}_e\underline{M}$ in this species are available from studies of the geographic distribution of electrophoretic alleles (76), and these can be deducted from the estimated values of β in order to obtain the contribution from excision and selection. Except for roo on the X chromosome, these values are all of the order of 10 or more (note that $\underline{N}_e$ for the X chromosome is three-fourths that for the autosomes, so that the difference between the X chromosome and the autosome data for roo probably largely reflects the difference in $\underline{N}_e$). If the expected mean number of elements per individual is unchanging over time, then from equation (d) $\beta \approx 4\underline{N}_e\underline{u}$. Estimates of $\underline{N}_e$ for East Coast D. melanogaster populations from the frequencies of allelism between recessive lethals suggest values of the order of $2\text{x}10^4$ (61). With a β value of 20,

Tab. 2. Estimates of the parameters of the probability distribution of element frequencies for Drosophila melanogaster elements from natural populations.

Element	Chromosome	α	β	Sample size
297	X	0.05	16.7	20
412	X	0	30.0	20
copia	X	0	40.0	20
roo	X	0.70	11.0	13
I	X	∞	∞	20
roo	3L	1.25	15.0	12
roo	3R	1.20	16.5	11

Note: $4N_eM$ is equal to 8.8 for autosomal loci and 6.5 for X-linked loci of East Coast D. melanogaster (76). The contributions of selection and excision to β can be estimated by deducting these values from the appropriate entries in the table.

Data for 297, 412, and copia are from Montgomery and Langley (59); data for roo X chromosomes are original data of B. Charlesworth; data for roo third chromosomes are from a re-examination by B. Charlesworth of material by Montgomery et al. (60); data for I are from Leigh Brown and Moss (47).

this yields an estimate of 2.5×10^{-4} for u. (The estimates of N_e are subject to considerable uncertainty, so that this value should not be taken too seriously.)

The differences among elements in values of β suggest that there are real differences between elements in rates of transposition, etc. Unfortunately, there are at present no reliable quantitative estimates of the rates of transposition for these elements that could be used to compare with these estimates. The magnitude of β in these data is so high that drift is clearly rather ineffective in relation to the deterministic forces responsible for reducing element frequencies; for element frequencies to be predominantly controlled by drift, β values of the order of one would be required (see, for example, Ref. 81).

Stochastic Loss of Elements

Charlesworth (12) and Kaplan et al. (38) have modeled the process of stochastic loss of elements from finite populations, suggested by Engels (17) as an explanation for the absence of the elements responsible for hybrid dysgenesis from old laboratory stocks of D. melanogaster, and described briefly above. As shown by Charlesworth (12), in a very small population the expected time to loss of all elements is of the order of at least $1/v$, unless the initial mean copy number per individual is very small; the time is substantially larger than $1/v$ in bigger populations. This is difficult to reconcile with the fact that laboratory strains lacking active P elements may be only 30 years old or less, corresponding to 750 generations at 25° (2). A high rate of excision of P elements from their chromosomal sites and a very small mean number of active elements per individual in the populations from which the laboratory strains in question were collected would be required to rescue the stochastic loss hypothesis for P elements. Since the mean copy number is of the order of 30 for natural populations (71), the latter condition is unlikely to be satisfied.

Other data relevant to the question of stochastic loss of P elements are reviewed by Engels (18). It seems most probable that the recent invasion hypothesis of Kidwell (39,40) is valid, particularly in view of the evidence for the presence of P elements in unrelated species and their absence from the closest relatives of D. melanogaster (15). It is conceivable, however, that the stochastic loss hypothesis is valid for the I element system of hybrid dysgenesis, since these have a lower mean copy number per individual than P elements (71) and the laboratory strains that lack active I elements are over 60 years old (2). They are also present in the relatives of D. melanogaster (6). Whether or not excision rates for I elements are sufficiently high to account for their loss over 1,500 generations is unclear.

A striking feature of P elements, and of some families of plant transposable elements, if that a substantial fraction of elements are defective and unable to produce the enzymes needed for their own transposition (21,64). They can, however, transpose in the presence of complete elements. Kaplan et al. (38) have constructed a model in which such defective elements are generated by mutation from functional elements. The equilibrium state of a large population is such that a large fraction of elements are defective, provided that the defective elements suffer no replicative disadvantage in host cells that contain a sufficient number of complete elements to supply them with transposase functions. It is even possible that defective elements may enjoy a selective advantage, due to faster replication. In a finite population, stochastic loss of complete elements may occur relatively fast (in terms of evolutionary time) because of their low equilibrium numbers. This could lead to the evolution of an element family, all of whose members are incapable of transposition. The ultimate fate of such a family would be extinction, unless excision is also impossible.

Muller's Ratchet: Theory

This is a process by which insertions of elements into genomic regions which lack crossing over can cause a gradual increase in copy number over time, as a result of the effects of genetic drift. The theory of Muller's ratchet has been worked out for the case of conventional deleterious mutations (22,29), based on an idea of Muller (62), but essentially the same principles apply to transposable elements provided that excision rates are sufficiently small. The principle is as follows. Consider first a segment of chromosome in an infinitely large population. If there is a rate of insertion of elements into this segment of U per generation, then (as we have seen) an equilibrium will be established under counter-selection or regulated transposition, such that there will be an approximately Poisson distribution between individuals of the numbers of copies of elements in this segment. The nature of this distribution will be similar for a freely-recombining segment of chromosome and a segment that lacks crossing over (see, for example, Ref. 29). If U is sufficiently large in relation to the pressure of selection, then the frequency of chromosomes lacking elements in the given segment will be low. In a finite population, this means that this class of chromosomes is vulnerable to loss by drift.

If there is crossing over in the segment in question, it is possible for the zero-element class to be regenerated by crossing over, and so the form of the distribution of element numbers will remain similar to that for an infinite population. If there is no crossing over, however, the zero class will be permanently lost in the absence of loss of elements by excision. Once it is lost, the class with one element in the region in question will be vulnerable to loss, and so on, so that the form of the distribution of copy number moves steadily to the right in a ratchet-like process. The speed of this process depends on the inverse of the product of the species population size and the frequency of the zero class (29), and so it will proceed very slowly in a natural population, especially if the size of the chromosome segment is small, so that the mean copy number at equilibrium is low.

In principle, the effect of the ratchet in producing a build-up of elements in regions where crossing over is suppressed can be distinguished from the equilibrium between unequal exchange and transposition discussed above by the fact that the ratchet produces an unlimited build-up of elements, given a sufficiently long time, whereas the latter process will only produce a limited increase, unless exchange is totally suppressed and there is no regulation of copy numbers. The ratchet also fails to operate in the presence of very low frequencies of recombination, unless the population size is very small (22), and will thus only produce an excess of elements in genomic regions that totally lack crossing over.

Muller's Ratchet: Data

It is therefore somewhat doubtful whether this process could account for the excess of elements near the base of *D. melanogaster* chromosomes described above, since this is a relatively small segment of the genome and crossing over is not completely absent (49). It is possible, though, that the concentration of *Drosophila* elements in the centric heterochromatin (23) is due to the ratchet, since exchange is virtually absent here (10). It may also be relevant to the huge build-up of repeated DNA in the chromosome 1 polymorphism of the newts of *Triturus cristatus* and *T. marmoratus* studied by MacGregor and colleagues (53). In this system, crossing over is suppressed between two chromosome rearrangements that are maintained permanently heterozygous as a result of recessive lethality. Since the polymorphism is found in several different species of *Triturus*, it is clearly of ancient origin, and so plenty of evolutionary time has been available for the ratchet to work.

Similarly, the excess of middle repetitive DNA in the neo-Y chromosome of *D. miranda*, reported by Steinemann (77), could be due to the operation of the ratchet. This chromosome is the result of a centric fusion between an autosome and the original Y chromosome, and so has been maintained strictly through the male line, without crossing over, since the establishment of the fusion. It is of sufficiently ancient origin that partial degeneration of the Y and partial dosage compensation of the corresponding neo-X chromosome have evolved.

Finally, the small size of laboratory populations means that the ratchet will operate much more rapidly than in large natural populations. The apparent build-up of elements near the breakpoints of balanced lethal inversions described by Montgomery et al. (60) may thus be due to the ratchet.

CONCLUSIONS

The main conclusion from the models and data presented above is that there is nothing that we know at present that is strikingly inconsistent with the view that transposable elements are maintained in populations as a result of transpositional increase in copy number. This view also has the merit of providing detailed explanations for a range of phenomena associated with transposable elements. In particular, it convincingly explains the fact that elements in bacteria, yeast, and *Drosophila* are usually present at low frequencies at individual chromosomal sites into which they can insert; this fact is almost impossible to accommodate on the hypothesis that elements persist as a result of favorable mutations associated with their transpositional activities. Indeed, this observation suggests that elements in these organisms rarely or never have a positive role in directly promoting evolutionary change in association with an increase in their own frequency. It is possible, however, that certain classes of mutation, such as chromosome rearrangements (16,19,26,58,69), could be generated as a result of excision of elements or exchange between elements at different locations, and could be fixed in populations by drift or selection without leading to an increase in element frequency at the sites in question. Thus, transposable elements could act indirectly as a source of mutational variation for evolutionary change, but this would not have much influence on their distribution within populations.

It should be emphasized that low frequencies of elements at individual sites are not necessarily predicted by the models; it is relatively easy to generate situations in which elements rise to high frequencies or even fixation at many chromosomal sites. There are, indeed, examples of such elements, the classic one being the very abundant mammalian *Alu* sequences (73). [The status of this class of sequence as a transposable element has recently been questioned; and it is possible that it is more akin to a family of processed pseudogenes (70)]. The data indicating low frequencies demand that some force or forces are actively opposing transpositional increase in copy number, at a rate that is high in comparison with the reciprocal of effective local population size (Tab. 2).

The nature of these forces is not clear; regulation of rate of transposition in relation to number of element copies per individual, selection against insertional mutations, and the induction of dominant lethal chromosome rearrangements by crossing over between elements at different sites may all be implicated. Experimental tests to distinguish between these possibilities are difficult to carry out, and rely on statistical predictions of the various models. At present, there is some weak evidence for an effect of insertional mutations on one out of three families of *Drosophila* retroviral elements (60). There is also evidence that some families of *Drosophila* elements tend to accumulate in regions of the genome where crossing over is suppressed, as would be expected on the last of the three modes of selection (44). Unfortunately, this effect is also expected from the operation of the stochastic process known as Muller's ratchet. Although this process seems somewhat unlikely to explain these data, it may be involved in some natural examples of a large build-up of middle-repetitive DNA sequences in chromosome regions where crossing over is suppressed.

The possible mechanisms involved in the stabilization of element frequencies are not mutually exclusive, which may account for some of the ambiguities of the presently available, rather fragmentary, evidence.

Indeed, it seems probable that the well-documented property of regulation of transposition is an acquired evolutionary response to the deleterious effects of transposable elements on host fitness (14). Further experiments and observations that take account of the predictions of the various models, for a wider range of species than has been studied up to now, are needed if these questions are to be resolved.

ACKNOWLEDGEMENTS

This work was supported by National Science Foundation grant BSR-8-16629 and Public Health Service grant 1-R01-GM 36405-01, and a grant from the Louis Block fund of The University of Chicago. I am grateful to Chuck Langley for several years of discussions and collaboration on this topic.

REFERENCES

1. Aquadro, C.F., S.F. Deese, M.M. Bland, C.H. Langley, and C.C. Laurie-Ahlberg (1986) Molecular population genetics of the alcohol dehydrogenase gene region of Drosophila melanogaster. Genetics 114:1165-1190.
2. Bregliano, J.-C., and M.G. Kidwell (1983) Hybrid dysgenesis determinants. In Mobile Genetic Elements, J.A. Shapiro, ed. Academic Press, New York, pp. 363-410.
3. Bridges, C.B. (1935) Salivary chromosome maps. J. Hered. 26:60-64.
4. Brookfield, J.F.Y. (1986) Population biology of transposable elements. Phil. Trans. Roy. Soc. Lond. B. 312:217-226.
5. Brown, A.H.D. (1979) Enzyme polymorphism in natural populations. Theor. Pop. Biol. 15:1-42.
6. Bucheton, A., M. Simonelig, C. Vaury, and M. Crozatier (1986) Sequences similar to the I transposable element involved in IR hybrid dysgenesis occur in other Drosophila species. Nature 322:650-652.
7. Bulmer, M.G. (1980) The Mathematical Theory of Quantitative Genetics. Oxford University Press, Oxford.
8. Cameron, J.R., E.Y. Loh, and R.W. Davis (1979) Evidence for transposition of dispersed repetitive DNA families in yeast. Cell 16:739-751.
9. Campbell, A. (1983) Transposons and their evolutionary significance. In Evolution of Genes and Proteins, M. Nei and R.K. Koehn, eds. Sinauer, Sunderland, Mass., pp. 258-279.
10. Carpenter, A.T.C., and B.S. Baker (1982) On the control of the distribution of meiotic exchange in Drosophila. Genetics 101:81-84.
11. Chao, L., C. Vargas, B.B. Spear, and E.C. Cox (1983) Transposable elements as mutator genes in evolution. Nature 303:633-635.
12. Charlesworth, B. (1985) The population genetics of transposable elements. In Population Genetics and Molecular Evolution, T. Ohta and K. Aoki, eds. Springer Verlag, Berlin, pp. 213-232.
13. Charlesworth, B., and D. Charlesworth (1983) The population dynamics of transposable elements. Genet. Res. 42:1-27.
14. Charlesworth, B., and C.H. Langley (1986) The evolution of self-regulated transposition of transposable elements. Genetics 112:359-383.
15. Daniels, S., L.D. Strasbaugh, L. Ehrman, and R. Armstrong (1984) Sequences homologous to P elements occur in Drosophila paulistorum. Proc. Natl. Acad. Sci., USA 81:6794-6797.

16. Davis, P.S., M.W. Shen, and B.H. Judd (1987) Asymmetrical pairings of transposons in and proximal to the white locus of Drosophila account for four classes of regularly occurring exchange products. Proc. Natl. Acad. Sci., USA 84:174-178.
16a. Doolittle, W.F., and C. Sapienza (1980) Selfish genes, the phenotype paradigm and genome evolution. Nature 284:601-607.
17. Engels, W.R. (1981) Hybrid dysgenesis in Drosophila and the stochastic loss hypothesis. Cold Spring Harb. Symp. Quant. Biol. 45:561-566.
18. Engels, W.R. (1986) On the evolution and population genetics of hybrid-dysgenesis causing transposable elements in Drosophila. Phil. Trans. Roy. Soc. Lond. B. 312:205-215.
19. Engels, W.R., and C.R. Preston (1984) Formation of chromosome rearrangements by P factors. Genetics 107:657-678.
20. Ewens, W.J. (1972) The sampling theory of selectively neutral alleles. Theor. Pop. Biol. 3:87-112.
21. Fedoroff, N.V. (1983) Controlling elements in maize. In Mobile Genetic Elements, J.A. Shapiro, ed. Academic Press, New York, pp. 1-63.
22. Felsenstein, J., and S. Yokoyama (1976) The evolutionary advantage of recombination. II. Individual selection for recombination. Genetics 83:845-859.
23. Finnegan, D.J., and D.H. Fawcett (1986) Transposable elements in Drosophila melanogaster. Oxf. Surv. Eukar. Genes 3:1-62.
24. Fitzpatrick, B.J., and J.A. Sved (1986) High level of fitness modifiers induced by hybrid dysgenesis in Drosophila melanogaster. Genet. Res. 48:89-94.
25. Ginzburg, L.R., P.M. Bingham, and S. Yoo (1984) On the theory of speciation induced by transposable elements. Genetics 107:331-341.
26. Goldberg, M.L., J.-Y. Shen, W.J. Gehring, and M.M. Green (1983) Unequal crossing-over associated with asymmetrical synapsis between nomadic elements of the Drosophila genome. Proc. Natl. Acad. Sci., USA 80:5017-5021.
27. Golding, G.B., C.F. Aquadro, and C.H. Langley (1986) Sequence evolution within populations under multiple types of mutation. Proc. Natl. Acad. Sci., USA 83:427-431.
28. Grindley, N.D.F., and R.R. Reed (1985) Transpositional recombination in prokaryotes. Ann. Rev. Biochem. 54:863-896.
29. Haigh, J. (1978) The accumulation of deleterious genes in a population--Muller's ratchet. Theor. Pop. Biol. 14:251-267.
30. Hartl, D.L., and D.E. Dykhuizen (1984) The population genetics of Escherichia coli. Ann. Rev. Genet. 18:31-68.
31. Hartl, D.L., and S.A. Sawyer (1988) Multiple correlations among insertion elements in the genome of natural isolates of Escherichia coli (in press).
32. Hartl, D.L., and S.A. Sawyer (1988) Why do unrelated insertion sequences occur together in the genome of Escherichia coli? (in press).
33. Hartl, D.L., M. Medhora, L. Green, and D.E. Dykhuizen (1986) The evolution of DNA sequences in Escherichia coli. Phil. Trans. Roy. Soc. Lond. B. 312:191-204.
34. Hickey, D.A. (1982) Selfish DNA: A sexually transmitted parasite. Genetics 101:519-531.
35. Hilliker, A.J., R. Appels, and A. Schalet (1980) The genetic analysis of D. melanogaster heterochromatin. Cell 21:607-619.
36. Kaplan, N.L., and J.F.Y. Brookfield (1983) The effect on homozygosity of selective differences between sites of transposable elements. Theor. Pop. Biol. 23:273-280.

37. Kaplan, N.L., and J.F.Y. Brookfield (1983) Transposable elements in Mendelian populations. III. Statistical results. Genetics 104:485-495.
38. Kaplan, N.L., T. Darden, and C.H. Langley (1985) Evolution and extinction of transposable elements in Mendelian populations. Genetics 109:459-480.
39. Kidwell, M.G. (1979) Hybrid dysgenesis in Drosophila melanogaster. The relationship between the P-M and I-R interaction systems. Genet. Res. 33:205-217.
40. Kidwell, M.G. (1983) Evolution of hybrid dysgenesis determinants in Drosophila melanogaster. Proc. Natl. Acad. Sci., USA 80:1655-1659.
41. Kleckner, N. (1981) Transposable elements in prokaryotes. Ann. Rev. Genet. 15:341-404.
42. Langley, C.H., and C.F. Aquadro (1987) Restriction map variation in natural populations of Drosophila melanogaster white locus region. Mol. Biol. Evol. 4:651-663.
43. Langley, C.H., E.A. Montgomery, and W.F. Quattlebaum (1982) Restriction map variation in the Adh region of Drosophila. Proc. Natl. Acad. Sci., USA 79:5631-5625.
44. Langley, C.H., J.F.Y. Brookfield, and N.L. Kaplan (1983) Transposable elements in Mendelian populations. I. A theory. Genetics 104:457-472.
45. Langley, C.H., E.A. Montgomery, R.H. Hudson, N.L. Kaplan, and B. Charlesworth (1988) On the role of unequal exchange in the containment of transposable element copy number (manuscript in preparation).
46. Leigh Brown, AJ. (1983) Variation at the 87A heat shock locus in Drosophila melanogaster. Proc. Natl. Acad. Sci., USA 80:5350-5354.
47. Leigh Brown, AJ., and J.E. Moss (1987) Transposition of the I element and copia in natural populations of Drosophila melanogaster. Genet. Res. 49:121-128.
48. Li, W.-H., C.-C. Luo, and C.-I. Wu (1985) Evolution of DNA sequences. In Molecular Evolutionary Genetics, R.J. MacIntyre, ed. Plenum Press, New York, pp. 1-94.
49. Lindsley, D.L., and L. Sandler (1977) The genetic analysis of meiosis in female Drosophila. Phil. Trans. Roy. Soc. B. 277:295-312.
50. Lucchesi, J.C. (1976) Inter-chromosomal effects. In The Genetics and Biology of Drosophila. 1a, M. Ashburner and E. Novitski, eds. Academic Press, New York, pp. 315-330.
51. Mackay, T.F.C. (1985) Transposable element-induced response to artificial selection in Drosophila melanogaster. Genetics 111:351-374.
52. Mackay, T.F.C. (1986) Transposable element-induced fitness mutations in Drosophila melanogaster. Genet. Res. 48:77-87.
53. MacGregor, H.C., and S.K. Sessions (1986) The biological significance of variation in satellite DNA and heterochromatin in newts of the genus Triturus: An evolutionary perspective. Phil. Trans. Roy. Soc. Lond. B. 312:243-259.
54. McClintock, B. (1956) Controlling elements and the gene. Cold Spring Harb. Symp. Quant. Biol. 21:197-216.
55. Maynard Smith, J. (1976) Group selection. Quart. Rev. Biol. 51:277-283.
56. Miklos, G.L.G. (1985) Localized highly repetitive DNA sequences in vertebrate and invertebrate genomes. In Molecular Evolutionary Genetics, R.J. MacIntyre, ed. Plenum Press, New York, pp. 240-321.
57. Miklos, G.L.G., M.J. Healy, P. Pain, A.J. Howells, and R.J. Russell (1984) Molecular genetic studies on the euchromatin-heterochromatin junction in the X chromosome of Drosophila

melanogaster. I. A cloned entry point near to the uncoordinated (unc) locus. Chromosoma 89:218-227.

58. Mikus, M.D., and T.D. Petes (1982) Recombination between genes located on nonhomologous chromosomes in Saccharomyces cerevisiae. Genetics 101:369-404.

59. Montgomery, E.A., and C.H. Langley (1983) Transposable elements in Mendelian populations. II. Distribution of three copia-like elements in a natural population. Genetics 104:473-483.

60. Montgomery, E.A., B. Charlesworth, and C.H. Langley (1987) A test for the role of natural selection in the stabilization of transposable element copy number in a population of Drosophila melanogaster. Genet. Res, 49:31-41.

61. Mukai, T., and O. Yamaguchi (1974) The genetic structure of natural populations of Drosophila. Genetics 82:63-82.

62. Muller, H.J. (1964) The relation of recombination to mutational advance. Mut. Res. 1:2-9.

63. Nevers, P., and H. Saedler (1977) Transposable genetic elements as agents of instability and chromosomal rearrangements. Nature 268:109-115.

64. O'Hare, K., and G.M. Rubin (1983) Structures of P transposable elements of Drosophila melanogaster and their sites of insertion and excision. Cell 34:25-35.

65. Ohta, T. (1981) Population genetics of selfish DNA. Nature 292:648-649.

66. Ohta, T. (1983) Theoretical study on the accumulation of selfish DNA. Genet. Res. 41:1-16.

67. Ohta, T., and M. Kimura (1981) Some calculations on the amount of selfish DNA. Proc. Natl. Acad. Sci., USA 79:1129-1132.

67a. Orgel, L.E., and F.H.C. Crick (1980) Selfish DNA: The ultimate parasite. Nature 284:604-607.

68. Potter, S.S., W.J. Brorein, P. Dunsmuir, and G.M. Rubin (1979) Transposition of elements of the 412, copia, and 297 gene families of Drosophila. Cell 17:415-427.

69. Roeder, G.S. (1983) Unequal crossing over between yeast transposable elements. Molec. Gen. Genet. 190:117-121.

70. Rogers, J. (1985) Origins of repeated DNA. Nature 317:765-766.

71. Ronsserray, S., and D. Anxolabehere (1986) Chromosomal distribution of P and I transposable elements in a natural population of Drosophila melanogaster. Chromosoma 94:433-440.

72. Sawyer, S.A., and D.K. Hartl (1986) Distribution of transposable elements in prokaryotes. Theor. Pop. Biol. 30:1-16.

72a. Sawyer, S.A., D.E. Dykhuizen, R.F. DuBose, L. Green, T. Mutangadura-Mhlanga, D.F. Wolczyk, and D.L. Hartl (1987) Distribution and abundance of insertion sequences among natural isolates of Escherichia coli. Genetics 115:51-63.

73. Schmid, C.W., and C.-K.J. Shen (1985) The evolution of interspersed repetitive DNA sequences in mammals and other vertebrates. In Molecular Evolutionary Genetics, R.J. MacIntyre, ed. Plenum Press, New York, pp. 323-358.

74. Shapiro, J.A. (1983) Mobile Genetic Elements. Academic Press, New York.

75. Simmons, M.J., and J.F. Crow (1977) Mutations affecting fitness in Drosophila populations. Ann. Rev. Genet. 11:49-78.

76. Singh, R.S., and L.R. Rhomberg (1987) A comprehensive study of genic variation in natural populations of Drosophila melanogaster. I. Estimates of gene flow from rare alleles. Genetics 115:313-322.

77. Steinemann, M. (1982) Multiple sex chromosomes in *Drosophila miranda*: A system to study the degeneration of a chromosome. *Chromosoma* 89:59-76.

78. Strobel, E., P. Dunsmuir, and G.M. Rubin (1979) Polymorphism in the locations of elements of the 412, *copia* and 297 dispersed repeated gene families in *Drosophila*. *Cell* 17:429-439.

78a. Syvanen, M. (1984) The evolutionary implication of mobile genetic elements. *Ann. Rev. Genet.* 18:271-293.

79. Uyenoyama, M.K. (1985) Quantitative models of hybrid dysgenesis: Rapid evolution under transposition, extrachromosomal inheritance and fertility selection. *Theor. Pop. Biol.* 27:176-201.

80. Wade, M.J. (1978) A critical review of the models of group selection. *Quart. Rev. Biol.* 3:101-114.

81. Wright, S. (1931) Evolution in Mendelian populations. *Genetics* 16:97-159.

82. Young, M. (1979) Middle repetitive DNA: A fluid component of the *Drosophila* genome. *Proc. Natl. Acad. Sci., USA* 76:6274-6278.

83. Yukuhiro, K., K. Harada, and T. Mukai (1985) Viability mutations induced by the *P* mutations in *Drosophila melanogaster*. *Jap. J. Genet.* 60:531-537.

DELETIONS AND BREAKS INVOLVING THE BORDERS OF THE Ac ELEMENT IN THE bz-m2(Ac) ALLELE OF MAIZE

Hugo K. Dooner, Edward Ralston, and James English

Advanced Genetic Sciences
Oakland, California 94608

INTRODUCTION

The corn Ac-Ds transposable element (te) system, the first one described by McClintock (15,16), consists of the autonomous element Ac (Activator) and the nonautonomous element Ds (Dissociation). That is, even though both elements can transpose, only Ac can induce its own transposition and that of Ds. There is extensive evidence in the literature that the te Ac can undergo and mediate a series of mutational changes. One type of change described by McClintock (17-20) is mutation of Ac to Ds. The genetic and molecular analysis of mutant Ds derivatives from Ac has helped to define regions of Ac that specify transposition functions in maize (8,12,22). With the recent demonstration that Ac transposes in tobacco (1), we can expect a finer mutational resolution of the transposase function of Ac. In this paper, we describe other types of mutations, the analysis of which has allowed us to obtain information on different transposition properties of Ac.

The maize mutable allele bz-m2 (bronze-mutable 2) originated by transposition of the 4.6-kb te Ac into the second exon of bronze (bz), a gene involved in anthocyanin biosynthesis (23). In bz-m2, depending on the tissue and the genetic constitution at other loci, purple spots or streaks are superimposed on a bronze background. These result from restoration of Bz function following transposition of Ac away from the bz locus during development of the tissue. Uniformly colored seeds, both purple and bronze, are found regularly in bz-m2(Ac) ears. They arise primarily from transpositions that occur during gamete formation and that either permanently restore or eliminate Bz function.

We have isolated and genetically characterized a series of phenotypically stable bronze derivatives (bz-s) from bz-m2 that retain an active, transposing Ac closely linked to bz. The molecular analysis of two such derivatives leads to the following conclusions: (a) The termini of Ac can be involved in the formation of deletions that extend either into the element itself, producing a fractured Ac, or into adjacent DNA. (b) A deletion that removes the 8-base pair (bp) direct repeat of target sequence DNA, generated upon Ac insertion, does not affect transposition. (c) Two separate but closely linked elements can cause deletion of interstitial DNA, as well as chromosome breaks. (d) The unusually high recombination (14 kb/cM), previously observed within the bz locus (6,7), does not extend to the region immediately upstream (proximal).

ISOLATION AND GENETIC CHARACTERIZATION OF DERIVATIVES

Isolation of Stable Bronze (bz-s) Derivatives from bz-m2(Ac)

The mutable allele bz-m2(Ac) conditions a spotted or variegated kernel phenotype. However, a few percent of the kernels in a cross between plants homozygous for bz-m2(Ac) and plants carrying a standard bz allele have a stable bronze phenotype (18). The high frequency with which these arise suggests that most of these exceptions, referred to hereafter as bz-s derivatives, originate from Ac-mediated events. Therefore, their analysis should provide information on the transposition behavior of Ac. Our intention at the beginning of this work was to identify bz-s derivatives that appeared, on the basis of genetic data, to retain Ac at or very close to bz and to determine genetically whether the cis- and trans-acting transposition properties of the Ac element were modified. We thus hoped to be able to isolate and examine jointly the donor and recipient sites of the same transposition event that left behind a functionally impaired bz gene and either a modified or unmodified Ac element.

The crossing scheme used to isolate bz-s derivatives was the following. Homozygous Sh bz-m2(Ac) Wx plants were pollinated with sh bz-R wx testers. The distal marker sh (sh, shrunken vs Sh, plump kernels) is located only three units from bz and is, therefore, a good indicator of the allelic state of the bz locus; wx (wx, waxy vs Wx, nonwaxy kernels) is 25 units proximal to bz; and bz-R is the stable reference allele of the bz locus. Most of the kernels from this cross are spotted. Nonspotted, bronze exceptions were selected, planted, and crossed to a Ds reporter stock (sh bz-m2(DII) wx) to determine if Ac was still present in the genome, and to an Ac stock [sh bz-R wx-m9(Ac)] to test for mutations of Ac to Ds. Selections that segregated mostly plump, spotted seeds and shrunken, bronze seeds in equal numbers from the first cross, and gave only bronze seeds in the second cross, were classified as cases of stable bronze derivatives from bz-m2(Ac) that still had Ac closely linked to the bz locus. The new, stable bronze derivatives have been designated bz-s, and each new allele has been identified by the four digits of the pedigree number under which it was first grown (e.g., bz-s:2106).

Genetic Characterization of bz-s(Ac) Derivatives

Close linkage between bz-s and Ac. The genetic distance between the stable bz mutation and Ac was estimated as follows. Heterozygous Sh (bz-s Ac) Wx / sh bz-m2(DII) wx plants were pollinated with sh bz-R wx testers. In this cross, Ac and Ds should segregate from each other, except when a crossover event or an Ac transposition event leads to their cosegregation. Therefore, among a majority of bronze kernels, there will be a few scattered spotted exceptions that, depending on their origin, will carry either a recombinant or parental arrangement of flanking markers. The exceptions can be subsequently tested to determine if Ac and the reporter allele bz-m2(DII) segregate independently or not. Thus, it is possible to obtain estimates of linkage between bz and a transposed Ac and of the frequency of Ac transposition to unlinked sites.

Tab. 1. Classification of spotted kernel exceptions from the cross Sh bz(Ac) Wx / sh bz-m2(DII) wx x sh bz-R wx.

bz(Ac) allele	Kernel population	Origin*	Outside markers Sh Wx	sh wx	Sh wx	sh Wx	N	Frequency** ($x 10^{-2}$)
bz-s:2114(Ac)	7,898	Transpositions	0	20	1	1	22	0.55
		Crossovers	0	0	0	0	0	0
bz-s:2094(fAc)	20,277	Transpositions	0	10	0	2	12	0.11
		Crossovers	0	0	0	5	5	0.05

*Determined from the segregation ratios observed in backcrosses of the spotted exceptions to the bz-R parent.

**$\frac{N \times 2 \times 100}{\text{population}}$.

Kernel progenies, representing at least 1,500 gametes from the above cross, were scored for each bz-s derivative. In most derivatives, Ac mapped close to bz at distances ranging from 0.3 to over 30 centiMorgans (cM). Southern blot analysis of these derivatives revealed that Ac had transposed away from the bz locus, leaving an "empty" site. The preferential transposition of Ac to closely linked sites has been well documented (14,16,29). Two derivatives, identified initially as bz-s:2094 and bz-s:2114, showed extremely low or no recombination between bz-s and Ac and, upon further analysis, proved to be uniquely informative. The results from the crosses involving these derivatives are summarized in Tab. 1.

Clearly, Ac is very tightly linked to bz in both derivatives. In bz-s:2114, the recombinant Sh wx and sh Wx spotted types are rare. Since, upon subsequent testing of these rare recombinants, Ac was found to segregate independently of bz-m2(DII), we conclude that these types must have arisen from coincidental transposition and recombination events. Thus, the Ac element in bz-s:2114(Ac) appears to be inseparable from the bz-s mutation and, within the limits of resolution of the present data, can be expected to be no farther than 0.09 cM away from bz. In bz-s:2094, five sh Wx recombinant spotted types were obtained out of more than 20,000 seeds. This result places Ac 0.05 cM proximal to bz.

Differences in Ac transposition frequencies. It is also evident from the data in Tab. 1 that the Ac element in the two derivatives does not transpose to unlinked sites at the same frequency. The difference in the frequency of transposition of Ac in bz-s:2114 (5.5×10^{-3}) and in bz-s:2094 (1.1×10^{-3}) is statistically significant at the 5% level (27). The sh wx spotted kernel class arose exclusively from transpositions of Ac to unlinked sites and not from double crossing-over, a result expected based on the very high chiasma interference in the region (6). At the same time, this class includes the vast majority of cases of transposition of Ac to sites unlinked to bz (30/34), so its frequency in crosses of the type bz(Ac)/-bz(Ds) x bz should provide a measurement (slight underestimate) of the transposition of Ac from bz or its vicinity to unlinked sites in the genome. Based on this, we estimated the frequency of transposition of Ac to unlinked sites in the bz-m2(Ac) progenitor allele to be 7.4×10^{-3} (14/1890). This frequency does not differ significantly from the corresponding frequency in bz-s:2114, but is six- to seven-fold higher than that in bz-s:2094. Since the spotted phenotypes conditioned by bz-s:2094/bz-m2(DII) and bz-s:2114/bz-m2(DII) heterozygotes are indistinguishable, the Ac element in bz-s:2094 is as competent as the one in bz-s:2114 in its capacity to promote the transposition of Ds. However, its ability to transpose to unlinked sites is highly impaired.

MOLECULAR CHARACTERIZATION OF DERIVATIVES

Genomic Southern Analysis

The bz-m2(Ac) and its wild-type progenitor allele Bz-McC have been cloned (7,13). In the restriction map of the bz-m2(Ac) allele shown in Fig. 1, the 4.6-kb Ac element is represented by the shaded part of the diagram. The direction of transcription and the extent of the bz transcribed region are indicated by the arrow. The distal marker sh is shown to the left (3') and the proximal marker wx to the right (5') of the bz restriction map. The bronze sequences in bz-m2(Ac) that are homologous to the probes pAGS 551, 528, and 526, obtained by subcloning restriction fragments from Bz-McC, are indicated above the map (7). Fragment 551 contains the 3' end of the transcript and fragment 526, the 5' end of the transcript [Ralston et al. (23)]. Fragment 528 contains the middle of the transcribed region, and includes the site of Ac insertion in bz-m2(Ac). It allows one to monitor changes that have taken place in the Ac element itself or in sequences immediately adjacent to Ac.

Deletions around the right border of Ac. In order to determine the nature of the changes around the Ac insertion site, various genomic digests of bz-s:2114 and bz-s:2094 DNA were probed with fragment 528. We found that enzymes that cut within Ac (EcoRI, PvuII, and HindIII: Ref. 12) gave two bands in bz-m2(Ac), but only one band in bz-s:2114. For example, EcoRI gave 8.8-kb and 6.0-kb bands in bz-m2(Ac), but only an 8.8-kb band in bz-s:2114 (Fig. 2a). The band corresponding to the left (centromere-distal) fragment remained intact, but the band corresponding to the right (centromere-proximal) fragment was missing. Enzymes that cut within Ac also gave two bands in bz-s:2094. Again, the band corresponding to the left fragment was of the same size as in bz-m2(Ac), but the band corresponding to the right fragment was a different size. As illustrated by the EcoRI digest of Fig. 2a, the right fragment in bz-s:2094 was about 2 kb smaller than that in bz-m2(Ac). These observations indicate that a deletion of bz sequences immediately to the right of the Ac insertion has occurred in bz-s:2114 and a deletion of sequences from the right end of Ac has occurred in bz-s:2094.

To establish the extent of the deletions, the above blots were reprobed with fragment 526, the 1.6-kb KpnI fragment to the right of 528 (Fig. 2b). As expected, this probe hybridized to a single band in bz-m2(Ac), corresponding in mobility to the right-hand fragment detected with the 528 probe. However, in every bz-s:2114 digest, the 526 probe detected a new band approximately 0.8 to 0.9 kb smaller than the band

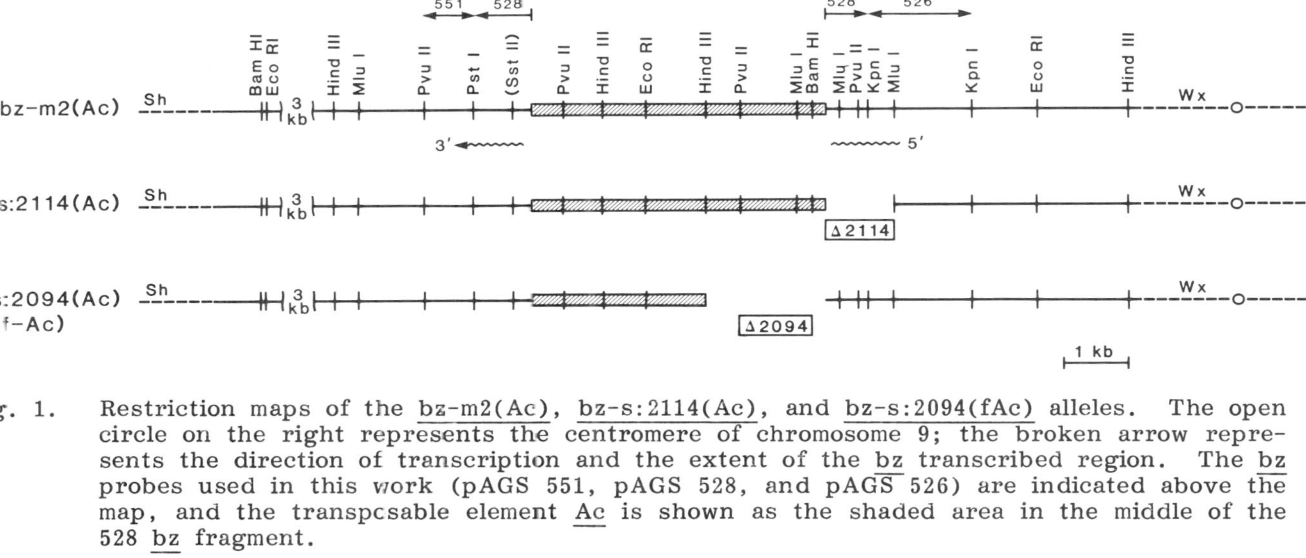

Fig. 1. Restriction maps of the bz-m2(Ac), bz-s:2114(Ac), and bz-s:2094(fAc) alleles. The open circle on the right represents the centromere of chromosome 9; the broken arrow represents the direction of transcription and the extent of the bz transcribed region. The bz probes used in this work (pAGS 551, pAGS 528, and pAGS 526) are indicated above the map, and the transpcsable element Ac is shown as the shaded area in the middle of the 528 bz fragment.

detected in bz-m2(Ac). Thus, the deletion appears to extend about 0.8 to 0.9 kb to the right of the Ac insertion element (Fig. 1). Based on more extensive Southern blot data (5), we concluded that an intact Ac element is present in bz-s:2114. Because an Ac element had been shown by both genetic and molecular criteria to reside at the bz locus, we renamed the mutation bz-s:2114(Ac). The bz-s:2094 digest showed only a 4-kb band corresponding in size to the right fragment detected with 528. This fragment also hybridized to an internal Ac probe (not shown), a result consistent with a 2-kb deletion of sequences from the right side of Ac in bz-s:2094. Other digests (not shown) established that the HindIII, PvuII, and BamHI sites from the right end of Ac had been lost (Fig. 1). Thus, in addition to a genetically active Ac 0.05 cM proximal to bz, the bz-s:2094 mutation appeared to carry a terminally deleted Ac element at the bz locus. We designated the element fractured Ac (fAc) and renamed the mutation bz-s:2094(fAc).

Nucleotide Sequence of the Ac-bz Junctions in bz-s:2114(Ac), bz-s:2094(fAc), and Their Progenitor Allele bz-m2(Ac)

In order to characterize the deletions in greater detail, the bz-s:2114(Ac) and bz-s:2094(fAc) mutations were isolated from their respective EMBL 3 genomic libraries (9,24). In both cases, the cloned DNA yielded fragments that hybridized to bz subclones and that were of the same size as predicted from earlier genomic Southern analyses (data not shown).

The adjacent deletion of bz-s:2114(Ac). To examine the right or proximal junctions between Ac and bz in bz-s:2114(Ac) and bz-m2(Ac),

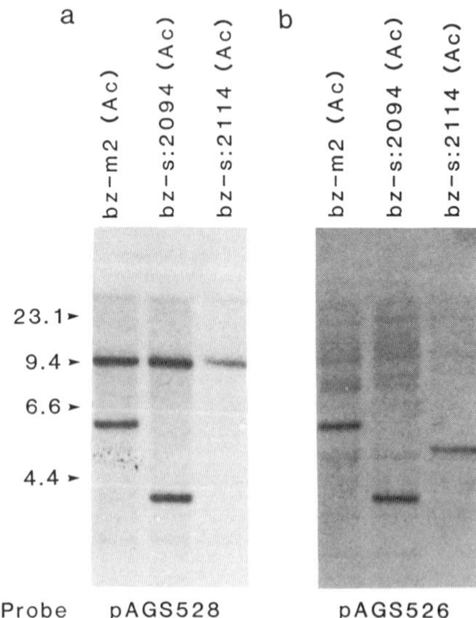

Fig. 2. Southern blot analysis of the bz-s:2114(Ac) and bz-s:2094(fAc) mutations. EcoRI digests of bz-m2(Ac), bz-s:2114(Ac), and bz-s:2094(fAc) DNA were electrophoresed, blotted onto nitrocellulose, and probed first with pAGS 528 (a) and subsequently with pAGS 526 (b).

MluI fragments spanning this junction were subcloned in M13 phage and sequenced (11). The sequences of the Ac-bz junctions in bz-m2(Ac) and bz-s:2114(Ac) are presented in Fig. 3. At the proximal (right) junction, 789 bp of upstream bz sequence, including the 8-bp direct repeat of the target sequence generated upon Ac insertion, have been deleted in bz-s:2114(Ac). The deletion removes part of the second exon, the single intron, and almost all of exon 1 of the bronze gene, terminating in a region of the gene corresponding to the 5' untranslated leader (23). This result shows that Ac can generate adjacent deletions and, when coupled with the genetic observations described above, that loss of the 8-bp target site repeat has no measurable effect on the transposition frequency of Ac. A deletion adjacent to the maize transposable element Mu-1 has also been reported in a stable null allele of Adh1 (28). The capacity to form adjacent deletions, demonstrated earlier in bacterial transposons (2), may be a general feature of maize transposable elements, although we know, from an analysis of several other bz-s derivatives, that such deletion mutations occur rarely in the progeny of bz-m2(Ac).

The distal junctions were also sequenced in order to determine whether the event that generated the deletion at the proximal or upstream junction might also have affected the distal or downstream junction between bz and Ac. A 2.3-kb SstII-EcoRI fragment from each allele (Fig. 1) was subcloned and sequenced from the SstII site in bz. No differences were detected in 267 nucleotides sequenced, corresponding to 67 bp of bz DNA and 200 bp of Ac DNA. We should point out that, while characterizing the bz-s:2114(Ac) derivative, we discovered two errors in the published Ac sequence (11,21,22).

Fractured Ac and the complex rearrangement of bz-s:2094(fAc). The sequence of the fAc-bz junction in the bz-s:2094(fAc) allele revealed an unusual rearrangement (Fig. 4). If the bz-s:2094(fAc) sequence is scanned from the left or distal (3') end, it appears identical to bz-m2(Ac). First, one finds the sequence of the left bz-Ac junction, including the left member of the 8-bp target site repeat, and then the left end of the Ac sequence. However, if one scans the bz-s:2094(fAc) sequence from the right or proximal (5') side, one finds, surprisingly, a typical "empty" site at the position that corresponds to the Ac insertion site in bz-m2(Ac)

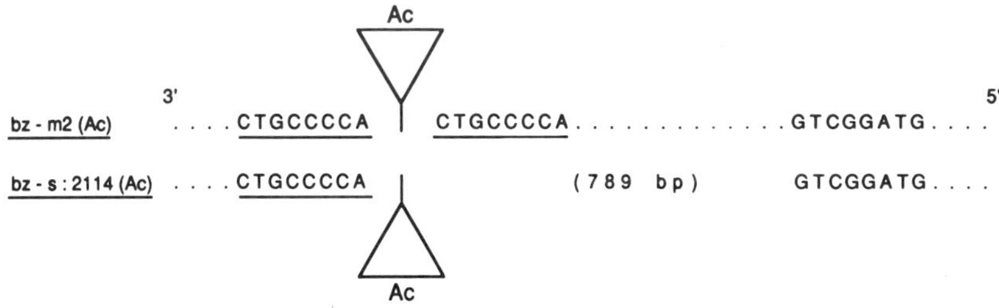

Fig. 3. DNA sequence of the bz-Ac distal (left) and proximal (right) junctions in bz-m2(Ac) and bz-s:2114(Ac), presented in the same order as the restriction map of Fig. 1. Ac is shown as an insert in the bronze target sequence. The 8-bp direct repeat flanking Ac in bz-m2(Ac) is underlined. The proximal (right) member of the repeat and 781 bp of bz sequence are deleted in the bz-s:2114(Ac) derivative.

NUCLEOTIDE SEQUENCE AROUND THE INSERTION SITES OF bz-m2(Ac) and bz-s:2094(fAc)

bz-m2(Ac)

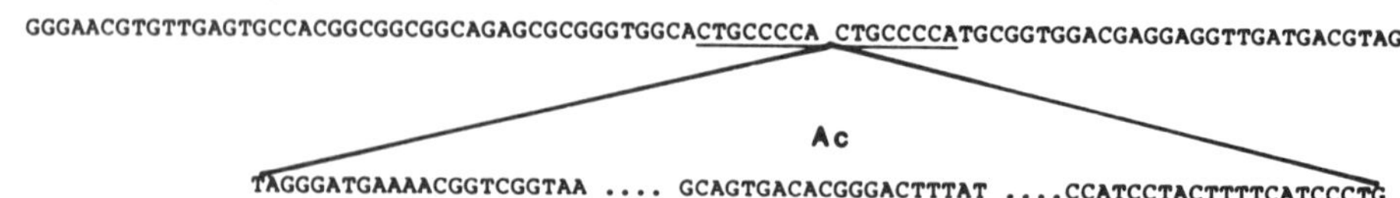

bz-s:2094(f-Ac)

GGGAACGTGTTGAGTGCCACGGCGGCGGCAGAGCGCGGGTGGCACTGCCCCA

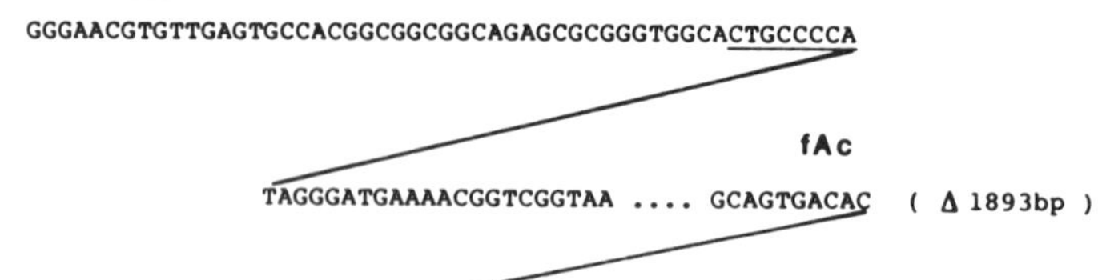

TGTTGAGTGCCACGGCGGCGGCAGAGCGCGGGTGGCACTGCCCCA TGCCCCATGCGGTGGACGAGGAGGTTGATGACGTAGT

Fig. 4. DNA sequence of the bz-Ac distal (left) and proximal (right) junctions in bz-s:2094(fAc). The 8-bp target site repeat is underlined in bz-m2(Ac), as is the 7+8 empty site in bz-s:2094-(fAc). The rightmost 1893 bp of Ac are deleted in the fractured Ac element, which is flanked by a repeat of the 45 bp of bz sequence immediately to the left of Ac in bz-m2(Ac).

(25). Ac has excised and left behind an imperfect 7+8 direct repeat at the previous insertion site. Continuing to the left, one finds 37 bp of bz sequence and then the fractured Ac element, which, based on the position of the break, appears to lack the rightmost 1,893 bases of the 4,565 bp of Ac (11,21,22). In other words, the 45 bp of bz sequence immediately to the left of the Ac insertion site in bz-m2(Ac) is present in bz-s:2094(fAc) at that position and also between the broken end of the fAc element and the empty site.

How this complex structure originated is hard to envision. We know that, in addition to the described rearrangement, there is a genetically active Ac 0.05 cM proximal to (to the right of) the empty site. Formally, the structure can be explained as the consequence of two separate events: a proximal transposition event that gave rise to the empty site and to the Ac element located to the right of bz; and an improbable illegitimate recombination event between bz sequences to the left of the empty site and the Ac element in the bz-m2(Ac) homolog. However, it is more plausible that the rearrangement is the product of a single, but complex and as yet unclear, Ac transposition event.

Derivatives of bz-s:2114(Ac) and bz-s:2094(fAc) Without Ac Activity

The Ac activity in bz-s:2114(Ac) resides at the bz locus. To confirm that the Ac activity detected genetically in bz-s:2114(Ac) corresponded to the Ac element detected molecularly, we proceeded to analyze derivatives from bz-s:2114(Ac) that had lost Ac activity. We pollinated the Ds reporter stock bz-m2(DI) wx with bz-s:2114(Ac) Wx and isolated uniformly bronze, nonwaxy exceptions from among a majority of spotted, nonwaxy seeds. These exceptions were progeny-tested to confirm that they had lost Ac activity and subjected to Southern blot analysis. Figure 5 shows a set of KpnI-PstI digests probed with fragment 528 (Fig. 1). In both representative bronze exceptions, there is a band corresponding to the bz-m2(DI) allele (4.5 kb) and a second band (2.0 kb) corresponding to the Ac empty site, i.e., to the fragment from bz-s:2114(Ac) that would be generated upon transposition of Ac from the bz locus. This is evidence that the only genetically active Ac element in the bz-s:2114(Ac) derivative resides at bz and can transpose.

Transposition interaction between Ac and the fAc element in bz-s:2094(fAc). The frequency of transposition of Ac to unlinked sites in bz-s:2094(fAc) was found to be several-fold lower than in bz-m2(Ac) (Tab. 1). The subsequent discovery of a fractured Ac element at bz, in addition to the Ac element 0.05 cM proximal to bz, raised the possibility of an interaction between these two elements. To investigate whether Ac interacts with fAc, a series of crosses was performed.

i) bz-m2(DI) wx x bz-s:2094(fAc) Wx. This cross is similar to the one described above for bz-s:2114(Ac). The intent, again, was to screen for losses of a genetically active Ac. Uniformly bronze, nonwaxy exceptions were selected, progeny-tested for loss of Ac, and analyzed by Southern blot. All 11 exceptions analyzed retained the fAc element at bz, but lacked Ac activity and the 2.6-kb PvuII band characteristic of an active Ac (3,26). That is, fAc was not structurally implicated in any of the cases of Ac loss. There is no indication from this result, obtained using the fAc stock as a male parent, that fAc interacts with Ac.

ii) bz-s:2094(fAc) x bz-m2(DI). This is the reciprocal cross of the one just presented and serves basically the same purpose. Losses of Ac can be selected as solid bronze seeds, except that in this case bz-s:2094(fAc) is used as the female parent. We have only limited data from this type of cross. However, one Ac exception (bz-s:3130.2) proved to be very interesting. Southern blot analysis indicates that fAc and the entire bz region proximal to fAc are deleted in this derivative (Fig. 6).

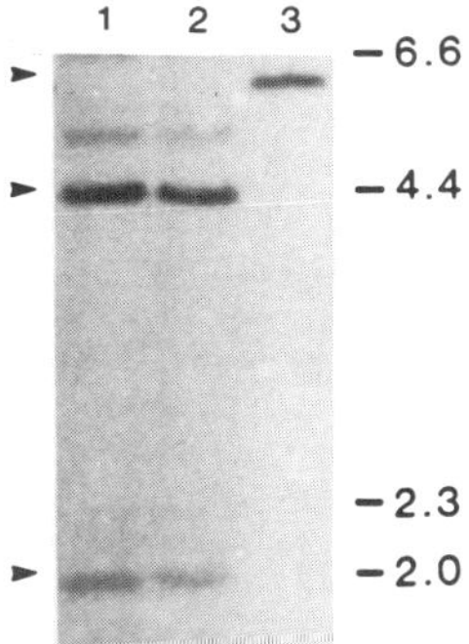

Fig. 5. Southern blot analysis of derivatives from bz-s:2114(Ac) without Ac activity. KpnI-PstI digests were hybridized to probe pAGS 528. Lanes 1 and 2: stable bronze selections from the cross bz-m2(DI) x bz-s:2114(Ac). Lane 3: bz-s:2114(Ac) control.

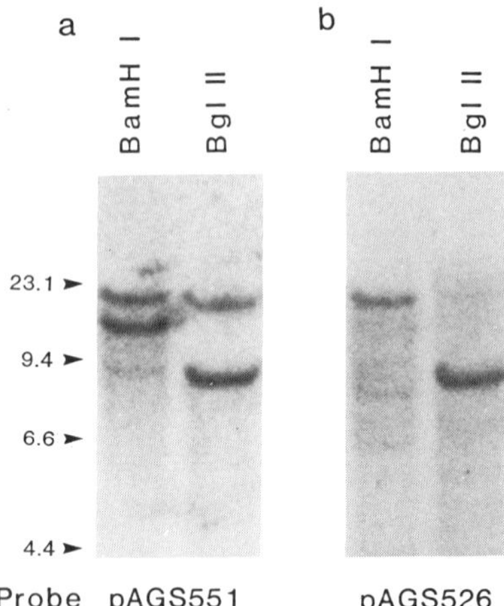

Fig. 6. Southern blot analysis of the bz-s:3130.2 deletion derivative from bz-s:2094(fAc) that has lost Ac, fAc, and the intervening region. DNA from bz-s:3130.2 / bz-m2(DI) heterozygotes was digested with BamHI and BglII. The blot was probed first with pAGS 528 (left) and subsequently with pAGS 526 (right).

The evidence is as follows. BamHI and BglII digests probed with 551 give two bands, one corresponding to bz-m2(DI) and one corresponding to the bz-s:3130.2 Ac$^-$ derivative. The latter band is of a different size from the one expected of bz-s:2094(fAc) (BamHI: 18.5 kb expected, .20 kb observed; BglII: 8.9 kb expected, 20 kb observed). This suggests that a change may have occurred in fAc. When the blot was reprobed with fragment 526 (to the right of fAc), the band corresponding to the Ac$^-$ derivative could not be detected, indicating a deletion of 526-homologous sequences in bz-s:3130.2. Other genomic digests (not presented) show that the 551 fragment is not affected by the deletion in the Ac$^-$ derivative. The origin of the deletion is most readily explained as an Ac-induced excision of a fragment extending from the left end of fAc to the right end of Ac. The type of transposon that would be generated is reminiscent of the complex, 30-kb Ds element in sh-m5933 (4).

The deletion causes the typical reduced male transmission of markers linked to it on the same chromosome arm (Tab. 2). As expected of a bz deletion, the effect is greater for sh than for wx, since sh is the closer marker. We suspect that our failure to recover a similar deletion, using bz-s:2094(fAc) as a pollen parent, may have been due partly to this reduced male transmission. It is conceivable, though, that the deletiogenic excision event does not occur with the same frequency in the male and female germ lines.

THE FRACTURED Ac REARRANGEMENT PROMOTES CHROMOSOME BREAKS

In the presence of an active Ac, Ds elements with a "double Ds" structure have the unique property of breaking chromosomes (10). In a double Ds, one 2-kb Ds element is inserted in opposite orientation into another. As a consequence, the 11-bp terminal inverted repeat is present

Tab. 2. Segregations showing reduced male transmission of the bz-s:3130.2 Ac derivative from bz-s:2094(fAc). Separate sh and wx testers were pollinated with Sh bz-s:3130.2 Wx / sh bz-R wx heterozygotes.

Tester	Segregation		Chi-square (1 df)
wx	Wx	357	6.2 *
	wx	430	
sh	Sh	580	63.8 **
	sh	885	

* P,0.05
** P,0.01

four times in a double Ds, both in direct and inverted orientation. In the fAc rearrangement--which contains 1½ Ac elements--there is also a pair of terminal repeats in direct orientation. The fractured Ac element is 0.05 cM away from an active Ac. We know that this distance represents at least 13 kb (E. Ralston, unpubl. results), a considerably larger distance than that separating the direct terminal repeats of a double Ds, yet possibly smaller than the distance separating the complex Ds ends in the 30-kb chromosome breaking insertion of sh-m5933.

We tested the ability of the fAc + Ac structure to break chromosomes by monitoring the loss of dominant markers located distal to this structure. A c sh Bz tester stock was pollinated with the fractured Ac stock C Sh bz-s:2094(fAc) Ac. If breaks occur at the bz locus during endosperm development, the distal markers C and Sh should be lost, resulting in colorless, shrunken sectors. C losses are easy to score; however, Sh losses result in a kernel indentation only if the sector is large. We found colorless sectors in practically all the kernels from the described cross, and shrunken sectors within some of the larger colorless sectors (Fig. 7).

The above result implicates the fAc + Ac structure in inducing chromosome breaks, but leaves open the possibility that fAc alone is a chromosome breaker. None of the several derivatives from bz-s:2094(fAc) that had lost Ac, but retained fAc at bz, showed breakage patterns when crossed into the c sh Bz tester line. Neither did the progenitor allele bz-m2(Ac) nor the bz-s:2114(Ac) derivative show breakage patterns, confirming that Ac alone cannot cause chromosome breaks. These results suggest that the breaks are due to an interaction between fAc and Ac. We are currently examining whether the interaction can occur in trans, as well as in cis, and are cloning the transposed Ac to determine its orientation relative to fAc. Döring et al. (10) have considered and discussed the evidence for several structures that could act as chromosome breakers: a single-element terminus, long-stem structures formed by elements in inverted orientation, and terminal repeats in direct orientation. Our current data are compatible with all of these models, but we are in the process of generating additional data that should help to distinguish among the simpler alternatives. Hopefully, the isolation of the transposed Ac will also provide a clue to the basis of the reduced transposition frequency of this element.

Fig. 7. Kernels from the cross c sh Bz x C Sh bz-s:2094(fAc) Ac showing colorless and shrunken endosperm sectors of various sizes, following breaks in 9S induced by the fAc + Ac structure.

THE RELATIONSHIP BETWEEN GENETIC RECOMBINATION AND PHYSICAL DISTANCE INSIDE AND OUTSIDE OF bz

We have reported (5,6) that genetic recombination within bz is more than 100 times higher than the average for the entire genome. Our estimate for the value of a cM within bronze is 14 kb. If this relationship between genetic and physical distance extended to the region immediately outside of bz, we would expect the Ac element 0.05 cM proximal to fAc in bz-s:2094(fAc) to be only 0.7 kb away. Yet, we do not find an Ac element within at least 13 kb of the fAc insertion site (E. Ralston, unpubl. data). We know that insertions within 1 kb of each other reduce intragenic recombination at bz (6). Possibly, insertions that are more than 13 kb from each other still interfere with recombination, but the effect is bound to be less pronounced. The discrepancy between the expected and the observed physical distance outside of bz is too high to be due to an insertion effect on recombination. This is particularly true since the intervening region was identical in the homologous chromosomes where recombination was measured, and most likely represents a true difference between recombination inside and immediately outside the bz gene. The analysis of other well-defined systems should provide information on the generality of this observation.

ACKNOWLEDGEMENT

We thank Diane Burgess for helpful suggestions on the manuscript. This paper No. 5-9 from AGS.

REFERENCES

1. Baker, B., J. Schell, H. Lorz, and N. Fedoroff (1986) Transposition of the maize transposable element Activator in tobacco. Proc. Natl. Acad. Sci., USA 83:4844-4848.

2. Calos, M., and J. Miller (1980) Molecular consequences of deletion formation mediated by the transposon Tn9. Nature 285:38-41.
3. Chomet, P., S. Wessler, and S. Dellaporta (1987) Inactivation of the transposable element Activator (Ac) is associated with its DNA modification. EMBO J. 6:295-302.
4. Courage-Tebbe, U., H.-P. Doring, N. Fedoroff, and P. Starlinger (1983) The controlling element Ds at the shrunken locus in Zea mays: Structure of the unstable sh-m5933 allele and several revertants. Cell 34:383-393.
5. Dooner, H.K. (1985) A deletion adjacent to the Ac insertion site in a stable derivative from bz-m2(Ac). In Plant Genetics, M. Freeling, ed. UCLA Symp. on Molec. and Cell. Biol., New Series, Vol. 35, pp. 561-573.
6. Dooner, H.K. (1986) Genetic fine structure of the bronze locus in maize. Genetics 113:1021-1036.
7. Dooner, H.K., E. Weck, S. Adams, E. Ralston, M. Favreau, and J. English (1985) A molecular genetic analysis of insertions in the bronze locus in maize. Molec. Gen. Genet. 200:240-246.
8. Dooner, H.K., J. English, E. Weck, and E. Ralston (1986) A single genetic unit specifies two transposition functions in the maize element Activator. Science 234:210-211.
9. Dooner, H.K., J. English, and E. Ralston (1988) The frequency of transposition of the maize element Activator is not affected by an adjacent deletion. Molec. Gen. Genet. Vol. 212 (in press).
10. Döring, H.-P., R. Garber, B. Nelsen, and E. Tillmann (1985) Transposable element Ds and chromosomal rearrangements. In Plant Genetics, M. Freeling, ed. UCLA Symp. on Molec. and Cell. Biol., New Series, Vol. 35, pp. 355-367.
11. English, J., E. Ralston, and H.K. Dooner (1987) Corrections in the nucleotide sequence of Activator. Maize Genet. Coop. Newslet. 61:81.
12. Fedoroff, N., S. Wessler, and M. Shure (1983) Isolation of the transposable maize controlling elements Ac and Ds. Cell 35:235-242.
13. Fedoroff, N., D. Furtek, and O.E. Nelson (1984) Cloning of the bronze locus in maize by a simple and generalizable procedure using the transposable element Activator (Ac). Proc. Natl. Acad. Sci., USA 81:3825-3829.
14. Greenblatt, I.M. (1984) A chromosome replication pattern deduced from pericarp phenotypes resulting from movements of the transposable element Modulator in maize. Genetics 108:471-485.
15. McClintock, B. (1948) Mutable loci in maize. Carnegie Institution of Washington Yearbook 47:155-169.
16. McClintock, B. (1952) Chromosome organization and gene expression. Cold Spring Harbor Symp. Quant. Biol. 16:13-47.
17. McClintock, B. (1955) Controlled mutation in maize. Carnegie Institution of Washington Yearbook 54:245-255.
18. McClintock, B. (1956) Mutation in maize. Carnegie Institution of Washington Yearbook 55:323-332.
19. McClintock, B. (1962) Topographical relations between elements of control systems in maize. Carnegie Institution of Washington Yearbook 61:448-461.
20. McClintock, B. (1963) Further studies of gene control systems in maize. Carnegie Institution of Washington Yearbook 62:486-493.
21. Mueller Neumann, M., J. Yoder, and P. Starlinger (1984) The DNA sequence of the transposable element Ac of Zea mays L. Molec. Gen. Genet. 198:19-24.
22. Pohlman, R., N. Fedoroff, and J. Messing (1984) The nucleotide sequence of the maize controlling element Activator. Cell 37:635-643.

23. Ralston, E., J. English, and H.K. Dooner (1988) Sequence of three bronze alleles of maize and correlation with the genetic fine structure. Genetics Vol. 119 (in press).
24. Ralston, E., J. English, and H.K. Dooner (1987) A fractured Ac element at the bz locus in maize: Origin, properties and molecular characterization. (In preparation.)
25. Sachs, M.M., W.J. Peacock, E.S. Dennis, and W.L. Gerlach (1983) Maize Ac/Ds controlling elements: A molecular viewpoint. Maydica 28:289-303.
26. Schwartz, D., and E. Dennis (1986) Transposase activity of the Ac controlling element in maize is regulated by its degree of methylation. Molec. Gen. Genet. 205:476-482.
27. Stevens, W.L. (1942) Accuracy of mutation rates. J. Genet. 43:301-306.
28. Taylor, L., and V. Walbot (1985) A deletion adjacent to the maize transposable element Mu1 accompanies loss of Adh1 expression. EMBO J. 4:869-876.
29. Van Schaik, N.W., and R.A. Brink (1959) Transpositions of Modulator, a component of the variegated pericarp allele in maize. Genetics 44:725-738.

TRANSPOSABLE ELEMENTS OF ANTIRRHINUM MAJUS

Hans Sommer, Reinhard Hehl, Enno Krebbers,
Ralf Piotrowiak, Wolf-Ekkehard Lönnig, and Heinz Saedler

Max-Planck-Institut für Züchtungsforschung
Egelsplad, D-5000
Köln 30, Federal Republic of Germany

INTRODUCTION

Genetic instability of loci affecting a directly observable property such as pigmentation or morphology is often responsible for the variegated appearance of many plant parts. Variegated plants of this kind can frequently produce phenotypically wild-type or nearly wild-type progeny due to reversion of the mutable allele in germinal tissue. In *Antirrhinum majus*, a large number of stable and unstable mutants have been isolated and genetically characterized, mainly in the first four decades of this century (for a compilation see Ref. 15,19).

The development of molecular techniques made it possible to analyze these mutants molecularly. Several unstable and stable alleles of the *nivea* (1,2,18,20) and the *pallida* loci (5,14) were characterized and shown to be due to the insertion of transposable elements.

In this article we report on the structural analysis of the three transposable elements *Tam1*, *Tam2*, and *Tam3* of *A. majus*, and a comparison of *Tam1* and *Tam3* with the transposable elements *En/Spm* and *Ac* of *Zea mays*, respectively. Furthermore, an interaction between the transposable elements *Tam1* and *Tam2*, leading to instability in a cross of the stable white lines 46 and 44, was molecularly analyzed. Analysis of paramutant F_1 plants from a cross of a paramutagenic and a paramutable line gives some hints on the nature of paramutation at the *nivea* locus in *A. majus*.

RESULTS

Structural Organization of the Transposable Elements Tam1, Tam2, and Tam3 of Antirrhinum majus

Structure of the transposable element Tam1. The transposable element *Tam1* was cloned from the highly unstable allele *niv*-53::*Tam1* (2) which was originally isolated by Kuckuck (12) and genetically characterized further by Harrison and Carpenter (7). Tam1 is residing in the promoter region of the chalcone synthase (chs) gene, 17 bp upstream of the TATA box (1), thereby blocking expression of the chs gene. Frequent somatic excision during development leads to highly variegated flowers, while excision in germinal cells gives rise to full red revertants. Sequencing across the ends of the element into flanking regions revealed that the ends of the element consist of 13-bp inverted repeats (IR), and that the element is flanked by 3-bp duplications generated from the target site upon integration (2).

In order to get further insight into the structural organization of Tam1, northern analysis was carried out with internal fragments of Tam1 as probes. The results indicate that at least two Tam1-specific mRNAs of respectively 2.4 kb and 5 kb in length are encoded by the element. The 2.4-kb mRNA represents about 0.01% of the poly(A)$^+$ fraction, while the 5-kb message is less abundant and more variable in amount than the 2.4-kb mRNA.

Tam1-specific cDNA clones were prepared and sequenced, and the sequences of the overlapping genomic clones Am63 and Am27 (1) were established. The genomic clones were sequenced single-stranded for most of the Tam1-specific parts, but the parts specifying coding regions [open reading frame (ORF) and exons; see Fig. 1] and about 1 kb of both termini were sequenced double-stranded. Comparison of cDNA and genomic sequences allowed us to derive the structural organization of the Tam1 element, as displayed in Fig. 1. The size of Tam1 is 14.9 kb. Transcription starts at the left end of Tam1, as shown by S1-mapping, at two distinct positions (a G at pos. 91 and an A at pos. 165, respectively). Transcription of Tam1 is in the opposite direction compared to transcription of the chs gene. The 3' end of the transcript is not well defined, since several poly(A) addition sites were found 570 to 610 bp before the right end of Tam1. Apparently almost the complete Tam1 element is transcribed into a long (14 kb) primary transcript which is then processed to yield the two final mRNAs of 2.4 kb and 5 kb seen in northern experiments. The open boxes in Fig. 1 represent sequences of about 2.2 kb in length which are present on the 2.4-kb mRNA. This mRNA has an ORF which could encode a protein of 497 amino acids. The ORF is preceded by a 5' untranslated region of either 364 bp or 439 bp, depending on which start site was used for transcription (see above). The 3' untranslated region is about 240 bp long.

Remarkably, a 1.4-kb long ORF is located in the fourth intron of Tam1 (Fig. 1). When this ORF is used as a radioactive probe in northern hybridization, only the 5-kb transcript lights up. So far we were not able to obtain cDNA clones derived from the 5-kb mRNA. In northern experiments with fragments from the left and right ends of Tam1, both mRNAs are lighting up, indicating that there is substantial homology between them. It is possible that the 5-kb transcript is still a precursor mRNA which is not processed completely to the final product.

Structure of the Tam2 element. The Tam2 element was isolated from the stable white-flowering allele niv-44 (genotype niv-44::Tam2), where it is inserted at the first exon/intron border of the chs gene (20). Sequence analysis of a genomic clone showed that the element is 5,187 bp long (11). In northern experiments with poly(A)$^+$ RNA from line 44, a Tam2-specific transcript of approximately 3 kb was detected which is not

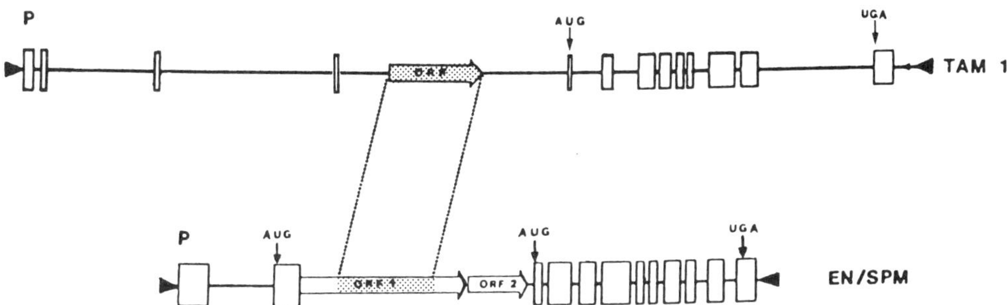

Fig. 1. Comparison of the structural organization of Tam1 (A. majus) and En/Spm (Zea mays). Open boxes represent exons; triangles, the terminal inverted repeats. Shaded regions are areas of homology between the elements. P = promoter.

found in other lines. Two overlapping cDNA clones comprising about 1.5 kb were obtained from RNA of line 44 and sequenced. Comparison of the genomic and cDNA sequences indicates that the message is spliced. Transcription of Tam2, assuming that the element is transcribed, is in the opposite direction with respect to transcription of the chs gene. The homology between cDNA and the genomic sequence is 100%, except for 86 bp at the 5' end of the cDNA which are not found on the genomic Tam2 sequence, suggesting that the 3-kb transcript probably originates from a different Tam2 copy elsewhere in the genome. This, and the fact that the largest ORF found on the cDNA is only 127 amino acids long, indicates that the Tam2 at the nivea locus is a nonautonomous, defective element, as already suspected from genetic analysis.

Homology between Tam1 and Tam2 had been realized earlier (20). Both elements have the endmost 13 bp in common and generate a 3-bp duplication of the target site upon integration. Comparison of the sequences of Tam1 and Tam2 revealed an additional homology between both elements: a 700-bp long segment extending in both elements from within the 3' untranslated region toward the ends shows 80% homology. The significance of this homology is unknown at present.

Structure of the Tam3 element. The Tam3 element was discovered in the highly variegated line niv-98 (genotype niv-98::Tam3) which was produced from a highly mutable pallida-recurrens line (8). The cloning and partial characterization of Tam3 was reported previously (18). Tam3 possesses 12-bp terminal inverted repeats and generates 8-bp duplications upon integration, except for one Tam3 copy cloned from an unstable pallida allele which was flanked by a 5-bp duplication (5). In niv-98, Tam3 is integrated 29 bp upstream of the TATA box in the promoter of the chs gene. Tam3 frequently excises in somatic and germinal cells, giving rise to variegated flowers and germinal revertants, respectively.

We have characterized the element further by sequence analysis of a genomic clone and of cDNA clones prepared from poly(A)$^{+}$ RNA of line 98. Tam3 is only 3,629 bp long, and is the smallest transposable element in A. majus discovered so far. It contains an ORF of 2.48 kb and apparently has no introns, since a 3.0-kb long cDNA clone is co-linear with the genomic clone. Tam3 is transcribed in the same direction as the chs gene.

We have no information yet on the exact size of the Tam3-specific transcript. In northern experiments only a 5.5-kb transcript was detected, which is very likely a chimeric mRNA derived from a different Tam3 copy inserted in a transcribed gene. Evidence for this comes from three chimeric cDNA clones which have 2.7 kb of their 3' end in common with Tam3, but which do not show any homology to Tam3 5' to position 698 of the Tam3 sequence. This indicates that these clones are either derived from a larger Tam3 element and that the Tam3 at the nivea locus is a defective element, or that the cDNA clones originate from a chimeric mRNA.

The latter explanation is supported by two lines of evidence. First, in genomic Southern blots of DNA restricted with BstNI (an enzyme which cuts in both ends of Tam3), the largest internal fragment observed with Tam3 as a probe is 3.2 kb in size, in agreement with that of the Tam3 copy at the nivea locus; and second, there is preliminary evidence that the cloned Tam3 transposes in transgenic tobacco plants, indicating that it is autonomous. The Tam3-internal transcription start has not been determined yet.

The first AUG codon is located 228 bases inside the ORF, reducing the coding capacity for a presumptive protein to 749 amino acids. The ORF is followed by a 3' untranslated region of 233 to 305 bases, depending on which of several possible poly(A) addition sites is used (cDNA clones showed that poly(A) addition occurs at several sites rather than at a distinct site). The 5' leader region of the transcript is about 450 to 500 bases long. Southern hybridization and copy number determination show that there are 15 to 25 Tam3 copies present in the haploid genome. The numbers are varying slightly between different lines.

Comparison of Tam1 and Tam3 with the Maize Transposable Elements En/Spm and Ac

Comparison of Tam1 and En/Spm. A comparison of the transposable elements Tam1 of *A. majus* and En/Spm of *Zea mays* (16) reveals some striking similarities. Both have inverted repeats at their termini which are identical, except for one nucleotide. Upon integration they generate 3-bp duplications of target site sequences (1,17). In northern hybridization both systems display a major transcript of 2.4 kb, but comparison of the sequences revealed no homology. Large primary transcripts (starting at the left end of the elements) seem to be transcribed from both elements (see Fig. 1), greater than 14 kb from Tam1 and about 8 kb from En/Spm, from which the mature mRNAs are derived by splicing. Both Tam1 and En/Spm have an ORF in a large intron (see Fig. 1). These ORFs show 56% homology at the DNA sequence level over a range of 1.3 kb. The homology is even more pronounced if the corresponding amino acid sequences are compared: regions of 50% to 70% homology are found, suggesting that these regions serve important function(s) in the transposition process. Altogether, both elements, Tam1 from the dicot plant *A. majus* and En/Spm from the monocot *Zea mays*, show a close structural relationship which could reflect an evolutionary relationship between the two elements.

Comparison of Tam3 and Ac/Ds. Recently we showed that similarities exist between Tam3 and the maize transposable element system Ac/Ds (18). Both Tam3 and Ac/Ds generate an 8-bp duplication of target sequences upon integration, and they have 7 bp of the endmost 11 bp IR in common.

A detailed analysis of the structural organization of Ac was reported recently (13).

Structural comparison of Tam3 and Ac reveals further homologies, in addition to those mentioned above. We estimate the size of the Tam3 transcript (based on a nearly full-size cDNA clone) to be approximately 3.1 kb [without the poly(A) tail], which is similar to the 3.3-kb long Ac transcript. Both transcripts have unusually long 5' untranslated regions of about 500 nucleotides (Tam3) and 600-700 nucleotides (Ac). The sizes of the respective proteins presumably encoded by the elements are very similar: 749 amino acids for Tam3 and 807 amino acids for Ac. Furthermore, these polypeptides have regions of fairly high homology (50-65%) in common (see Fig. 2). The conservation of these regions probably indicates a functional importance of these sequences for transposition, as well as a close evolutionary relationship of the two elements.

The main difference between the two elements are the four introns of Ac vs. no introns of Tam3. If these were removed from Ac, the sizes of the two elements would be comparable: Ac 3.9 kb and Tam3 3.6 kb; hence, Tam3 could be considered a streamlined Ac.

Possible Interactions Between Transposable Elements of Antirrhinum majus

Interaction between Tam1 and Tam2. The stable white-flowering line 44 contains a defective Tam2 element inserted at the first exon/intron border of the chs gene (see above). Tam2 has terminal inverted repeats identical to those of the Tam1 element. Since the IRs at the ends of transposable elements are thought to be the substrates for the transposase, the identity of the IRs suggests that complementation of the defective Tam2 by a functional Tam1 element might be possible. This was indeed observed in a cross of line 44 with line 46 (genotype niv-46::Tam1-46), a stable white flowering Tam1 allele derived from the highly unstable allele niv-53::Tam1 (B.J. Harrison and R. Carpenter, pers. commun). Niv-46 carries an excision-defective Tam1 derivative (Tam1-46) in the chs promoter (integrated at the same position as the original Tam1 in niv-53), which has a 5-bp deletion in the left IR (9), but which in all other respects seems to be identical with the original Tam1 element. A cross of line 44 with line 46 resulted in heterogeneous F_1 phenotypes ranging from

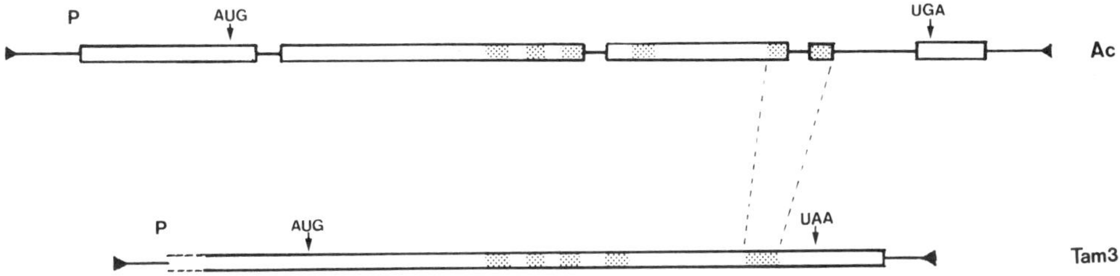

Fig. 2. Comparison of the structures of the transposable elements Tam3 (A. majus) and Ac (Zea mays). Open boxes are exons; triangles represent the terminal inverted repeats. Shaded areas are regions of homology. P = promoter (the promoter of Tam3 has not yet been mapped precisely).

white to flush rose with or without red spots, while a control cross of line 45 (another stable white line) with line 46 yielded only the expected stable white F_1 progeny (9). About half of the F_1 plants of the cross 44 X 46 were variegated, while the other half displayed a stable nonspotted phenotype. Genomic southern blots confirmed that the F_1 plants were heterozygous niv-46::Tam1-46/niv-44::Tam2. The variegation disappeared in the F_2 generation, and also when cuttings from variegated F_1 plants were grown.

In order to determine which of the two elements is responsible for the instability observed, two germinal revertants were cloned and the relevant regions sequenced. The results clearly demonstrate that in both cases the Tam2 element has been excised, suggesting, but not proving, that the excision-defective Tam1-46 element provides a functional transposase for excision of Tam2. An alternative explanation would be that an autonomous Tam2 in the background of line 46 is responsible for excision of the Tam2 at the nivea locus. But so far there is no evidence for an active Tam2 element in line 46. A similar activation of Tam2 was reported by Hudson et al. (10).

As mentioned above, the instability of the F_1 plants is not heritable and disappears in the F_2 generation. In addition, only about 50% of the F_1 plants show instability. How can these observations be explained? If a postulated Tam1 transposase is responsible for Tam1 excision in the cross, then a trans-acting function in the background of line 44 can be imagined which suppresses the Tam1 transposase. In the F_1, both activities are present, and it seems to be a matter of timing or dosage when the suppressing function overcomes the transposase activity. This explanation is supported by the observation that, in a few plants, escape from suppression was observed. When a variegated F_1 plant was selfed, most of the F_2 progenies were not flaked; only a few plants showed variegation, indicating segregation of the suppressing factor.

Paramutation in Antirrhinum majus. Paramutation is defined as the interaction of two alleles of a locus leading to a directed, heritable change of one (the paramutable) allele; the other allele, which induces the change, is called paramutagenic (3). This phenomenon has been most extensively studied at the R locus of *Zea mays*, but has also been observed in other plants (for review, see Ref. 3 and 6).

In *A. majus*, paramutation was observed at the nivea locus (7) in crosses of the highly variegated line 53 (carrying the transposable element Tam1 in the chs promoter region) with the stable white line 44 (containing the Tam2 element at the first exon/intron border). Instead of the expected highly variegated phenotype, the F_1 plants show a flush or granulated phenotype, in some plants with a variegated pattern superimposed upon it, but this disappears in the F_2 generation, in which the phenotype is much more stable. No variegated plants are observed to segregate in a Mendelian fashion through at least three generations of selfings. According to the definition (see above), line 44 is the paramutagenic and line 53 is the paramutable line. The effect is specific for line 44, as a similar cross between the stable white line 45 and the line 53 results in variegated F_1 plants only, as expected.

In order to get insight into the underlying molecular events causing the paramutation phenomenon, we have carried out three series of crosses of lines 44 and 53 (11). The results are in agreement with those reported by Harrison and Carpenter (7). Of 150 plants, 67% were some shade of flush or flush with a superimposed variegation, 23% were variegated to some degree on a white background, and the rest were white or red. The red plants are probably revertants of the niv-53 allele, while the white plants are assumed to be either an extreme form of paramutation or to have undergone unrelated mutations, either at the nivea locus or at other pigment loci.

Brink et al. (4) have proposed models involving chromosomal rearrangements in the case of paramutation at the R locus of maize. To test these models, genomic DNA isolated from various F_1 and F_2 plants was subjected to Southern blot analysis using an appropriate fragment of the chs gene as a probe (for experimental details, see Ref. 11). In most of the paramutant F_1 plants, both nivea alleles are present, apparently unchanged, as indicated by the characteristic junction fragments of Tam1 (4.7 and 4.4 kb) and Tam2 (9.0 and 1.9 kb) observed in the Southern blots. But there was one exception, plant Q, in which the Tam2 junction fragments were missing and a 5.7-kb wild-type band appeared (discussed later). The results demonstrate that paramutation at the nivea locus of A. majus involves no major rearrangements. They do not exclude minor sequence changes, but do rule out recombination or gene conversion events involving the two nivea alleles.

A second question concerning paramutation at the nivea locus is whether the presence of Tam2 at the nivea locus is required for paramutagenicity to occur in line 44. The answer comes from analysis of plant Q. Plant Q has a paramutant phenotype indistinguishable from other paramutant F_1 plants. Southern analysis indicated that Tam2 had been excised in plant Q. Sequencing across the site of integration confirmed the loss of Tam2 and 28 bp of exon sequences, rendering the chs gene inactive. The allele is designated niv-4432. When plants homozygous for the allele niv-4432 were crossed with the paramutable line 53, the F_1 offspring were indistinguishable from those of the paramutant cross 44 X 53 (see above), thus demonstrating that the homozygous line 4,432 is still paramutagenic. Therefore, the Tam2 element at the nivea locus is not required for paramutagenicity; rather, factor(s) in the background of line 44 must be responsible for the phenomenon.

CONCLUSIONS

The structural analysis of the transposable elements Tam1 and Tam3 of Antirrhinum majus has revealed striking similarities with the maize elements En/Spm and Ac. The homologies between Tam1 and En/Spm (and between Tam3 and Ac) suggest a common progenitor element from which the elements have evolved. Furthermore, the conservation of certain internal regions points to these regions being important for the transposition process. Since the conserved regions are located in ORFs, it is very likely that they represent domains of the respective transposases. Mutagenesis of these homologous areas could help to identify functions exerted by the transposase during transposition. The homologies also raise the question whether complementation is possible between Tam1 and En/Spm, or between Tam3 and Ac. Experiments to test this have been initiated.

The analysis of paramutant F_1 and F_2 plants of the cross line 44 X line 53 revealed that no gross rearrangements occurred in the paramutable allele niv-53::Tam1, in contrast to models proposed by Brink et al. (4) for paramutation at the R locus in maize. Furthermore, the paramutagenic property of line 44 is not associated with the allele niv-44::Tam2, but, rather, factor(s) in the background of line 44 seem to be responsible for the paramutation phenomenon. What are these factors? They could be Tam1 and/or Tam2 elements that produce a repressor or defective transposases which inhibit excision of the Tam1 element. We hope that further analysis will reveal the nature of these factors.

REFERENCES

1. Bonas, U., H. Sommer, and H. Saedler (1984) The 17-kb Tam1 element of Antirrhinum majus induces a 3-bp duplication upon integration into the chalcone synthase gene. EMBO J. 3:1015-1019.
2. Bonas, U., H. Sommer, B.J. Harrison, and H. Saedler (1984) The transposable element Tam1 of Antirrhinum majus is 17 kb long. Molec. Gen. Genet. 194:138-143.
3. Brink, R.A. (1973) Paramutation. Annu. Rev. Genet. 7:129-152.
4. Brink, R.A., E.D. Styles, and J.D. Axtell (1968) Paramutation: Directed genetic change. Science 159:161-170.
5. Coen, E.S., R. Carpenter, and C. Martin (1986) Transposable elements generate novel spatial patterns of gene expression in Antirrhinum majus. Cell 47:285-296.
6. Hagemann, R., and W. Berg (1977) Vergleichende Analyse der Paramutationssysteme bei höheren Pflanzen. Biol Zentralblatt 96:257-301.
7. Harrison, B.J., and R. Carpenter (1973) A comparison of the instabilities at the nivea and pallida loci in Antirrhinum majus. Heredity 31:309-323.
8. Harrison, B.J., and R. Carpenter (1978) John Innes Annual Report, Norwich, England.
9. Hehl, R., H. Sommer, and H. Saedler (1987) Interaction between the Tam1 and Tam2 transposable elements of Antirrhinum majus. Molec. Gen. Genet. 207:47-53.
10. Hudson, A., R. Carpenter, and E.S. Coen (1987) De novo activation of the transposable element Tam2 of Antirrhinum majus. Molec. Gen. Genet. 207:54-59.
11. Krebbers, E., R. Hehl, R. Piotrowiak, W-E. Lönnig, H. Sommer, and H. Saedler (1987) Molecular analysis of paramutant plants of Antirrhinum majus and the involvement of transposable elements. Molec. Gen. Genet. (in press).
12. Kuckuck, H. (1936) Über vier neue Serien multipler Allele bei Antirrhinum majus. Zeitschrift für Inductive Abstammungs - und Vererbungslehre 71:429-440.
13. Kunze, R., U. Stochaj, J. Laufs, and P. Starlinger (1987) Transcription of transposable element Activator (Ac) of Zea mays L. EMBO J. 6:1555-1563.
14. Martin, C., R. Carpenter, H. Sommer, H. Saedler, and E.S. Coen (1985) Molecular analysis of instability in flower pigmentation of Antirrhinum majus, following isolation of the pallida locus by transposon tagging. EMBO J. 7:1625-1630.
15. Nevers, P., N.S. Shepherd, and H. Saedler (1986) Plant transposable elements. Adv. in Bot. Res. 12:103-203.
16. Pereira, A., H. Cuypers, A. Gierl, Zs. Schwarz-Sommer, and H. Saedler (1986) Molecular analysis of the En/Spm transposable element system of Zea mays. EMBO J. 5:835-841.

17. Schwarz-Sommer, Zs., A. Gierl, R.B. Klösgen, U. Wienand, P.A. Peterson, and H. Saedler (1984) The Spm (En) transposable element controls the excision of a 2-kb DNA insert at the wxm-8 allele of Zea mays. EMBO J. 3:1021-1028.
18. Sommer, H., R. Carpenter, B.J. Harrison, and H. Saedler (1985) The transposable element Tam3 of Antirrhinum majus generates a novel type of sequence alterations upon excision. Molec. Gen. Genet. 199:225-231.
19. Stubbe, H. (1966) Genetic und Zytologie von Antirrhinum L. sect. Antirrhinum. Fischer, Jena GDR.
20. Upadhyaya, K.C., H. Sommer, E. Krebbers, and H. Saedler (1985) The paramutagenic line niv-44 has a 5 kb insert, Tam2, in the chalcone synthase gene of Antirrhinum majus. Molec. Gen. Genet. 199:201-207.

THE STATE OF DNA MODIFICATION WITHIN AND FLANKING MAIZE TRANSPOSABLE ELEMENTS

Jeffrey L. Bennetzen, Willis E. Brown, and Patricia S. Springer

Department of Biological Sciences
Purdue University
West Lafayette, Indiana 47907

ABSTRACT

Modification of the *Mu1*-like transposable elements, primarily or exclusively 5-methylcytosine *in* the sequences 5'-CG-3' and 5'-CNG-3', is strongly correlated with the loss of *Mutator* activity. Chromosomal copies of *Mu1*-like elements and the maize retrotransposon *Bs1* are fully digested by 5-methylcytosine-sensitive restriction enzymes in DNA from plants in which element transposition has been detected. In some cases, *Mutator* activity may be lost through a process not associated with *Mu* element modification. Under other circumstances, particularly inbreeding for the *Mutator* trait, *Mu* element modification and loss of *Mutator* activity are concurrent. When *Mu* element modification occurs, HinfI and MluI sites in and near the *Mu1* terminal inverted repeats are often modified first. Modification at these HinfI sites is not seen in some lines that had lost *Mutator* activity one or more generations earlier, and are fully modified at sites within the *Mu* element. The *Mu1*-internal NotI and EcoRII sites are always and exclusively modified in any *Mutator*-derived line that has lost activity through a modification-associated process. Extensive modification within a *Mu* element does not lead to detected modification in flanking sequences.

In crosses between *Mutator*-loss plants with modified *Mu* elements and active *Mutator* plants with unmodified *Mu* elements, the progeny initially contain a mixture of modified and unmodified elements. During development of the plant, the ratio of modified to unmodified elements progressively increases. This gradual dominance of modified elements is approximately equivalent regardless of whether the *Mutator*-loss plant was the male or female parent.

INTRODUCTION

In animal systems, changes in the modification pattern of particular DNA sequences have been correlated with the developmental activation of tissue-specific genes (20,35). Generally, demethylation of distinctive 5-methylcytosine residues within or adjacent to an animal gene promoter is seen to coincide with transcriptional activation of that gene.

The nuclear DNA of higher plants is extensively methylated, primarily as 5-methylcytosine in the sequences 5'-CG-3' and 5'-CNG-3' (25,43). In wheat, cytosines are 5-methylated in over 90% of all 5'-CG-3' and in over 80% of all 5'-CNG-3' sequences in the nuclear genome (25). Similar results are observed with maize and other plant species (unpubl. observ.). An unambiguous role for this wide-ranging modification of plant DNA has not been demonstrated. In our limited experiments with the Adh1, Bz, Sh, and Wx loci of maize, we have found no 5-methylcytosine within these genes or in nearby single-copy sequences. This lack of genic modification is observed in DNA from somatic tissues, whether or not they detectably express the gene analyzed (12). Conversely, the highly repetitive sequences that are found a few kilobase pairs distal to single-copy maize genes are fully modified in most cases (11,12). By these criteria, we have proposed that extensive 5-methylation of cytosine differentiates the highly repetitive majority of maize nuclear DNA (29) from the unmodified domain of low-copy number DNA sequences that have the potential for genetic expression (12).

Two lines of evidence directly suggest that 5-methylation of cytosine plays a role in the control of gene expression in plants. First, after nuclear integration, exogenously transformed sequences that are weakly expressed have been shown to be more resistant to digestion with 5-methylcytosine-sensitive restriction enzymes than are highly expressed integrants of the same DNA construct (26). These poorly expressed sequences can also be genetically reactivated by treatment of the transformed plant cells with 5-azacytidine (47). Second, restriction enzymes that are sensitive to 5-methylcytosine digest active transposable elements more efficiently than quiescent or inactive transposable elements (5,6,15,17-19, 42,48).

The Mutator transposable element system of maize (36) is particularly well suited to studies of modification, since the element involved in most mutations (14,34,46), Mu1, has been cloned (4), sequenced (2), and found to become extensively modified when germinal transposition activity (5-7) or somatic mutability (15,48) is lost. Moreover, loss of Mutator activity is frequent (5,9,15,36,48,49), and can be routinely initiated by interbreeding of Mutator stocks (1,6,7,38). Finally, although the modified state of Mu elements is generally and eventually dominant (7,9,39,49), the modification-associated loss of Mutator-derived somatic mutability can occasionally be reversed by crossing an inactive line to an active line (9,49).

The experiments reported here extend our studies on modification of Mutator transposable elements and on the plant retroviral transposable element Bs1 (27,28). The results presented demonstrate that: (a) active Mu and Bs1 elements are unmodified, (b) Mu elements are associated with unmodified flanking sequences, (c) significant variation exists in the modification of particular nucleotides within inactivated Mu elements, (d) modification of a Mu1-like element at bronze does not lead to detected modification of sequences flanking the element, (e) modified Mu elements are "dominant" to unmodified elements, (f) modified Mu-element dominance is efficient whether the Mu-inactivated parent is the male or female, and (g) modified Mu-element dominance is progressive through most or all somatic tissues of the developing plant.

MATERIALS AND METHODS

DNA Sources and Preparation

All Mutator-derived stocks were advanced generations of material, kindly provided by D.S. Robertson. The presence or absence of Mutator activity in each of these lines was scored by mutagenic activity (36) or, where applicable, by the phenotypic reversion pattern of an unstable, Mutator-induced allele of bronze, bzMum4 (5). Maize line 1s2p was generously provided by M. Freeling. Bronze and Bs1 hybridizational probes were derived from plasmids, kindly provided by D. Furtek and O. Nelson (23), and by M.A. Johns and M. Freeling (28), respectively.

Maize genomic DNA was purified either as previously described (4) or by the technique of Shure et al. (44).

Gel Blot Hybridization Analysis

Eight to ten micrograms of total maize DNA were digested with a two-fold excess of the appropriate restriction enzyme(s), according to the manufacturers' specifications, and size-fractionated on horizontal agarose gels. The resultant gels were replica-blotted onto nitrocellulose or nylon filters. These filters were hybridized and washed according to our standard protocol (3). The Mu hybridization probe used was a purified 1.0 kb Tth111I fragment of Mu1 (2,3), labeled by either nick translation or random primer extension (22). A purified 842-bp PstI fragment, homologous to the 3' end of the bronze locus (21,23), was labeled by random primer extension and used as probe in analysis of bzMum4. Filters testing Mu-element modification were washed free of bound Mu probe, and rehybridized to either maize ribosomal DNA (6) or mitochondrial DNA (kindly provided by K. Newton) to test for restriction enzyme digestion efficiency.

RESULTS

Mu1-like elements have numerous sites for 5-methylcytosine-sensitive restriction enzymes (Fig. 1). Multiple analyses with AvaI, AvaII, BglI, EcoRII, HinfI, HpaII, MluI, MspI, NotI, and SstII indicate that these 5-methylcytosine-sensitive enzymes fully digest most or all Mu1-like elements in an active Mutator line (5,6,8,15,48, and unpubl. observ.).

Fig. 1. Restriction map of the maize transposable element Mu1. With the exception of TaqI, Tth111I, and BstNI, all of the restriction enzymes whose sites are noted are sensitive to 5-methylation of cytosine in the sequences 5'-CG-3' and/or 5'CNG-3' (6,30). Sites presented are from the sequence of Barker et al. (2). Arrows indicate the terminal inverted repeats of Mu1. A = AvaI, B = BstNI, E = EcoRII, H = HinfI, M = MluI, N = NotI, Q = TaqI, S = SstII, T = Tth111I.

Similar analysis of the maize retroviral transposable element Bs1 (27,28) demonstrates that the two endogenous Bs1 elements in the maize line 1s2p (28) are unmodified at internal PstI and SalI sites (Fig. 2 and data not shown). Hence, both Mutator and Bs1 elements are unmodified, unlike the majority of plant nuclear DNA (25,43), in plants in which their transposition has been detected (1,7,28).

When Mutator-associated mutagenic, transpositional, or somatic mutability activities are lost, many restriction sites within the whole population of Mu1-like elements in a plant are modified (5,6,8,15,48). Some restriction sites (e.g., EcoRII) appear to be more commonly modified than others (e.g., AvaI) (6). The two HinfI sites in the Mu1 terminal inverted repeats are commonly utilized for analysis of modification in Mu elements (15,48). In many lines that have lost somatic mutability of the Mutator-derived bronze mutation, bzMum4 (5,14), most of the Mu elements are resistant to digestion with HinfI (Fig. 3). However, in some other mutability-loss lines, most Mu1-like elements are fully digested at one or both HinfI sites (Fig. 3, lane 6). Moreover, in some lines that have lost Mutator mutagenic and transpositional activity upon intercrossing of diverse Mutator lines (5-7,38), virtually none of the Mu1-like elements are resistant to HinfI cleavage (5) (Fig. 4). Hence, in our studies, HinfI is one of the less dependable enzymes in predicting the activity and modification status of a Mutator-derived line.

Robertson originally derived Mutator-loss stocks from approximately 10% of his outcrosses to standard (non-Mutator) maize backgrounds (36) or in 100% of the progeny from four generations or more of intercrossing of separate Mutator lines (38). Although the Mu elements in Mutator intercross-loss (ML_i) lines are heavily modified at the EcoRII and other internal restriction sites (5,6,8) (Fig. 4), we have not seen modification at any restriction site in Robertson's Mutator outcross-loss (ML_o) lines (6) (Fig. 4).

In cases where a population of Mu elements vary in their degree of modification, analysis of the activity and modification of a single Mu element would be most informative. The unstable bronze mutation, bzMum4 (5,14), contains the insertion of a 1.4-kb Mu1-like element in the middle of

Fig. 2. Restriction map of the two preretrovirus-like Bs1 elements in the maize inbred 1s2p (28) derived from gel blot hybridization analysis (not shown). The probe used in these hybridization analyses was a mixture of three purified DNA fragments from a PstI/-BamHI digest of Bs1, yielding 2494 bp from the center of the element (27). Enzymes sensitive to 5-methylation of cytosine at 5'-CNG-3' and/or 5'-CG-3' are indicated by asterisks. The restriction map derived from gel blot hybridization analysis of the endogenous Bs1-like elements in 1s2p did not differ in any way from the map predicted by the sequence (27) of the Bs1 element that was cloned from Adh1 mutant S5446 by Johns et al. (28). Arrows indicate the direct terminal repeats of Bs1. B = BamHI, E = BstEII, H = HindIII, P* = PstI, S* = SalI, T = SstI.

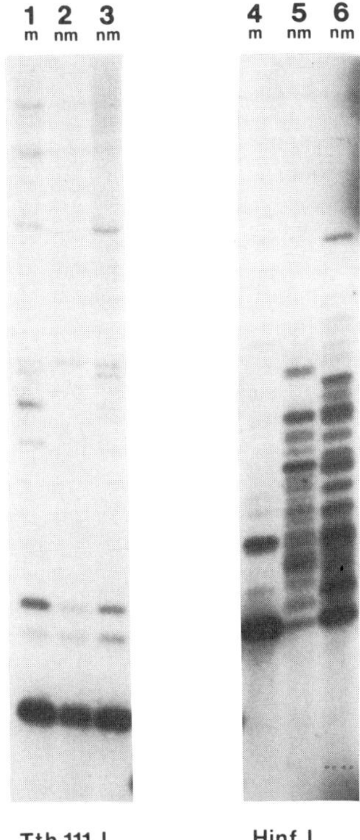

Fig. 3. Restriction enzyme digestion patterns of Mu elements from maize lines differing in mutability at the Mutator-induced allele bzMum4. Total maize DNA digested with Tth111I (lanes 1-3) or HinfI (lanes 4-6) was size fractionated on a 20 cm, 0.8% agarose gel. This gel was filter replicated and hybridized to a Mu probe labeled by random hexamer priming (22), as described in the section on Materials and Methods. Lanes 1 and 4 contain DNA derived from a plant that grew from a single bzMum4 mutable ("m") kernel. Lanes 2, 3, 5, and 6 contain DNA derived from two different plants that grew from two kernels that had independently lost bzMum4 mutability and were, therefore, termed nonmutable ("nm") (5,9).

the bronze structural gene (Fig. 5). Germinal reversion of this mutation, although rare, does occur via a relatively precise excision of the Mu element (14). For several years we have followed the somatic mutability of bzMum4 as an indicator of Mutator activity (5,8,9). Gel blot hybridization analysis of bzMum4 (Fig. 6) indicates that this bronze allele, like all other single-copy maize genes analyzed (12,33), is not extensively modified with 5-methylcytosine. In cases where bzMum4 retains somatic mutability, the Mu1-like element is seen to be completely unmodified at the internal NotI site, but can be partially or fully modified at the three MluI sites near the ends of the element (Fig. 5 and 6). Interestingly, in both of the mutable lines in which some MluI modification is observed, the two nearby MluI sites at one end of Mu1 are fully resistant to digestion, while the single MluI site at the other end of Mu1 is less than 50% digested (Fig. 6 and data not shown).

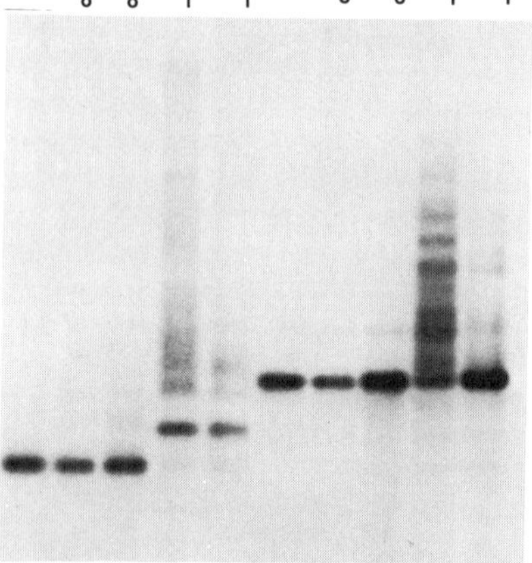

Fig. 4. Modification status of EcoRII and HinfI sites in various Mutator-derived lines. Lanes 1-5 contain DNAs double digested with Tth111I and EcoRII. Lanes 6-10 contain the same DNAs, in the same order, digested with HinfI. The cleaved DNA was size fractionated on a 15 cm, 0.9% agarose gel. The resultant gel was filter replicated and hybridized to a nick-translated Mu probe, as described in Materials and Methods. Lanes 1 and 6: DNA from an active Mu line (Mu). Lanes 2, 3, 7, and 8: DNA from a line that had lost Mutator mutagenic and transpositional activity on outcrossing (ML_o) (7,36). Lanes 4, 5, 9, and 10: DNA from two separate lines that had lost Mutator mutagenic and transpositional activity on intercrossing (ML_i) (7,38). The enzyme BstNI, an isoschizomer of EcoRII that is not sensitive to cytosine methylation at 5'-CNG-3' or 5'-CG-3' (6,30), completely digested all of the Mu elements in these lines (6, and data not shown). Rehybridrization of this filter with a maize ribosomal DNA probe (6) indicated that the EcoRII and HinfI digestions went to completion in each sample, but that the Tth111I digestion was incomplete in lanes 4 and 5 (data not shown).

Gel blot hybridization analysis of two lines that had lost mutability at bzMum4 (5,8,9) indicates that the internal NotI and MluI sites had all become resistant to digestion (Fig. 5 and 6). Nine other sites for 5-methylcytosine-sensitive restriction enzymes within or flanking the bronze structural gene remain fully digestible when mutability is lost, and the Mu element is modified (Fig. 5 and 6). The SmaI site employed in this analysis is probably less than 100 bp downstream from the Mu1 insertion site. Hence, the internal modification of the Mu element at bzMum4 does not lead to any detected modification of sequences flanking the element.

In a third line that had lost mutability, analysis with several restriction enzymes seems to indicate a complex rearrangement of the locus (Fig. 6, lane 13). Deletions and other forms of rearrangement initiated at one end of a transposable element have been commonly observed with many plant and bacterial systems (31,32), including Mutator (45).

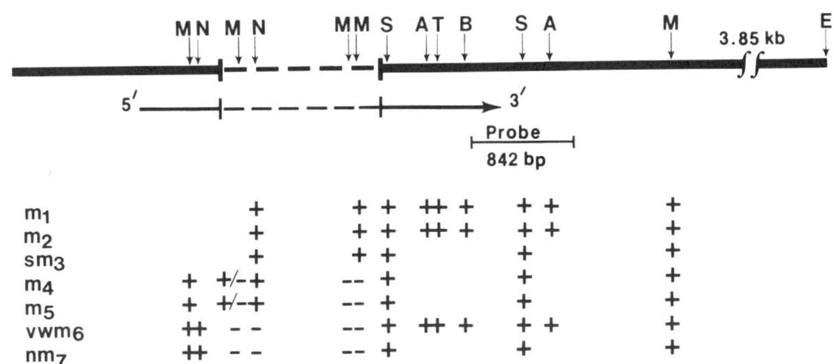

Fig. 5. Modification of the *bzMum4* locus as a function of its somatic instability. The mutant *bzMum4* contains an insertion of a 1.4 kb *Mu1*-like element (indicated by the dashed line) between the central *NotI* and *SmaI* sites in *Bronze* (23). + = complete digestion, - = no digestion, and +/- = partial digestion by the enzyme directly above the symbol on the restriction map. All restriction sites positioned on this map, except *EcoRI*, are for enzymes that are sensitive to 5-methylation of cytosine in 5'-CG-3' and/or 5'-CNG-3'. sm = strongly mutable [over 200 phenotypic reversion events (PRE) per kernel], m = mutable (100-200 PRE per kernel), vwm = very weak mutable (1-20 PRE per kernel), and nm = nonmutable (zero PRE per kernel) (5). The gel blot hybridization analyses giving rise to this map are from Fig. 6 and data not shown. sm_3 through vwm_6 were kernels from the same outcross of a mutable *bzMum4* line to *bronze-shrunken* deletion inbred 1302 (5,9). All other kernels were from separate ears of a similar outcross. A = *AvaI*, B = *BglI*, E = *EcoRI*, M = *MluI*, N = *NotI*, S = *SmaI*, T = *SstII*.

The 5-methylcytosine-sensitive restriction enzymes *PstI* and *PvuII* digest plant nuclear DNA poorly (Fig. 7). Less than 20% of the stained DNA in a *PstI* or *PvuII* digestion migrates at a molecular weight of less than 20 kb, and much of this is presumably due to contamination with organellar DNA. A *Mu* element hybridizational probe, however, identifies over 80% of all *Mu* elements as migrating at molecular weights of less than 20 kb in these digests, despite the fact that there are no *PstI* or *PvuII* sites in these *Mu1*-like elements. This result suggests that *Mu* elements are at least four-fold more likely to be found inserted in the unmodified DNA, characteristic of potentially active, low-copy number genes, than they are in modified, generally highly repetitive sequences (11,12). Since the *Mutator*-derived plant investigated in this study was an intercross-loss *Mutator* line with fully modified *Mu* elements, this experiment also demonstrates that modification of *Mu* elements does not generally lead to extensive modification of flanking sequences.

In all cases where careful studies have been performed, chiefly in bacteria (31) but including *Ac* in plants (24), transposable elements have been shown to exhibit some degree of sequence or regional insertion specificity. It remains to be determined whether *Mu* elements recognize a specific sequence or sequences associated with low-copy number DNA, or are transposing to these regions primarily because they are open and accessible (unlike the heterochromatic, condensed regions that contain primarily repetitive DNA). In previous studies from our laboratory (3,7),

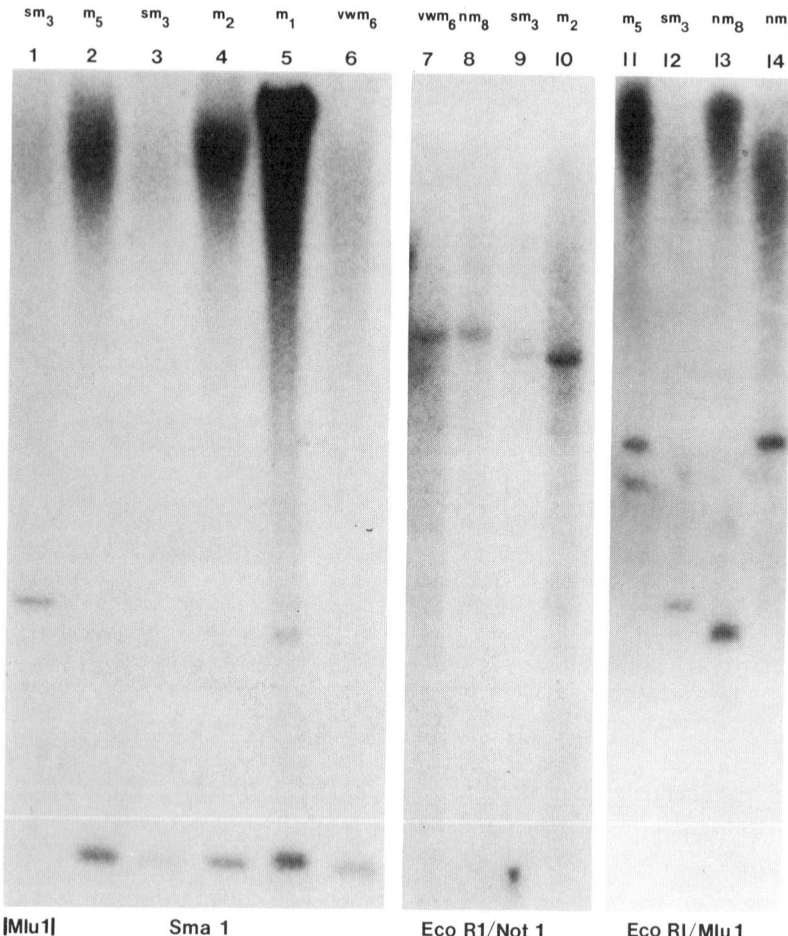

Fig. 6. Gel blot hybridization analysis of modification at bzMum4. The three gels shown were all 26 cm long and 0.8% agarose. The random primer extended Bronze probe employed in these hybridizations was an 842-bp PstI restriction fragment from the 3' end of the locus (see Fig. 5). m_1 - nm_7 are all as described in the legend to Fig. 5. nm_8 (lanes 8, 13) was a nonmutable kernel derived from a separate bzMum4 outcross to 1302 (5,9) and appears to be a complex rearrangement of the locus. Lane 1: an MluI digest. Lanes 2-6: SmaI digests. Lanes 7-10: EcoRI plus NotI double digestions. Lanes 11-14: EcoRI plus MluI double digestions.

the exceptional mutagenic activity (36,37) of Mutator was explained as a likely outcome of a high frequency of transposition per Mu element, a gene-specific insertion preference for Mu elements, or both. Since Mu elements can transpose at frequencies averaging nearly one transposition per element per sexual generation (1,7) and also are found preferentially in undermodified maize DNA, it seems likely that both factors play an important role.

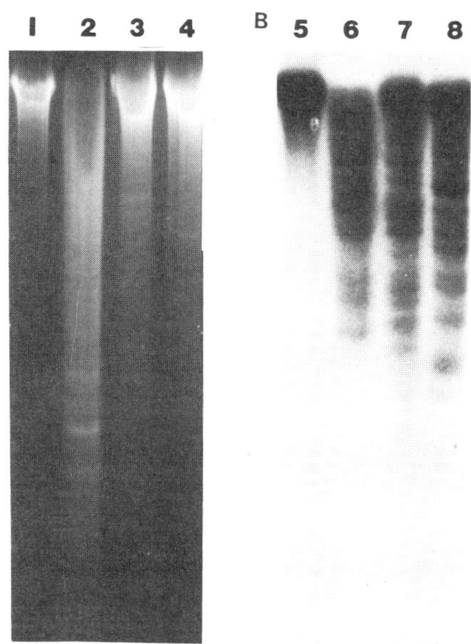

Fig. 7. Modification status of sequences flanking Mu elements in a modified Mutator stock. Lanes 1-4 contain the ethidium bromide stain pattern, and lanes 5-8, the Mu element hybridization profile of uncut (lanes 1, 5), HindIII digested (lanes 2, 6), PstI restricted (lanes 3, 7), and PvuII digested (lanes 4, 8) DNA from a Mutator intercross loss line containing 100% modified Mu elements. None of these restriction enzymes have sites within Mu1-like elements. Of the three enzymes employed in this experiment, only HindIII cleaves maize nuclear DNA efficiently, due to its insensitivity to cytosine methylation. The results shown are for a 15 cm, 0.7% agarose gel filter replicated and hybridized to a nick-translated Mu probe, as described in the section on Materials and Methods.

In crosses between Mutator-derived maize lines with multiple unmodified Mu elements and lines with numerous modified Mu elements, we have observed that modification appears to be progressively dominant (Fig. 8). In younger tissues of a plant derived from such a cross, both modified and unmodified elements are about equally abundant (Fig. 8A, lanes 1-3, and 8B, lanes 1-3 and 7-9). Leaves higher up on the same plant progressively attain a higher percentage of modified elements (Fig. 8A, lanes 4-9, and 8B, lanes 4-6 and 10-12). Finally, the tassel, silk, and ear tissues often have only modified Mu elements (Fig. 8A, lanes 10-15). This gradually dominant process of modification was observed when the modified Mu elements were derived from either the male or female parent (compare Fig. 8A, lanes 1-9, to Fig. 8B, lanes 1-12).

DISCUSSION

Numerous studies have now shown that inactivity of Mutator and other plant transposable elements often correlates with modification of the elements involved (5,6,15,17-19,42,48). That this is not the only mode in

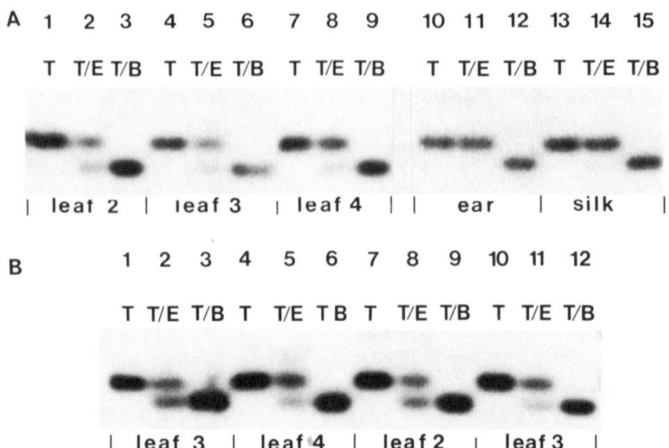

Fig. 8. Progressive dominance of Mu element modification. A: Progressive modification in various tissues from a single plant derived from the cross of a female parent with modifed Mu elements to a male parent with unmodified Mu elements. B: Progressive modification in leaves from two plants (one in lanes 1-6, the other in lanes 7-12) from a cross of a female parent with unmodifed Mu elements to a male parent with modified Mu elements. Each sample was digested with the 5-methylcytosine-insensitive enzyme TaqI (T) for one lane and then double digested with TaqI and EcoRII (T/E), and TaqI and BstNI (T/B). BstNI and EcoRII recognize the same restriction sites in Mu elements, but EcoRII cleavage is blocked by 5-methylcytosine in the sequence 5'-CNG-3'. BstNI is 5-methylcytosine insensitive. Ear and silk DNA preparations were from the same unfertilized ear. These three gels were all 15 cm in length and 1.5% agarose. Hybridization was to the central, 1.0 kb Tth111I fragment of Mu1 labeled by random hexamer priming (22).

which Mutator is regulated is exemplified by the observation that Robertson's original Mutator outcross-loss lines (36) contain multiple Mu elements that are unmodified, but are mutagenically and transpositionally quiescent (6,7). Although a somewhat lowered Mu element copy number in these Mutator outcross-loss lines could perhaps account for the loss of Mutator mutagenic activity (6,15), it cannot account for the drastic decrease in the average Mu element transposition frequency (7). More likely, the decreased Mutator activity in these outcross lines is due to the segregational loss of one or more positive factors (e.g., a transposase) necessary for transposition (3,6,32,36).

In most of our crosses following mutability at bzMum4 (5,9), we have found that outcross loss of mutability is associated with Mu element modification (e.g., Fig. 3). Moreover, in most cases (8,41), loss of mutability in these and similar lines is correlated with a loss of Mutator mutagenic activity. It is not at all clear why outcross loss of Mutator activity may be commonly associated with Mu-element modification in some crossing programs (15,48) and rare or nonexistent in others (6,36). One possible clue to this mystery is the variable effect of genetic background on the retention of Mutator activity (8,40). Some backgrounds may induce or prevent modification of Mu elements, for instance, more efficiently than others. Alternatively, different populations of Mu elements might vary in their ratio of activating to modifying components.

Generally, multiple 5'-CNG-3' and 5'-CG-3' cytosines are methylated when a Mu element is inactivated. However, little or no cytosine methylation is concurrently propagated in sequences flanking the element. This suggests that the activity(ies) responsible for recognition of the Mu element and initiation of the modification process must also delimit the methylation activity of Mutator sequences. This extensive and sharply delineated DNA modification process has not been described in any other biological system.

Cytosine methylation of Mu elements is extensive, discrete, and relatively predictable, at least upon repetitive intercrossing of active Mutator lines or upon crossing of a modified Mutator-derived line to an unmodified Mutator line. The actual sites modified under any given circumstance do seem to vary, however. In our work, modification of internal EcoRII and NotI sites seems to be consistently predictive of Mutator activity, while the HinfI and MluI sites in and near the Mu1 terminal inverted repeats are not particularly dependable. In studies of the low-copy number, endogenous Mu1-like elements found in some non-Mutator maize inbreds (6,7,10,16), we have found that these elements contain modified HinfI sites, but are unmodified at internal EcoRII and HpaII sites (10). To date, our data suggest that terminal modification at HinfI and MluI sites may occur early in an inactivation process, but may be lost through subsequent crosses in an inactive line or through reactivation by crossing to an active line (9). Modification at internal Mu element sites like EcoRII and NotI may occur later in an inactivation process, but may be both more stable and more inhibitory to Mutator activity.

In the progeny of a cross between an active Mutator line with unmodified Mu elements and an inactive Mutator-derived line with modified Mutator elements, we see a gradual increase in the ratio of modified to unmodified elements during development of the plant. Maize tassels and ears seem to have the highest proportion of modifed Mu elements, in some cases reaching 100%. This effect may or may not be a sign of a tissue-specific enhancement of modified Mu element dominance in the gametophyte. Alternatively, the greater developmental "age" (in terms of both time and the number of mitotic divisions removed from the original zygote) of tassels and ear shoots may account for their preponderance of modified Mu elements, as appears to be true for upper leaves.

Modified element dominance seems to pass efficiently through both male and female gametes, in accordance with recent genetic studies showing ilttle or no unique maternal or paternal contribution to inactivation of a mutable bzMum4 allele by lines containing a nonmutable bzMum4 allele (9). A recent study from our laboratory (6) reporting somewhat increased modification dominance through the female must, in light of the results presented here, be more carefully interpreted. Several DNA preparations from F_1 plants of crosses where the modified Mu elements were contributed by the female parent were isolated from immature ear shoots, while, in this same study, most DNA preparations of F_1 plants from crosses where the unmodified Mutator parent was the female were made from young leaves. In addition, the partial female dominance in retention and loss of Mutator-derived mutability at bronze2 reported by Walbot (49) appears, in our experiments with bzMum4 (9), to be a reflection of aleurone dosage differences and not true female dominance.

Obviously, much more work will be necessary to discover the determinants and effects of Mu element modification and inactivation. In particular, the significance and nature of the variation in Mu element modification patterns observed remain a mystery. Also, the inactivational dominance of modified Mu elements, even by lines that had become modified and lost Mutator activity one or more generations previously (6,39), requires a greater degree of communication between different Mu elements, or between Mu elements and endogenous "modulator" genes (8,32), than we have come to expect. DNA modification-associated negative regulation appears to be a common phenomenon with maize transposable elements, and may also be responsible for other genetic phenomena [for instance, paramutation (13)] not yet molecularly detailed. Hence, future studies of the biology of Mu element modification and Mutator regulation may be particularly fruitful.

ACKNOWLEDGEMENTS

We wish to thank C. Carter and R.P. Fracasso for excellent technical assistance, and M. Freeling, D. Furtek, R. Johns, O. Nelson, and K. Newton for graciously providing several plasmids used in this study. This research was supported by grants 86-CRCR-1-2121 from the U.S. Department of Agriculture, 8552557A1-DCB from the National Science Foundation, and, from the McKnight Foundation.

REFERENCES

1. Alleman, M., and M. Freeling (1986) The Mu transposable elements of maize: Evidence for transposition and copy number regulation during development. Genetics 112:107-118.
2. Barker, R.F., D.V. Thompson, D.R. Talbot, J. Swanson, and J.L. Bennetzen (1984) Nucleotide sequence of the maize transposable element Mu1. Nucl. Acids Res. 12:5955-5967.
3. Bennetzen, J.L. (1984) Transposable element Mu1 is found in multiple copies only in Robertson's Mutator maize lines. J. Molec. Appl. Genet. 2:519-524.
4. Bennetzen, J.L., J. Swanson, W.C. Taylor, and M. Freeling (1984) DNA insertion in the first intron of maize Adh1 affects message levels: Cloning of progenitor and mutant Adh1 alleles. Proc. Natl. Acad. Sci., USA 81:4125-4128.
5. Bennetzen, J.L. (1985) The regulation of Mutator function and Mu1 transposition. UCLA Symposium on Molecular and Cellular Biology 35:343-354.
6. Bennetzen, J.L. (1987) Covalent DNA modification and the regulation of Mutator element transposition in maize. Molec. Gen. Genet. 208:45-51.
7. Bennetzen, J.L., R.P. Fracasso, D.W. Morris, D.S. Robertson, and M.J. Skogen-Hagenson (1987) Concomitant regulation of Mu1 transposition and Mutator activity in maize. Molec. Gen. Genet. 108:57-62.
8. Bennetzen, J.L., A. Cresse, W.E. Brown, and L. Lee (1987) Molecular cloning of maize genes by transposon tagging with Mutator. UCLA Symposium on Molecular and Cellular Biology 62:183-204.
9. Bennetzen, J.L. (1988) Inactivation and reactivation of a Mutator-derived bronze mutable allele in maize (submitted for publication).
10. Bennetzen, J.L., R.P. Fracasso, D. Horvath, and C. Carter (1988) Variable presence and modification of sequences related to transpos-

able element Mu1 in non-Mutator maize lines (manuscript in preparation).

11. Bennetzen, J.L., and K.S. Schrick (1988) Active, single copy maize genes are flanked by diverse highly repetitive DNA sequences (manuscript in preparation).
12. Bennetzen, J.L., and K.S. Schrick (1988) Low copy number and highly repetitive DNA sequences in Zea mays are organized in respective modified and unmodified domains (manuscript in preparation).
13. Brink, R.A. (1973) Paramutation. Ann. Rev. Genet. 7:129-152.
14. Brown, W.E., D.S. Robertson, and J.L. Bennetzen (1987) Molecular analysis of multiple Mutator-derived alleles of the bronze locus of maize (submitted for publication).
15. Chandler, V.L., and V. Walbot (1986) DNA modification of a maize transposable element correlates with loss of activity. Proc. Natl. Acad. Sci., USA 83:1767-1771.
16. Chandler, V.L., C. Rivin, and V. Walbot (1986) Stable non-Mutator stocks of maize have sequences homologous to the Mu1 transposable element. Genetics 114:1007-1021.
17. Chomet, P.S., S. Wessler, and S.L. Dellaporta (1987) Inactivation of the maize transposable element Activator (Ac) is associated with its DNA modification. EMBO J. 6:295-302.
18. Cone, K.C., F.A. Burr, and B. Burr (1986) Molecular analysis of the maize anthocyanin regulatory locus C1. Proc. Natl. Acad. Sci., USA 83:9631-9635.
19. Dellaporta, S.L., and P.S. Chomet (1985) The activation of maize controlling elements. In Genetic Flux in Plants, B. Hohn and E.S. Dennis, eds. Springer Verlag, Wien/New York, pp. 169-216.
20. Doerfler, W. (1983) DNA methylation and gene activity. Ann. Rev. Biochem. 52:93-124.
21. Fedoroff, N.V., D.B. Furtek, and O.E. Nelson, Jr. (1984) Cloning of the bronze locus in maize by a simple and generalizable procedure using the transposable controlling element Activator (Ac). Proc. Natl. Acad. Sci., USA 81:3825-3829.
22. Feinberg, A., and B. Vogelstein (1983) A technique for radio-labelling DNA restriction endonuclease fragments to high specific activity. Anal. Biochem. 132:6-13.
23. Furtek, D.B. (1986) Cloning and sequence analysis of the bronze1 locus of maize: Location of a transposable element insertion and comparison of two wild-type alleles. Ph.D. Dissertation, University of Wisconsin at Madison.
24. Greenblatt, I.M. (1984) A chromosomal replication pattern deduced from pericarp phenotypes resulting from movements of the transposable element, Modulator, in maize. Genetics 108:471-485.
25. Gruenbaum, Y., T. Naveh-Many, H. Cedar, and A. Razin (1981) Sequence specificity of methylation in higher plant DNA. Nature 292:860-862.
26. Hepburn, A.G., L.E. Clarke, L. Pearson, and J. White (1983) The role of cytosine methylation in the control of nopaline synthase gene expression in a plant tumor. J. Molec. Appl. Genet. 2:315-329.
27. Jin, Y.K., and J.L. Bennetzen (1988) Structure and coding properties of Bs1, a maize retrovirus-like transposon (submitted for publication).
28. Johns, M.A., J. Mottinger, and M. Freeling (1985) A low copy number, copia-like transposon in maize. EMBO J. 4:1093-1102.
29. Hake, S., and V. Walbot (1980) The genome of Zea mays, its organization and homology to related grasses. Chromosoma 79:251-270.

30. Kessler, C., P.S. Neumeier, and W. Wolf (1985) Recognition sequences of restriction endonucleases and methylases - A review. Gene 33:1-102.
31. Kleckner, N. (1981) Transposable elements in prokaryotes. Ann. Rev. Genet. 15:341-404.
32. Nevers, P., N.S. Shepherd, and H. Saedler (1986) Plant transposable elements. Adv. Bot. Res. 12:103-203.
33. Nick, H., B. Bowen, R.J. Ferl, and W. Gilbert (1986) Detection of cytosine methylation in the maize alcohol dehydrogenase gene by genomic sequencing. Nature 319:243-246.
34. O'Reilly, C., N.S. Shepherd, A. Pereira, Z. Schwarz-Sommer, I. Bertram, D.S. Robertson, P.A. Peterson, and H. Saedler (1985) Molecular cloning of the al locus of Zea mays using the transposable elements En and Mu. EMBO J. 4:877-882.
35. Riggs, A.D., and P.A. Jones (1983) 5-Methylcytosine, gene regulation, and cancer. Adv. Cancer Res. 40:1-30.
36. Robertson, D.S. (1978) Characterization of a mutator system in maize. Mutat. Res. 51:21-28.
37. Robertson, D.S., and P.N. Mascia (1981) Tests of four controlling element systems of maize for mutator activity and their interaction with Mu mutator. Mutat. Res. 84:283-289.
38. Robertson, D.S. (1983) A possible dose-dependent inactivation of Mutator (Mu) in maize. Molec. Gen. Genet. 191:86-90.
39. Robertson, D.S. (1986) Genetic studies of the loss of Mu mutator activity in maize. Genetics 113:765-773.
40. Robertson, D.S. (1986) Loss of Mu mutator activity when active Mu systems are transferred to inbred lines. In Maize Genetics Cooperative Newsletter 60:10.
41. Robertson, D.S. (1987) Additional evidence on the correlation of somatic mutability and germinal mutator activity in Mu-induced aleurone mutants. In Maize Genetics Cooperative Newsletter 61:13-14.
42. Schwartz, D., and E. Dennis (1986) Transposase activity of the Ac controlling element in maize is regulated by its degree of methylation. Molec. Gen. Genet. 205:476-482.
43. Shapiro, H.S. (1976) Distribution of purines and pyrimidines in deoxyribonucleic acids. In CRC Handbook of Biochemistry and Molecular Biology, G.D. Fasman, ed. CRC Press, Cleveland, Ohio, pp. 241-283.
44. Shure, M., S. Wessler, and N. Fedoroff (1983) Molecular identification and isolation of the Waxy locus in maize. Cell 35:225-233.
45. Taylor, L.P., and V. Walbot (1985) A deletion adjacent to the maize transposable element Mu-1 accompanies loss of Adh1 expression. EMBO J. 4:869-876.
46. Taylor, L.P., V.L. Chandler, and V. Walbot (1986) Insertion of 1.4 kb and 1.7 kb Mu elements into the Bronze-1 gene of Zea mays. Maydica 31:31-46.
47. Van Slogteren, G.M.S., P.J.J. Hooykaas, and R.A. Schilperoort (1984) Silent T-DNA genes in plant lines transformed by Agrobacterium tumefaciens are activated by grafting and 5-azacytidine treatment. Plant Molec. Biol. 3:333-336.
48. Walbot, V., V. Chandler, and L. Taylor (1985) Alterations in the Mutator transposable element family of Zea mays. UCLA Symposium on Molecular and Cellular Biology 35:333-342.
49. Walbot, V. (1986) Inheritance of Mutator activity in Zea mays as assayed by somatic instability of the bz2-mu1 allele. Genetics 114:1293-1312.

EXTRACHROMOSOMAL *Mu*

Venkatesan Sundaresan

Cold Spring Harbor Laboratory
Cold Spring Harbor, New York 11724

ABSTRACT

Maize lines known as Robertson's *Mutator* (*Mu*) lines generate unstable recessive mutations at high rates. These lines carry actively transposing copies of the elements *Mu1* and *Mu1.7*. We have discovered extrachromosomal forms of the *Mu1* and *Mu1.7* transposons. We have shown that these molecules are most likely covalently closed circular DNA (ccDNA). Further, we find that their occurrence is correlated with *Mu* transposition so that they are probably generated during *Mu* transposition as transposition intermediates, abortive intermediates, or excision products. Our data suggest that *Mu1* and *Mu1.7* circles account for a significant fraction of the total ccDNA found in *Mutator* lines, but that other *Mu*-homologous circles are also present in these lines.

INTRODUCTION

The *Mutator* (*Mu*) system of maize was identified by Robertson in a maize line that generated recessive mutants at high frequencies (18). Molecular analysis of *Mu*-induced mutants has shown that these mutants are associated with the insertion of a 1.4-kilobase (kb) element, *Mu1* (26), or a larger but homologous 1.7-kb element, *Mu1.7* (28). Transposons *Mu1* and *Mu1.7* are found in *Mu* lines, usually at 20-30 copies and 0-5 copies per cell, respectively (for review, see Ref. 14). *Mu1* has been sequenced and found to have 200-base-pair (bp) terminal inverted repeats, 100-bp internal direct repeats, and four open reading frames (ORF) of 300-500 bp (2). Although the *Mu* elements are unusually active transposons, with high rates of forward transposition that can result in up to one-half of the *Mu1* elements at new locations in the outcross progeny (1), little is known about the mechanism of transposition and as yet no mRNA or transposase has been identified. In an attempt to understand the mechanism of *Mu* transposition, we have searched for, and found, extrachromosomal molecules of *Mu1* and *Mu1.7* DNA that are associated with *Mu* activity.

MATERIALS AND METHODS

Maize Lines

Active maize lines, abbreviated Mu:1s2p, were derived by back-crossing Robertson's Mutator lines into the Freeling 1s2p inbred background (1). These Mu-active families displayed clonal striping of leaves and no modification of the HinfI sites in Mu1 and Mu1.7, which is indicative of actively transposing lines (6), and carried 20-25 copies of Mu1. Samples were either from leaves and stems of one-month-old plants or from whole tassels in which the average sporocyte was in prophase I. Other active Mu lines examined--standard Mu (DR84-1254) and purple aleurone Mu (DR79-9027X8028), and two inactive Mu lines (DR77-1070 and DR79-9562X8563)--were gifts from D. Robertson (Iowa State University).

The methods used for isolation and detection of extrachromosomal DNA, Southern blotting, electron microscopy, etc., have been described elsewhere (Ref. 27).

RESULTS

Detection of Extrachromosomal Mu

When DNA from leaf or stem tissue of an active Mu plant is fractionated on a CsCl-EtBr gradient, and the fractions are electrophoresed on an agarose gel, blotted, and hybridized to a Mu1 probe, hybridization to the ccDNA fractions can be detected as a faint band, as shown in Fig. 1A. Using ccDNA size markers, the size of the DNA in this band was estimated to be 1.4 kb, which is consistent with the size of Mu1. Using copy number controls, we estimate that it is present at approximately 0.1 copy/cell. This extrachromosomal Mu1 can be detected at similar copy numbers in many different Mu lines with different genetic backgrounds.

Mu lines can spontaneously lose activity on outcrossing or inbreeding, giving inactive lines carrying Mu1 sequences that do not transpose (20,21). Such inactive Mu lines do not have any extrachromosomal copies of Mu1, as shown by the example in Fig. 1B. Therefore the chromosomal Mu1 elements do not generate extrachromosomal Mu1 unless they are actively transposing, and the extrachromosomal Mu1 is associated with Mutator activity rather than with the presence of Mu1 sequences.

Genetic data on the timing of Mu activity suggest that Mu is most active developmentally at the premeoitic or meoitic stage (19). When DNA from immature tassels corresponding to this developmental stage was examined by fractionation, the copy number of the extrachromosomal Mu1 was found to be significantly higher than in leaf and stem tissue (Fig. 1C vs 1A). The higher concentration of the circles in this tissue (0.3-0.6 copy/cell) permitted the detection of extrachromosomal Mu1.7, a 1.7-kb variant of Mu1 that is present in fewer copies than Mu1 (28).

Structure of the Extrachromosomal Mu

The extrachromosomal Mu1 and Mu1.7 were detected in the fractions corresponding to ccDNA. To confirm that the extrachromosomal Mu is in fact ccDNA, the corresponding fractions have been subjected to a variety of treatments prior to electrophoresis. The extrachromosomal Mu1 and Mu1.7 are resistant to NaOH and proteinase K treatment, but not to DNase 1 (27). Restriction enzyme digestions show that these molecules have circular restriction maps (Fig. 2). The circles could be generated as expected from the linear maps of the integrated Mu elements by circularization at the ends without any major rearrangement or deletions. The DNA sequence at the junction is not known. The DNA sequence at the 5' end of Mu1 is GAG... and at the 3' end is ...CTC. If the ends were joined directly without any insertion or deletion, we would expect to generate the sequence ...CTCGAG... at the junction, which is the recognition sequence for XhoI. The circles are resistant to XhoI digestion, suggesting that this is not the case (27), and that the junction carries some alteration of the Mu1 sequence, e.g., the 9-bp host sequence duplication.

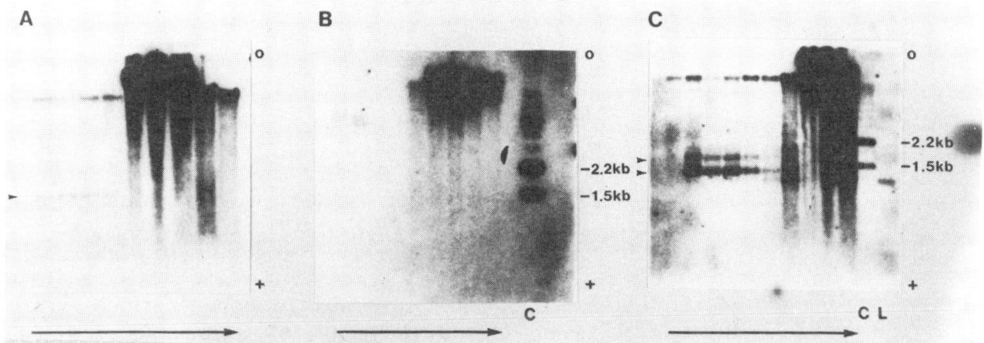

Fig. 1. Autoradiograms of Southern blots of uncut DNA fractions from CsCl/EtBr equilibrium gradients hybridized to Mu1-specific, radiolabeled RNA probe. Fifty to 100 micrograms of total uncut DNA are centrifuged to equilibrium in a CsCl/EtBr gradient, and fractions are collected from the gradient, electrophoresed on a 1% agarose gel, and blotted to nitrocellulose filters. The probe used was a Mu1 RNA, synthesized using SP6 polymerase, and contained one terminal inverted repeat plus most of the internal sequence of Mu1. Long arrows at the bottom indicate concentrations of CsCl from high to low; ccDNA fractions are on the left, and linear DNA fractions are to the right. O, Origin of the gel electrophoresis; +, anode. (A) DNA from leaf and stem of a Mu:1s2p line (MF83*5). Arrowhead, 1.4-kb band. (B) DNA from leaf and stem of an inactive Mu line (DR77-1070). (Faint band in the ccDNA fractions at 4 kb corresponds to a tracer amount of exogenously added E. coli plasmid carrying a Mu insertion.) Lane C contains ccDNA size markers, including plasmids p ED1 (2.2 kb) and p ED2-1 (1.5 kb) that carry Mu1 insertions. (C) DNA from meiotic tassels of Mu:1s2p line (MF83*5). Arrowheads, 1.4- and 1.7-kb bands. Supercoiled DNA markers at 2.2 kb and 1.5 kb in lane C are from plasmids p ED1 and p ED2.1; hybridization to p ED2-1 represents 0.4 pg of Mu1 DNA. Lane L contains linear DNA size markers, including DNA digested with Hind111; the hybridizing band here is a 1.1-kb Mu1 fragment.

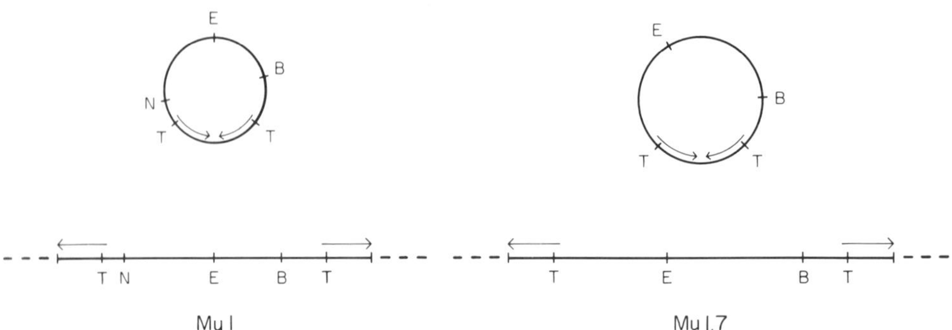

Fig. 2. Restriction map of Mu1 and Mu1.7 ccDNA with the map of linear integrated Mu1 and Mu1.7 shown below for reference. B, BstN1; E, BstE11; N, Not1; and T, Tth111-1. The Not1 site is absent in Mu1.7 ccDNA, as in genomic Mu1.7 (4).

Electron microscopy was used to study the DNA in the closed circular fractions to determine if a large number of circular molecules other than Mu1 are present in the nucleus of a Mutator maize plant. Both supercoiled and relaxed circles of 1.4 kb could be observed (Fig. 3a, b). A size distribution study showed that most of the molecules observed in a Mu:1s2p line were in the size class of 1.2-1.4 kb (Fig. 3c), with additional peaks at 0.8 kb and 1.6-1.8 kb. The sizes of Mu1 and Mu1.7 are 1.4 kb and 1.7 kb, respectively. The concentrations of the molecules in the 1.2-1.4 kb class and the 1.6-1.8 kb class, as estimated by the EM study, were in the same range as the estimates of Mu1 and Mu1.7 concentrations, based on the intensity of hybridization on Southern blots. This suggests that the peaks at 1.2-1.4 kb and 1.6-1.8 kb represent Mu1 and Mu1.7 circles. When the number of circles that are clearly not Mu1 or Mu1.7, by size, was estimated by the EM study, they accounted for less than one-third of the total circles.

Identification of Other Mu Circles

Recently, a new element, Mu3, has been identified as an insertion into the Adh1 gene of a Robertson Mu line (7). This element is 1.85 kb in size and has 200-bp inverted repeats homologous to Mu1 and Mu1.7, but the internal sequence is not homologous to Mu1 or Mu1.7. Subsequently, other elements have been identified that carry terminal inverted repeats homologous to Mu1 or Mu1.7; these elements are also believed to transpose in Mu lines (Chandler et al., this Volume). In the Mu:1s2p line used for the studies in Fig. 1 and Fig. 3, the only circles detected were Mu1 and Mu1.7. However, in other Mu lines this is not the case, and larger circles hybridizing to the Mu1 probe could be seen. For example, when the DNA from a Robertson purple aleurone Mu line was examined for Mu circles, we could see, in addition to Mu1 and Mu1.7 circles, bands corresponding to 1.85-kb and 2.0-kb circles (Fig. 4). These probably represent circular forms of other Mu elements, such as Mu3, found in this line. Further, a library of the circular fractions from this line has been constructed and clones have been identified that contain the Mu1 termini, but with internal sequences not homologous to Mu1. One of these clones is the same size as Mu3, and hybridizes to a Mu3-specific probe. Therefore it is likely that all the elements of the Mu family of transposons can exist as extrachromosomal circles.

DISCUSSION

We have shown that active Mu lines carry extrachromosomal copies of the Mu1 and Mu1.7 transposons. These extrachromosomal Mu sequences appear to be closed circular DNA molecules. Restriction mapping data suggest that they correspond to circularized forms of Mu1 and Mu1.7, joined at their terminal inverted repeats. The exact sequence at the junction is presently unknown, but the data suggest that the inverted repeats are not simply blunt-ended together. The extrachromosomal Mu circles are correlated with mutator activity. Maize lines that carry nontransposing Mu sequences do not have any extrachromosomal Mu circles. Further, there is an overall correlation between increased copy number and increased mutator activity in the developing tassel. Therefore, the Mu circles are probably generated by Mu transposon activity, either by excision or by replication of the integrated chromosomal Mu elements.

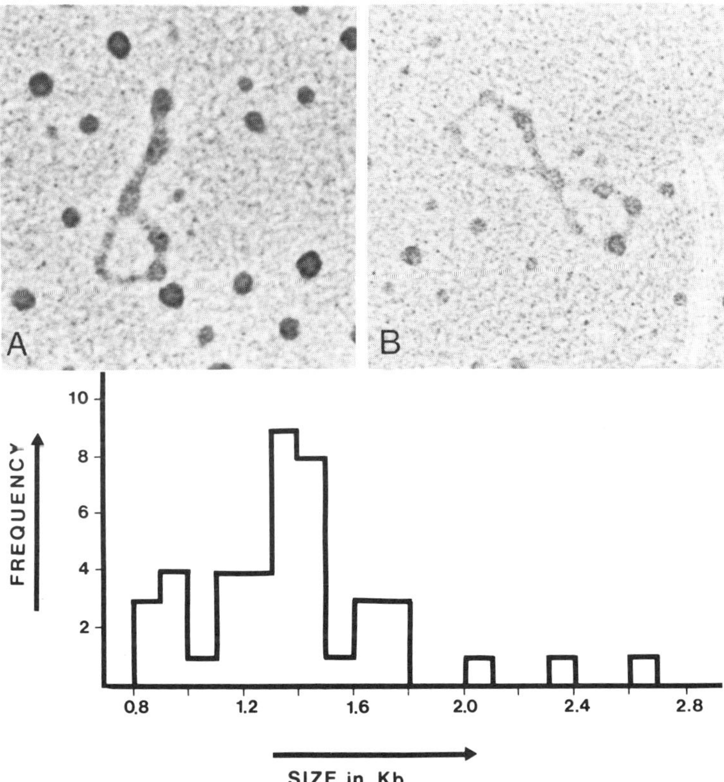

Fig. 3. Electron micrographs of purified ccDNA from an active Mu line, showing 1.4-kb (A) supercoiled and (B) relaxed circular molecules. (C) Size distribution of the circular molecules in the ccDNA fraction. Each bar represents the number of individual molecules within a 0.1-kb size range.

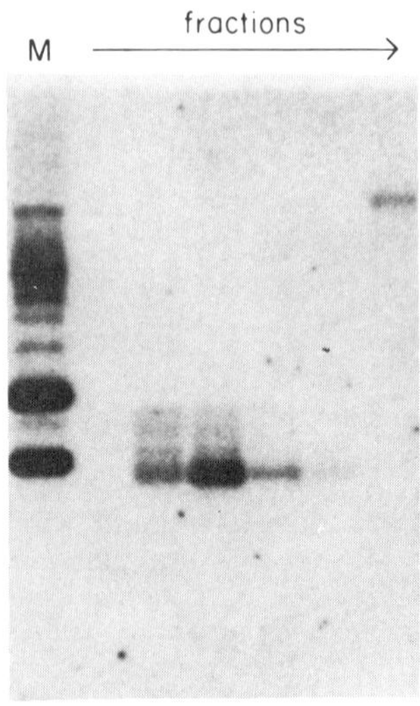

Fig. 4. Autogradiogram of unit DNA fractions from a Robertson purple aleurone Mu line (DR79-9027x8028), fractionated on a CsCl/EtBr gradient, showing at least four circular DNA species including Mu1 hybridizing to a Mu1-specific probe. Arrow on top indicates concentrations of CsCl from high to low, with linear fractions on the right. M, closed circular DNA size markers.

Mu-induced mutations are unstable and revert somatically, as manifested by spotted kernels, striped leaves, etc. (14). This reversion must be due to excision of the Mu insertion from the gene into which it is inserted. The excision of a Mu element from the chromosome could be followed by ligation of both the Mu ends and of the chromosome, restoring the gene function and generating a Mu circle. A possible mechanism for this process is outlined in Fig. 5. A mechanism of excision of a plant transposable element by base-pairing between the inverted repeats has already been suggested by Saedler and Nevers (24). Ligation of the Mu ends of this cruciform structure would ultimately lead to a circular molecule with the inverted repeats juxtaposed as in Fig. 5. The DNA sequence of the junction of the inverted repeats has yet to be determined, but it should lend insights into the mechanism of Mu excision. The excised Mu element may then be capable of reintegration at a new chromosomal location, as suggested for Ac element.

At present it is not known if the Mu circles are capable of reintegration. This question could be answered by transforming non-Mu maize with the Mu circles and looking for integration, but it is not presently possible to obtain transformed maize plants. It has been suggested that Mu transposition may be replicative (1), as in the case of bacterial transposons, e.g., Tn3 and Tn501, that transpose through a replicative mechanism proposed by Shapiro (25). However, the bacterial transposons such as Tn3 differ from Mu in at least one important respect; that is, they do not show transposase-dependent excision. Further, no extrachromosomal forms of these transposons have been identified. Therefore, the mechanism of Mu transposition is likely to be different from these in at least

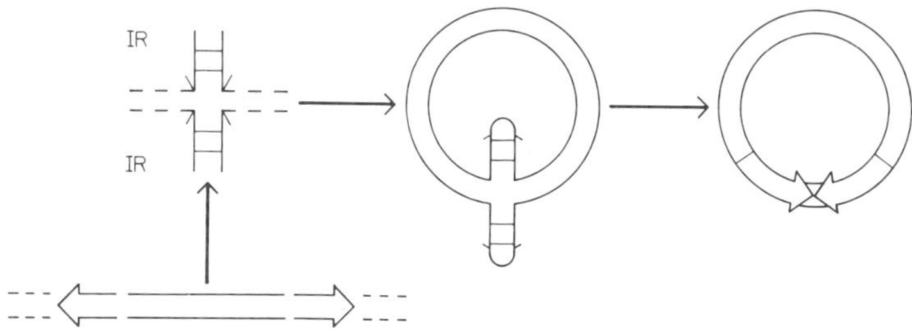

Fig. 5. A possible mechanism by which an extrachromosomal circular Mu1 molecule might be generated by excision from an integrated chromosomal Mu1. The first step shown follows from the model of Saedler and Nevers (24).

some important aspects. If Mu transposition is replicative, the extrachromosomal circles may also represent real or abortive replication intermediates that are generated by a mechanism different from the classical Shapiro model for replicative transposition.

While no other plant transposable element has been shown to exist extrachromosomally, there are other examples of extrachromosomal forms of transposons. The best-studied case is that of the retrotransposons, which include copia and Ty1. The transposons of this class are thought to transpose through extrachromosomal DNA intermediates following reverse transcription. In the case of copia, transposon circles believed to be intermediates have been identified and studied (8, 9). However, we believe that Mu1 is unlikely to be a retrotransposon because it does not have either long terminal direct repeats or the structure of the DIRS element (5), and it does not encode any protein with homology to reverse transcriptase. Therefore, the existence of the circles is likely to be due to a process other than reverse transcription.

In eukaryotic systems, the only other nonretroviral-like transposon besides Mu for which an extrachromosomal form has been identified is the transposon Tc1 of the nematode Caenorhabditis elegans (22, 23). Tc1 shares some other features with Mu1: it is small (1.6 kb), has terminal inverted repeats (54 bp), is found as a homogeneous, multicopy family, and appears not to encode any transposase. The extrachromosomal Tc1 differs from Mu1 in that 90% of Tc1 is found as a linear species, and only 10% of it is circular. In the case of Mu1, we have not detected any linear extrachromosomal species. However, in view of their similarities, TC1 and Mu1 might belong to a class of transposons that can be distinguished from both the retroviral-like elements such as copia and Ty and the two-element systems such as the P element in Drosophila and the controlling elements (e.g., Ac-Ds) in maize. If so, the common feature of an extrachromosomal form of the transposon could be a clue to a distinctly different mechanism of transposition and, perhaps, transmission.

A third example of an extrachromosomal transposon involves the bacterial transposon Tn10. When Tn10 transposase is overexpressed in a plasmid, Tn10 circles are generated by excision (15). The same observation can be reproduced in vitro using purified Tn10 transposase (16). Tn10 differs from the other bacterial transposons such as Tn3 in that it appears to transpose by excision and reintegration rather than by replication (3). It is not presently known whether the observed *Tn10* circles present real intermediates or abortive intermediates in transposition.

Extrachromosomal ccDNAs have been found in *Drosophila* tissue culture cells and in *Drosophila* embryos; these DNAs hybridize to middle-repetitive DNA sequences (17). Because a large fraction of the maize genome is middle-repetitive DNA, which has been considered to be at least potentially transposable (for reviews, see Refs. 10 and 29), and because a *copia*-like transposon has been bound in an insertional mutant in maize (11), we questioned whether a variety of transposons or plasmids were represented in the ccDNA fractions from our gradients. Our EM analysis indicated that most of the molecules, based on size and occurrence frequency within our sample, were *Mu1*. Although it is possible that some circles seen in EM are of mitochondrial origin, the circles seen in mitochondria of male-fertile N-type maize have been reported to be 1.94 kb or larger (12, 13). Most of the transposon systems identified in *Drosophila* are retrotransposons, in contrast to maize. Therefore it is possible that these results reflect a difference in the abundance of different classes of transposons in the maize vs *Drosophila* genomes. In this context it is worth noting that the only known maize retrotransposon, *Bs1*, was identified after infection of maize plants with Barley Stripe Mosaic Virus (11). It is possible that there are many retrotransposon families in maize that are normally quiescent but could become activated by stresses, as suggested by the "genome shock" hypothesis of McClintock.

In conclusion, we have found that the elements of the *Mu* family of transposons in Robertson's *Mutator* lines of maize can exist as extrachromosomal closed circular DNA molecules. While the presence of these circles is dependent on *Mu* transposition activity, the exact role of these circles in the transposition process is as yet unknown. Regardless of their role in transposition, the existence of extrachromosomal *Mu* elements is of interest in that it demonstrates that transposable elements can be found as stable DNA molecules independent of the host DNA. This fact is well established for retrotransposons, and it is not surprising given their close relationship with retroviruses. The *E. coli* bacteriophage *Mu* can exist as a virus or as a transposon integrated in the host DNA. The observations with *Mu1* and *Tc1* extend the idea that the dividing line between viruses and transposable elements may not be sharply defined in general, and may have interesting evolutionary implications for the horizontal transmission of genetic information.

REFERENCES

1. Alleman, M., and M. Freeling (1986) *Genetics* 112:107-119.
2. Barker, R.F., D.V. Thompson, D.R. Talbot, J. Swanson, and J. Bennetzen (1984) *Nucl. Acids Res.* 12:5955-5967.
3. Bender, J., and N. Kleckner (1986) *Cell* 45:801-815.
4. Bennetzen, J.L. (1985) In *Plant Genetics: UCLA Symposia in Molecular Biology, New Series*, M. Freeling, ed. Liss, New York, Vol. 35, pp. 343-354.
5. Cappello, J., K. Handelsman, and H.F. Lodish (1985) *Cell* 43:105-115.
6. Chandler, V.L., and V. Walbot (1986) *Proc. Natl. Acad. Sci., USA* 83:1767-1771.

7. Chen, C.H., K.K. Oishi, B. Kloeckener-Gruissem, and M. Freeling (1987) Genetics 116:469-477.
8. Flavell, A.J., and C. Brierley (1986) Nucl. Acids Res. 14:3659-3669.
9. Flavell, A.J., and D. Isch-Horowicz (1981) Nature, London 292:591-595.
10. Freeling, M. (1984) An. Rev. Plant Physiol. 35:277-298.
11. Johns, M.A., J. Mottinger, and M. Freeling (1985) EMBO J. 4:1093-1102.
12. Kemble, R.J., and J.R. Bedbrook (1980) Nature London 284:565-566.
13. Levings, C.S., III, D.M. Shah, W.W.L. Hu, D.R. Pring, and D.H. Timothy (1979) In Extrachromosomal DNA, ICN-UCLA Symposium on Molecular and Cellular Biology, D.J. Cummings, P. Borst, I.G. David, and S.M. Weissman, eds. Academic Press, New York, Vol. 16, pp. 63-73.
14. Lillis, M., and M. Freeling (1986) Trends in Genet. 2:183-188.
15. Morisato, D., and N. Kleckner (1984) Cell 39:181-190.
16. Morisato, D., and N. Kleckner (1987) Cell (in press).
17. Mossie, K.G., M.W. Young, and H.E. Varmus (1985) J. Molec. Biol. 182:31-48.
18. Robertson, D.S. (1978) Mutat. Res. 51:21-28.
19. Robertson, D.S. (1980) Genetics 94:969-978.
20. Robertson, D.S. (1983) Molec. Gen. Genet. 191:86-90.
21. Robertson, D.S. (1986) Genetics 113:765-773.
22. Rose, A.M., and T.P. Snutch (1984) Nature, London 311:485-486.
23. Ruan, K.S., and S.W. Emmons (1984) Proc. Natl. Acad. Sci., USA 81:4018-4022.
24. Saedler, H., and P. Nevers (1985) EMBO J. 4:585-590.
25. Shapiro, J.A. (1979) Proc. Natl. Acad. Sci., USA 76:1933-1937.
26. Strommer, J.N., S. Hake, J. Bennetzen, W.C. Taylor, and M. Freeling (1982) Nature, London 300:542-544.
27. Sundaresan, V., and M. Freeling (1987) Proc. Natl. Acad. Sci., USA 84:4924-4928.
28. Taylor, L.P., V.L. Chandler, and V. Walbot (1986) Maydica 31:31-45.
29. Walbot, V., and C.A. Cullis (1985) An. Rev. Plant Physiol. 36:367-396.

MOLECULAR CHARACTERIZATION OF SUPPRESSOR-MUTATOR (Spm)-INDUCED MUTATIONS AT THE bronze-1 LOCUS IN MAIZE: THE bz-m13 ALLELES

John W. Schiefelbein, Victor Raboy,
Hwa-Yeong Kim, and Oliver E. Nelson

Laboratory of Genetics
University of Wisconsin
Madison, Wisconsin 53706

ABSTRACT

The bz-m13 allele of maize contains a defective Suppressor-mutator (dSpm) transposable element and gives rise to a variety of stable and unstable derivatives in the presence of an autonomous Suppressor-mutator (Spm) element. The dSpm-13 element of bz-m13 consists of 2,241 base pairs (bp) and is located within the second exon of the bronze-1 (bz) gene. A number of the stable derivatives, both functional and nonfunctional, derived from bz-m13 were characterized molecularly. Results from genomic DNA blotting experiments indicate that the dSpm-13 element had excised from the locus in each stable derivative analyzed. The unstable derivatives bz-m13CS9 and bz-m13CS6 contain dSpm elements in the same position and orientation as the dSpm-13 element, but they differ in the length of the element. The dSpm-13CS9 element is 902 bp and arose via a deletion between two 5-bp direct repeats within the dSpm-13 element. The dSpm-13CS6 element is 2,239 bp and only differs from dSpm-13 by a 2-bp deletion at the end of one of the 13-bp terminal inverted repeats. The effect of these deletions on the frequency and timing of Spm-induced excision is discussed herein. In the absence of Spm, each of the bz-m13 alleles conditions a nonmutant phenotype despite the presence of the insertions in the second exon. The role of RNA splicing in this phenomenon and the recent finding of an acceptor splice site within the terminal inverted repeat are also discussed.

INTRODUCTION

One of the most intensively studied transposable element systems in maize is the Suppressor-mutator (Spm, also known as Enhancer-Inhibitor, En-I) system. Many of the basic genetic aspects of the system were elucidated by McClintock and Peterson, respectively, in the 1950s (17,18,20,

21,26,27,28). More recently, studies at the molecular level have become feasible and, as a result, a more complete understanding of the transposable element system exists (1,3,9,12,14,23-25,32-37,39).

The Spm system is made up of two basic components. One is the autonomous or transposition-competent member, Spm (or En), so named because of its two recognizable trans-acting functions, suppression and mutation. The other component of the system is a heterogeneous group of nonautonomous or defective Spm (dSpm) elements that are only able to transpose or excise in response to an autonomous Spm.

Our laboratory has been interested in studying the elements of the Spm family at the bronze-1 (bz) locus of maize. The bz locus is useful because it affects pigment deposition in the maize plant and seed and, therefore, transposable element mutations can be easily identified. Additionally, a molecular clone of the locus has been isolated (8), and the gene product, UDPglucose:flavonoid glucosyltransferase (UFGT), has been identified and characterized (5,16). To date, we have focused on one dSpm insertion mutant at the bz locus and its Spm-induced derivatives. This mutant, bz-m13, and several of its derivatives have been described in previous publications (1,9,14,15,22,32,33). In this report, we review the previous studies and present new molecular data concerning the bz-m13 alleles.

MATERIALS AND METHODS

Plant Materials

The isolation of the various bz-m13 alleles has been reported (22,32). A brief description of the relevant genotypes follows:

Bz-McC2. The wild-type allele contained in McClintock's c-m5(Spm) Sh Bz wx-m8(dSpm) line which gave rise to the bx-m13 allele.

bz-m13. The initial dSpm insertion mutation harboring a 2.2-kb element within the second exon of the bz gene. In this chapter, bz-m13 refers to those sublines displaying a high germinal mutation rate, such as bz-m13Ref4 or bz-m13Ref5 (22).

bz-m13CS1, bz-m13CS3, bz-m13CS5, etc. Unstable derivative alleles (states) that differ from bz-m13 in the frequency and timing of somatic reversion in the presence of Spm.

Bz'(m13)-1, -2, -3, etc. Stable, colored derivatives (revertants) that do not respond to Spm.

bz'(m13)-1, -2, -3, etc. Stable, bronze derivatives (recessives) that do not respond to Spm.

Genomic DNA Cloning

The isolation of genomic clones from the bz-m13, Bz-McC2, Bz'(m13)-3, and bz-m13CS9 alleles has been described (9,14,32). A genomic DNA clone from the bz-m13CS6 allele was isolated by digesting DNA with BglII and cloning into the BamHI site of the λEMBL3 vector (10).

DNA Sequencing

The Sanger dideoxy method (31) was used to sequence both M13mp (44) clones and pUC plasmid (41) subclones. Deletion subclones of the plasmids were created according to the method of Henikoff (13).

DNA and RNA Blot Hybridizations

DNA and RNA isolation and blotting, nick translations, hybridizations, and other routine procedures were performed as previously described (14,32).

RESULTS AND DISCUSSION

Structure and Sequence of the bz-m13 Allele

Isolation of bz-m13. Spm-controlled bz mutant alleles were isolated by making use of a line carrying the wx-m8 allele (dSpm at the wx locus) and the c-m5 allele (Spm at the cl locus) (22). In this line, the bz locus is flanked by members of the Spm family on the same chromosome. One of the mutants isolated from this study was an unstable allele, bz-m13, that harbors a dSpm element at the locus. In the absence of an autonomous Spm element in the genome, bz-m13 conditions full anthocyanin pigmentation, indistinguishable from nonmutant alleles (Fig. 1). In the presence of Spm, bz-m13 displays large pigmented sectors of tissue on a bronze (recessive) background (Fig. 1). This phenotypic response of bz-m13 to Spm demonstrates the two trans-acting functions of Spm, suppression and mutation, as originally described by McClintock (17,18). The expression of bz-m13 that occurs in the absence of Spm is eliminated by the suppressor function while the mutator or excision function induces early reversion events during development.

Sequence of the dSpm-13 element. From a genomic clone of the bz-m13 allele (9), the DNA sequence of the dSpm element at the locus (dSpm-13) has been determined. It contains 2,241 bp and is identical to the reported sequence of the dSpm (Spm-I8) element present at the wx-m8 allele (12), as modified by the elimination of a thymine residue at position 821 (25). An interesting possibility suggested by the DNA sequences is that the element present at the wx-m8 allele may have transposed into the Bz-McC2 allele to create the bz-m13 allele.

Based on the sequence data, the dSpm-13 element can be interpreted as a deletion derivative of the autonomous Spm (En) element. Using the orientation and coordinates given in the published sequence of En-1 (25), the sequence from position 861 to 6906 (inclusive) of En-1 is not represented in the dSpm-13 element (Fig. 2). However, the highly structured sequences located within 200-300 bp of each end of En-1 are present in dSpm-13. We will refer to these sequences as the subterminal repetitive regions. These regions consist of a complex array of short repetitive sequence elements (25,34). Their importance in determining the frequency and timing of excision will be discussed in a later section. The dSpm-13 element also has two features common to all members of the Spm family: 13-bp perfect terminal inverted repeats and the production of a 3-bp duplication of host DNA at the insertion site (Fig. 3).

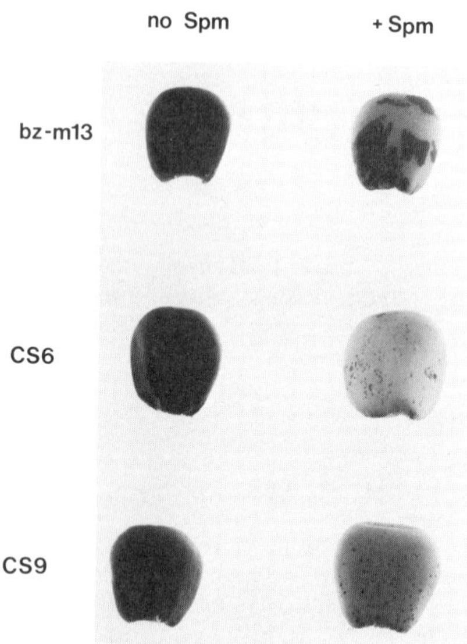

Fig. 1. Aleurone pigmentation in typical kernels from bz-m13 and derivatives in the presence and absence of a standard, active Spm. Each indicated allele is present in one copy, e.g., bz-m13 means bz/bz/bz-m13.

The dSpm-13 element is located in the second exon of the bz gene, 38 nucleotides away from the single intron. The element is oriented such that the promoter for the gene 1 (major) En transcript (25) would direct transcription opposite to the direction of bz-promoted transcription (Fig. 2). This orientation is opposite that of the identical Spm-I8 element in the wx-m8 allele and, as discussed in a later section, leads to interesting differences in host gene expression in the absence of Spm.

Stable Derivatives of bz-m13

One of the exceptional attributes of the bz-m13 allele is its high rate of germinal mutation in the presence of Spm. Typically, this rate is 70% or higher, meaning that out of every 100 gametes from a plant carrying bz-m13 and a standard, active Spm element, approximately 70 gametes will have undergone some sort of change at the bz locus and will no longer carry the original bz-m13 allele (22). This high rate of germinal mutation parallels the early and frequent somatic mutations that occur in bz-m13. This sort of correlation between germinal and somatic phenotypes has been recognized and reported by McClintock for other Spm-controlled mutations (18).

Functional derivatives. At a frequency reported to be approximately 25 to 30%, the bz-m13 allele "reverts" to a stable "Bz" form that conditions fully pigmented tissues in the presence or absence of Spm (22). Enzyme analyses of several of these Bz'(m13) derivatives (abbreviated Bz') demonstrated that although these alleles condition identical visual phenotypes, they vary in the level of UFGT activity produced (22). Alterations in the enzymatic properties of proteins, encoded by functional derivatives of transposable element mutations, have been observed by others (6,7,40,42).

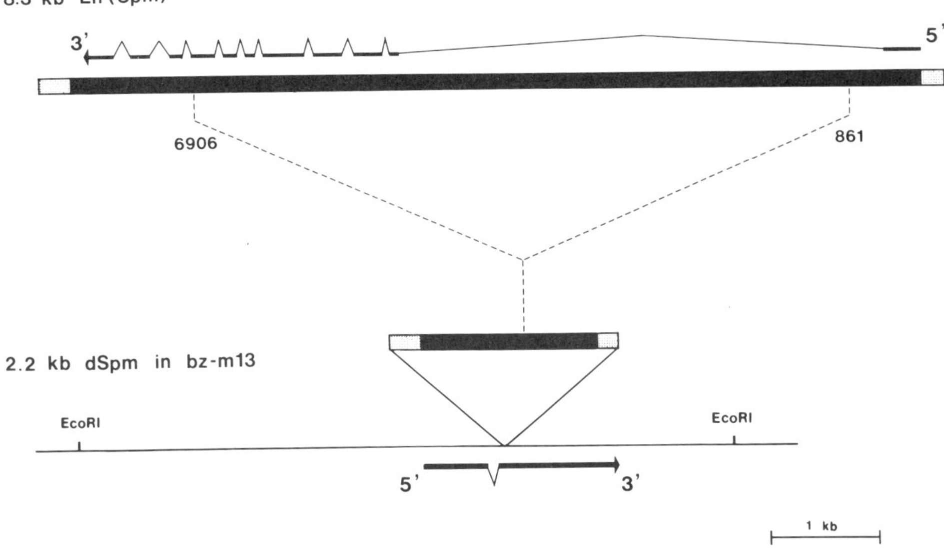

Fig. 2. Structural comparison of the autonomous En (Spm) element and the dSpm element in bz-m13. The portion of the En-1 element deleted in the dSpm-13 element is noted by the dashed lines; the coordinates are those from the En-1 element (25). The subterminal repetitive regions are indicated by the stippled portions of the bars. The structure of the gene 1 En transcript, as reported by Pereira et al. (25), is shown above the En (Spm) element. The structure of the Bz transcript shown below the Bz gene structure is from Furtek (11). Note that the direction of transcription from the Bz promoter is opposite to the direction from the En (Spm) promoter.

The Bz' derivative that conditioned the highest level of UFGT activity in the Nelson and Klein study (22), Bz'-3, has been cloned (32) and sequenced (33). This allele no longer contains the dSpm-13 element and also does not contain the 3-bp duplication created upon insertion of dSpm-13 (Fig. 3). Thus, the nucleotide sequence of Bz'-3 is identical to that of the Bz-McC2 progenitor allele and, consequently, has been used as a control in many of our studies.

DNA blot hybridization analysis has been carried out on eight Bz' derivatives, including the six derivatives in the Nelson and Klein study (22). The results indicate that, in each case, the dSpm-13 element has been excised from the locus and that, within the limits of resolution of this analysis (30 bp), the wild-type structure of the locus has been restored (Fig. 4A). This suggests that the alterations in UFGT activity, measured in the various Bz' derivatives (from 20 to 100% of wild-type) (22), are due to small changes in nucleotide sequence at the dSpm-13 insertion site following excision. These nucleotide changes are believed to occur in multiples of three (e.g., loss or gain of three or six nucleotides), so that the translational reading frame is not altered in the second exon. Presumably, a difference in the number or type of amino acids encoded by nucleotides at this site is responsible for the enzymatic alterations. An examination of

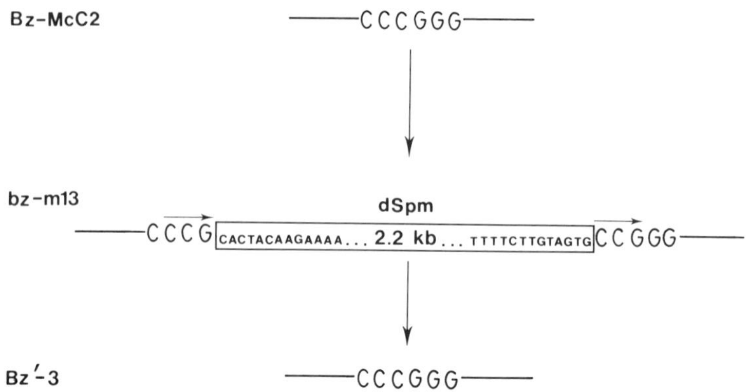

Fig. 3. DNA sequence of the bz-m13 progenitor, bz-m13, and a bz-m13 revertant at the dSpm-13 insertion site. The boxed region represents the dSpm-13 element, including the 13-bp terminal inverted repeats shown. The 3-bp sequence duplicated upon insertion of the dSpm-13 element is indicated by an arrow. Note that the insertion occurred into and eliminated an SmaI restriction enzyme site (CCCGGG).

the nucleotide sequences in these Bz' derivatives may reveal the importance of this site for UFGT activity and function. A study of this type has been reported by Wessler et al. (42) for revertants of the Ds mutant, wx-m1, that have altered enzyme activities.

Nonfunctional derivatives. The most frequent class of Spm-induced germinal mutations of bz-m13 are those which produce a recessive "bz" allele that does not respond to Spm. These occur at the rate of 45% to 50% (22) and are designated bz'(m13) alleles (abbreviated bz'). These have not been examined biochemically, but their visual phenotype indicates that they are unable to produce a functional UFGT. At the molecular level, DNA blot hybridization analyses show that the structure of the bz locus in the majority of the bz' derivatives is indistinguishable from the wild type at this level of analysis (Fig. 4B). These alleles are probably the result of excisions of the dSpm-13 element that alter the DNA sequence and disrupt the normal reading frame (i.e., gain or loss of 1, 2, 4, 5, etc., nucleotides). Somatic and germinal excision products, isolated and sequenced from the wx-m8 allele, demonstrate that these sorts of nucleotide changes do occur when a dSpm element is excised (35).

Of the 13 bz' derivatives examined, only one is detectably different from wild type on the basis of genomic DNA blotting. This derivative, bz'-10, can be interpreted as an excision of the dSpm-13 element and a deletion of an additional 40 to 90 nucleotides (Fig. 4B). An event of this type has been reported for a null derivative of the Ds mutant, Adh1-Fm335, in which the excision of the Ds element is accompanied by an adjacent deletion of 77 bp (4).

Unstable Derivatives of bz-m13

In the presence of Spm, the bz-m13 allele gives rise to new unstable derivatives. These are recognized by the altered pattern of somatic

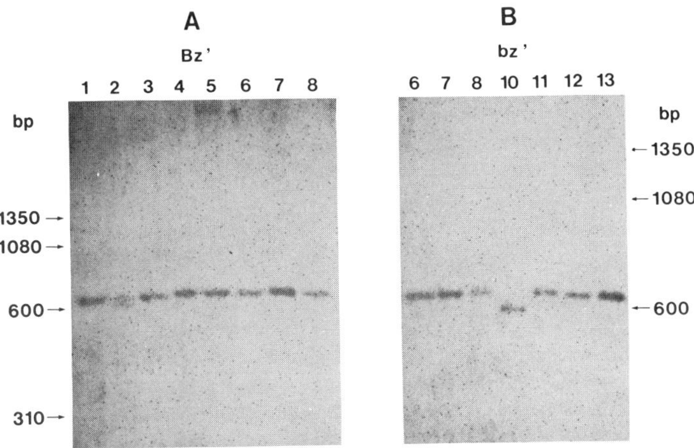

Fig. 4. Blot hybridization analysis of genomic maize DNAs digested with BstEII/SacI. DNAs were isolated from plants homozygous for the indicated alleles (e.g., the first lane of blot A contains DNA from the Bz'-1 allele). After digestion, DNAs were fractionated on agarose gels, denatured, and transferred to nitrocellulose filters. The filters were hybridized to ^{32}P-labeled pD3MS9, which contains a 630-bp MluI/SacI fragment of unique Bz sequence (32). This probe almost completely covers the 650-bp BstEII/SacI fragment produced by the Bz-McC2 allele and expected by any near-perfect excision of the dSpm-13 from the bz locus. Size standards from λ DNA included in each gel are shown.

reversion (pigmented spots or sectors) in the aleurone. McClintock characterized these types of derivatives from Spm-induced mutations at the al locus as "changes in state" of the original mutation (18). Thus, these kinds of bz-m13 derivatives are designated bz-m13CS1, bz-m13CS2, etc. (abbreviated as CS1, CS2, etc.). Two of these new derivatives that have been analyzed in detail, CS6 and CS9, will be emphasized in this report. Their phenotypes are illustrated in Fig. 1. The small colored sectors in the presence of Spm indicate that, in the derivatives, somatic reversion events occur later during endosperm development. Therefore, the dSpm elements in the derivatives are altered in their ability to respond to the mutator (excision) function of Spm.

Structure of dSpm elements in derivatives. The structure of the dSpm elements present in six independent unstable derivatives of bz-m13 has been reported (32), and the most recent results are presented in Fig. 5. This study and others (1,36,39) have demonstrated that the majority of these derivatives arise by deletions occurring within the original dSpm element at the locus. Furthermore, each of the identified deletions extends into one of the subterminal repetitive regions located within 200-300 bp of the end of the element. This suggests that the subterminal repetitive regions are important in determining the excision properties of dSpm elements, perhaps by forming a particular secondary structure or by providing a substrate for the Spm transposase or both. In the remainder of this report, we will focus on the CS6 and CS9 derivatives for which we have obtained DNA sequence data.

Somatic excision is impaired in the CS6 and CS9 derivatives. In order to further characterize the response of dSpm elements to Spm, an altered Spm (Spm-weak; Ref. 19 and 21), as well as an altered En (En-low; Ref. 29), was introduced into the bz-m13, CS6, and CS9 lines. Both Spm-weak and En-low are known to reduce the frequency and delay the timing of somatic reversion in dSpm elements as compared to the standard Spm or En elements (19,21,29). When bz-m13, CS6, and CS9 were present in combination with Spm-weak or En-low, a reduction in the degree of somatic reversion was observed in each case. The relative differences among the bz-m13 alleles, however, were maintained. This experiment indicates that the somatic reversion observed in each allele is due to events (e.g., excision and/or transposition) induced by the trans-

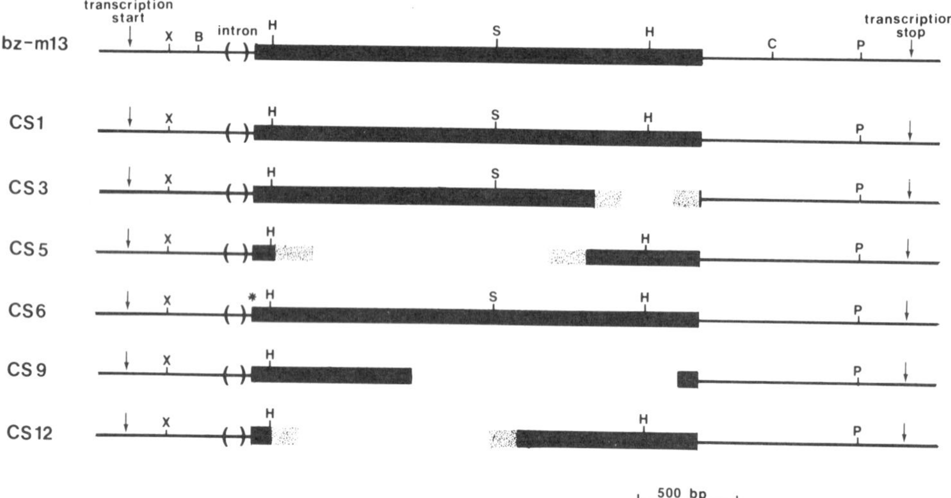

Fig. 5. Structure of bz-m13 and derivative alleles. These restriction endonuclease cleavage site maps indicate the regions of the 2.2-kb dSpm-13 element determined to be present (solid bars) and absent (open areas) in each of the derivatives, based upon direct DNA sequence analysis for the bz-m13, CS6, and CS9 alleles, and based upon the results of Schiefelbein et al. (32), for the CS1, CS3, CS5, and CS12 alleles. The lightly shaded bars in CS3, CS5, and CS12 represent regions of the dSpm elements where uncertainty exists over the exact deletion end points. The asterisk near the left border of the CS6 element indicates the position of the 2-bp terminal deletion in that allele. The transcription start and stop positions, as well as the location and size of the Bz intron, are taken from Furtek (11). The orientation of the restriction maps has been reversed from the previously published version (32), in light of the direction of Bz transcription. Restriction sites: B, BstEII; C, SacI; H, HincII, P, PstI; X, XmaI (SmaI).

acting factor(s) of Spm (En), rather than by inactivation of the autonomous element in some cell lineages.

A molecular characterization of excision, occurring in bz-m13 and these derivatives in the presence of Spm, has also been conducted. DNA was isolated from a bulked group of three-week-old plants carrying one of the bz-m13 alleles and a standard, active Spm. When digested with a restriction enzyme (EcoRI) cutting outside the bz locus and probed with a ^{32}P-labeled Bz fragment, DNA from plants carrying the bz-m13 allele and Spm produced an abundant amount of a hybridizing fragment of wild-type size (Fig. 6). This indicates a high level of somatic excision in these plants. On the other hand, DNA from plants carrying either the CS6 or CS9 alleles and Spm produced no detectable wild-type-sized fragment (Fig. 6). The absence of an excision band from the derivatives may be due to a very low level of excision or may reflect the later developmental timing of excision in the derivatives. This experiment demonstrates, once again, that the dSpm elements in the CS6 and CS9 alleles are impaired in their ability to respond to the trans-acting function(s) of Spm involved in excision and/or transposition.

CS9 deletion occurs between 5-bp direct repeats. DNA sequence analysis of the CS9 dSpm element (designated dSpm-13CS9) shows that it consists of 902 bp and is inserted at the identical position within the bz locus as its parent element, dSpm-13 (Fig. 7). These results are consistent with those of the previous mapping studies (32). The 1,339-bp deletion of dSpm sequence extends approximately 80 bp into the subterminal repetitive region at the right end of the element (with respect to the bz gene). This eliminates the potential to form a portion of the proposed stem and loop structure (34,36). In Fig. 7, a stem represents the potential interactions that may occur between the sequences of the subterminal repetitive regions at each end during the excision or transposition process. Obviously, the deletion in CS9 will alter the interactions that are possible between the sequences of the subterminal repetitive region. This may be the cause of the altered response of CS9 to Spm. Figure 7 also indicates that the deletion event occurred between two 5-bp direct repeats. These particular repeats, are at En-1 positions 106-110 and 7491-7495, and at dSpm-13 positions 106-110 and 1444-1448. A portion of this 5-bp repeat is also a part of the consensus 13-bp repeat ACCGACACTCTTA (12) present many times in the subterminal repetitive region at each end of the element.

The event that created the dSpm-13 element from Spm apparently also involves a deletion between two direct repeats. In this case, 3-bp repeats were involved. These are located at En-1 positions 861-863 and 6907-6909 (25). A comparison of sequences located at the deletion breakpoints of both dSpm-13 and dSpm-13CS9 is given in Fig. 8. There is no homology between the repeats used in these deletion events, although a common 6-bp sequence is present at the 5'-deletion breakpoint in each case (underlined with dashes in Fig. 8). The sequences of two other deletion events involving an element the same as dSpm-13, the Spm-I6078 element from a1-m1, have been reported (36,39). No homology between the deletion breakpoints was uncovered in those cases, although one of these deletion events has the 6-bp common sequence at its 5'-deletion endpoint (Spm-I1112) (39). It appears that direct repeats may be involved or may even facilitate these types of deletions by Spm, but they are not required. Sequence characterization of additional deleted elements will be necessary to fully understand the deletion process.

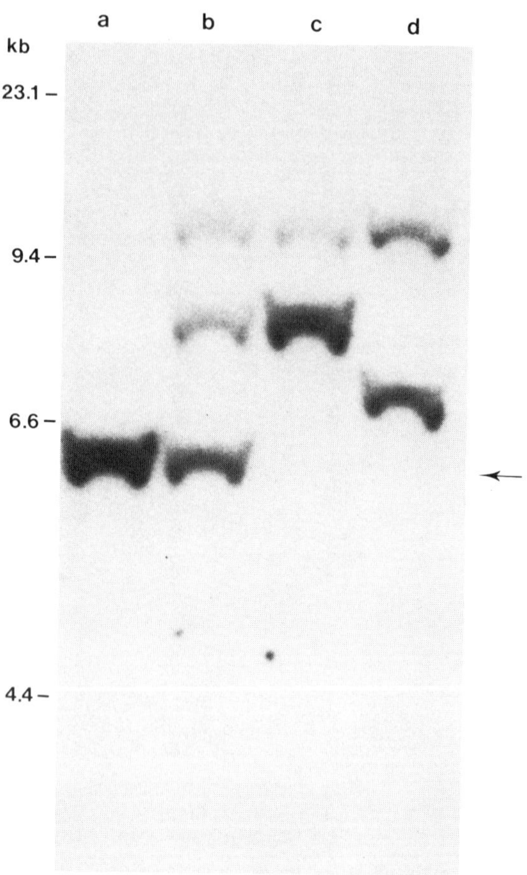

Fig. 6. Blot hybridization analysis of genomic maize DNAs digested with EcoRI. DNAs were isolated from a group of three-week-old plants of the following genotypes: (a) Bz'-3/Bz'-3; (b) bz-m13/bz, Spm present; (c) CS6/bz, Spm present; and (d) CS9/bz, Spm present. After digestion, DNAs were fractionated on agarose gels, denatured, and transferred to nitrocellulose filters. The filters were hybridized to a ^{32}P-labeled Bz-specific 850-bp Mbo I DNA fragment adjacent to the dSpm-13 insertion site. The wild-type EcoRI fragment size (i.e., expected excision band size) is marked by an arrow. The hybridizing fragment of approximately 10 kb present in lanes b, c, and d is due to the EcoRI fragment produced by the recessive bz allele. Size standards taken from λ DNA included in the gel are shown.

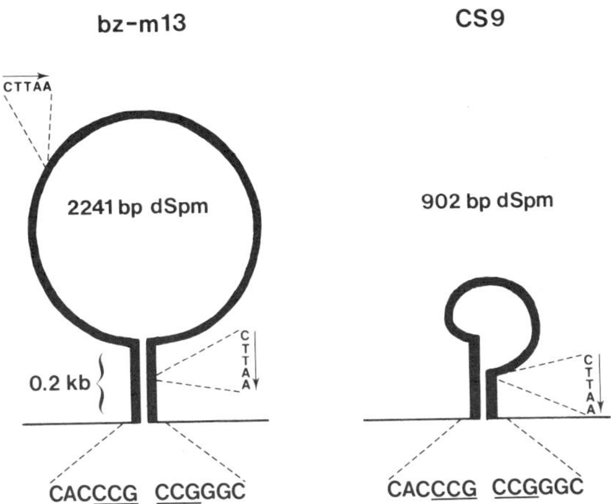

Fig. 7. Structural comparison of the dSpm elements in bz-m13 and CS9. The thick bars represent dSpm sequences and the thin lines represent bz locus sequences. The portions of the dSpm elements containing the subterminal repetitive sequences that are discussed in the text are shown together to indicate the possibility of interactions between the ends. The 5-bp repeats involved in the deletion event creating CS9 are indicated by arrows. The sequence given at the bottom of each figure is the bz sequence adjacent to each insert, with the 3-bp host duplication underlined. All sequences are given from 5' to 3' with respect to Bz transcription.

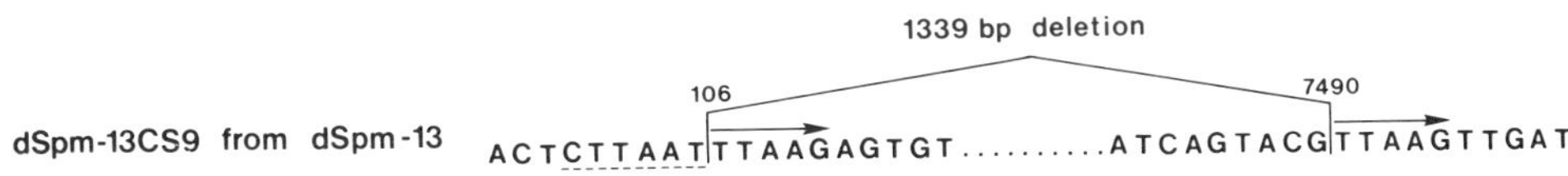

Fig. 8. DNA sequences from the deletion endpoints giving rise to the dSpm-13 and the dSpm-13CS9 elements. Arrows indicate the short direct repeats present at the ends of each deletion. Endpoint positions are given with respect to the numbering of the En-1 element (25). Note that because of the repeats at each end, the actual deletion point is uncertain. Also note that the sequence is given from 5' to 3' with respect to the En major transcript, so it is complementary to the sequence presented in Fig. 7. A common 6-bp sequence present at the 5'-deletion endpoint is underscored with a dashed line.

The dSpm-13CS6 element has a 2-bp deletion at the end of one terminal inverted repeat. Nearly all of the deletions affecting the 2.2-kb dSpm elements at the bz and al loci, thus far examined, range from 0.4 to 1.5 kb, with most in the 1.2 to 1.5-kb range (32,36,39). Also, these deletions all extend into the subterminal repetitive region but not the 13-bp terminal inverted repeats of the element. In these two respects, the deletion event giving rise to the dSpm-13CS6 element in the CS6 allele differs. DNA sequencing of the dSpm-13CS6 element shows that it is 2,239 bp in length. It is identical in its site of insertion and in its sequence to dSpm-13, except for a 2-bp deletion at the 3'-most end of the element (at the 5' end with respect to the bz gene; Fig. 5 and 9). Since the terminal inverted repeats are considered to be important or even essential for transposition in most mobile genetic elements (38), it is not surprising that this small deletion has been able to dramatically affect the excision of this element by Spm.

Although the dSpm element in the CS6 allele has a 2-bp deletion at the end of a terminal inverted repeat, the change may instead be considered as a single base substitution mutation, since the two adjacent nucleotides from the bz gene may now be considered as a part of the inverted

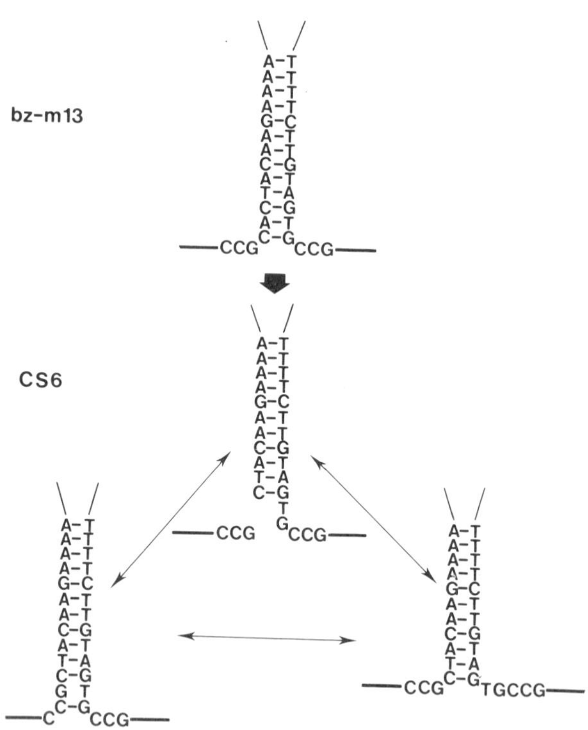

Fig. 9. Nucleotide sequence of the terminal inverted repeats and the target site duplication present in the bz-m13 and CS6 alleles. The terminal inverted repeats are shown in a paired configuration that may or may not reflect the situation during excision and/or transposition. Possible alternative pairing configurations for the dSpm-13CS6 element are given in the lower left and lower right portions of the figure.

repeat (Fig. 9). Therefore, the altered excision properties of this mutant may be considered to be caused by an A to G transition in the second nucleotide of the inverted repeat. However, this sort of configuration alters the 3-bp duplication. Thus, if the 3-bp duplication is important for proper and efficient excision of dSpm elements, then this change may contribute to the difference in the somatic reversion observed. Finally, with the alteration at the end of the dSpm in CS6, its excision products would be of interest to characterize. The types and frequencies of various excision products may be different from those produced by the original dSpm-13 simply because of the altered nucleotide sequence at the end. Perhaps this type of study would provide a new insight into the mechanism of Spm-induced excision/transposition (30).

Effects of dSpm Elements on bz Gene Expression

One of the major goals of our laboratory over the past several years has been to understand the ways that transposable elements are able to alter maize gene expression. In the bz-m13 alleles, gene expression is still permitted in the absence of Spm as demonstrated by the observation of fully pigmented kernels (Fig. 1). Since the dSpm insertions in each of these are located in an exon, the obvious question concerned the ability of the gene to be expressed with a foreign DNA segment of 0.9 to 2.2 kb within an exon. We have examined this question intensively with one of these alleles, the CS9 allele, and have recently reported the results of this study (14). In the remainder of this chapter, we will review the main findings of this investigation.

RNA processing uses an acceptor splice site in the terminal inverted repeat to remove the dSpm sequence from CS9 transcripts. In the absence of an autonomous Spm element, a high level of gene expression is detected in CS9, with approximately 40% to 50% of wild-type levels of UFGT activity measured in crude extracts (14). RNA blot hybridization experiments, carried out with poly$(A)^+$ RNA prepared from husks of CS9 plants, reveal two RNA species that hybridize to Bz probes (Fig. 10). The larger molecular weight species corresponds to a transcript including both Bz and dSpm sequences. The smaller molecular weight species is approximately equivalent in size to the nonmutant Bz transcript and does not hybridize to a probe consisting of only dSpm sequences (Fig. 10). These results suggest that transcription proceeds completely through the dSpm and ends at the usual Bz termination signals. The dSpm portion of the primary transcript is then removed by RNA processing to create a nearly wild-type-sized RNA species.

In order to characterize the wild-type-sized RNA, cDNA was prepared from CS9 RNA. Sequence analysis of a cDNA clone corresponding to a wild-type-sized RNA revealed an aberrant RNA splicing event. RNA sequences from the 5' side of the normal Bz intron had been spliced to RNA from the 3'-most inverted repeat of the dSpm (Fig. 11). The 3' end of the RNA removed from the pre-mRNA ends with an AG dinucleotide and conforms to the plant consensus acceptor splice site sequence (2). Thus, the dSpm and, therefore, Spm (En) contain within their terminal inverted repeats a sequence that can act as an acceptor splice site (14). This mechanism may be one explanation for the observation of gene expression in the absence of Spm for many dSpm mutations (17,21). Apparently, splicing dSpm sequences from the primary transcript is possible when elements are in this particular orientation; whereas, the reported premature polyadenylation occurring in the wx-m8 allele (12) may be the rule when in

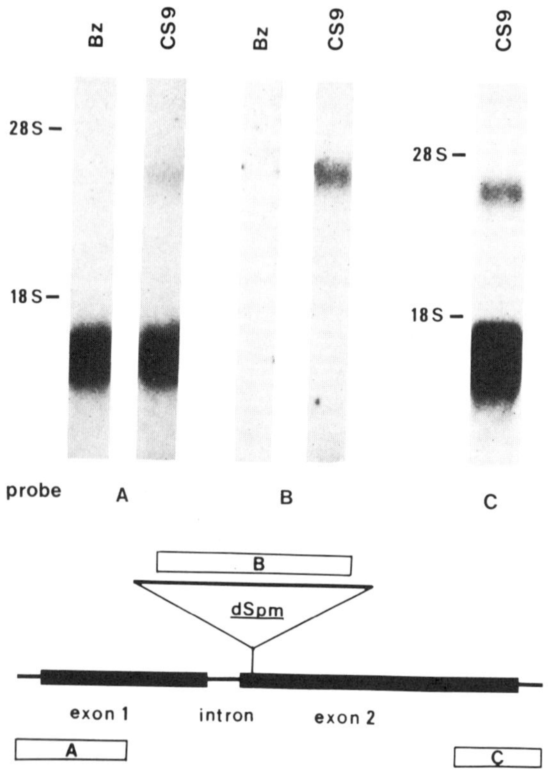

Fig. 10. Northern blot analyses of poly(A)$^+$ RNAs with Bz and dSpm probes. Poly(A)$^+$ RNAs were isolated from the husk tissue of plants homozygous for the indicated alleles at the bz locus in the absence of Spm; Bz refers to Bz'-3. Poly(A)$^+$ RNAs were fractionated in 1.0% formaldehyde-agarose gels, transferred to nitrocellulose filters, and probed with the indicated ^{32}P-labeled DNA probes. The diagram below the autoradiogram describes the DNA structure of CS9. Probes are indicated by open boxes; probe A contains a Bz sequence 5' to the dSpm, probe B contains most of the dSpm, and probe C contains a 3' Bz sequence. Blots probed with A and B were made by fractionating the poly(A)$^+$ RNAs of Bz'-3 and CS9 in duplicate lanes in the same gel, transferring to a nitrocellulose filter, and dividing the filter into two parts, each containing the RNAs from the two genotypes.

the opposite orientation. Recently, Wessler et al. (43) have reported that Ds transposable elements can be spliced from RNA utilizing multiple 5'-donor sites near the termini of these elements. Therefore, RNA splicing may be a general mechanism by which transposable element sequences can be eliminated from maize gene transcripts.

Aberrant splice maintains the proper reading frame and produces an altered polypeptide. The splicing event revealed by the cDNA analysis of CS9 would maintain the proper translational reading frame, since the 38

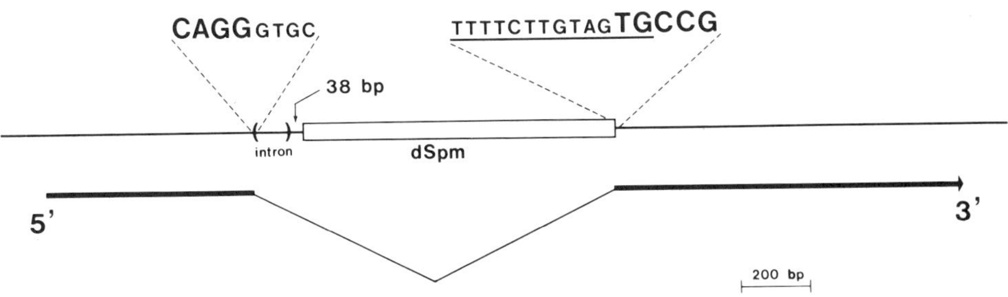

Fig. 11. Splicing of the CS9 transcript based on cDNA analysis. The structure of the spliced CS9 transcript is indicated by the thick bar and the dSpm element is represented by an open box. Nucleotides remaining in the spliced transcript are represented by large capital letters. Nucleotides removed from the primary transcript are represented by small capital letters. The 13-bp terminal inverted repeat is underlined. Note that 2 bp from the right end of the dSpm remain in the spliced transcript.

nucleotides between the intron and the dSpm are removed and two nucleotides from the dSpm element and three nucleotides of host duplication are added (i.e., net loss of 33 nucleotides from the normal Bz mRNA; Fig. 11). However, this event would result in a loss of 11 amino acids from the predicted polypeptide and might be expected to affect the gene product. In fact, UFGT extracted from CS9 is less thermally stable than UFGT extracted from a nonmutant Bz allele (14). Therefore, the CS9 allele essentially contains a new intron, consisting, in part, of transposable element sequences, that leads to the production of an altered polypeptide.

SUMMARY

The exceptionally high rate of mutation induced by the Spm element upon the bz-m13 allele has produced an array of stable and unstable alleles at the bz locus. The genetic, biochemical, and molecular characterization of these mutants has provided a new insight into the Spm family of transposable elements and their effects on gene structure and function. Further characterization of these bz-m13 derivatives, as well as the identification and analyses of new dSpm insertions in the bz gene, promises to uncover additional facets of transposable element action and plant gene expression.

ACKNOWLEDGEMENTS

We thank Douglas Furtek, Nina Fedoroff, and Barbara McClintock for helpful discussions of these results. This research was supported by the College of Agriculture and Life Sciences, University of Wisconsin, by National Science Foundation Grant DCB-8507895, and by National Institutes of Health Predoctoral Training Grant T32-GM07133. J.W.S. was supported by a predoctoral fellowship from the Agrigenetics Corporation. H.-Y.K. was supported by a fellowship from the Rockefeller Foundation. This is paper No. 2954 from the Laboratory of Genetics, University of Wisconsin-Madison.

REFERENCES

1. Banks, J., J. Kingsbury, V. Raboy, J.W. Schiefelbein, O. Nelson, Jr., and N. Fedoroff (1985) The Ac and Spm controlling element families in maize. Cold Spring Harbor Symp. Quant. Biol. 50:307-311.
2. Brown, J.W.S., G. Feix, and D. Frendewey (1986) Accurate in vitro splicing of two pre-mRNA plant introns in a HeLa cell nuclear extract. EMBO J. 5:2749-2758.
3. Cone, K.C., F.A. Burr, and B. Burr (1986) Molecular analysis of the maize anthocyanin regulatory locus C1. Proc. Natl. Acad. Sci., USA 83:9631-9635.
4. Dennis, E.S., W.L. Gerlach, W.J. Peacock, and D. Schwartz (1986) Excision of the Ds controlling element from the Adh1 gene of maize. Maydica 31:47-57.
5. Dooner, H.K., and O.E. Nelson (1976) Genetic control of UDPglucose: flavonol 3-O-glucosyltransferase in the endosperm of maize. Biochem. Genet. 15:509-519.
6. Dooner, H.K., and O.E. Nelson (1979) Heterogeneous flavonoid glucosyltransferases in purple derivatives from a controlling element-suppressed bronze mutant in maize. Proc. Natl. Acad. Sci., USA 76:2369-2371.
7. Echt, C.S., and D. Schwartz (1981) Evidence for the inclusion of controlling elements within the structural gene at the waxy locus in maize. Genetics 99:275-284.
8. Fedoroff, N.V., D.B. Furtek, and O.E. Nelson (1984) Cloning of the bronze locus in maize by a simple and generalizable procedure using the transposable controlling element Ac. Proc. Natl. Acad. Sci., USA 81:3825-3829.
9. Fedoroff, N., M. Shure, S. Kelly, M. Johns, D. Furtek, J. Schiefelbein, and O. Nelson, Jr. (1984) Isolation of Spm controlling elements from maize. Cold Spring Harbor Symp. Quant. Biol. 49:339-345.
10. Frischauf, A., H. Lehrach, A. Poustka, and N. Murray (1983) Lambda replacement vectors carrying polylinker sequences. J. Mol. Biol. 170:827-842.
11. Furtek, D. (1986), Cloning and sequence analysis of the bronze locus of maize: Location of transposable element insertions and comparison of two wild type alleles. Ph.D. Thesis, University of Wisconsin, Madison.
12. Gierl, A., Z. Schwarz-Sommer, and H. Saedler (1985) Molecular interactions between the components of the En-I transposable element system of Zea mays. EMBO J. 4:579-583.
13. Henikoff, S. (1984) Unidirectional digestion with exonuclease III creates targeted breakpoints for DNA sequencing. Gene 28:351-359.
14. Kim, H.-Y., J.W. Schiefelbein, V. Raboy, D.B. Furtek, and O.E. Nelson (1987) RNA splicing permits expression of a maize gene with a defective Suppressor-mutator transposable element insertion in an exon. Proc. Natl. Acad. Sci., USA 84:5863-5867.
15. Klein, A.S., and O.E. Nelson, Jr. (1983) Biochemical consequences of the insertion of a Suppressor-mutator (Spm) receptor at the bronze-1 locus in maize. Proc. Natl. Acad. Sci., USA 80:7591-7595.
16. Larson, R.L., and E.H. Coe, Jr. (1977) Gene-dependent flavonoid glucosyl transferase in maize. Biochem. Genet. 15:153-156.
17. McClintock, B. (1954) Mutations in maize and chromosomal aberrations in Neurospora. Carnegie Inst. of Washington Yearbook 53:254-260.

18. McClintock, B. (1955) Controlled mutation in maize. Carnegie Inst. of Washington Yearbook 54:245-255.
19. McClintock, B. (1956) Mutation in maize. Carnegie Inst. of Washington Yearbook 55:323-332.
20. McClintock, B. (1961) Further studies of the Suppressor-mutator system of control of gene action in maize. Carnegie Inst. of Washington Yearbook 60:469-476.
21. McClintock, B. (1965) The control of gene action in maize. Brookhaven Symp. Biol. 18:162-184.
22. Nelson, O.E., Jr., and A.S. Klein (1984) Characterization of an Spm-controlled bronze-mutable allele in maize. Genetics 106:769-779.
23. Paz-Ares, J., U. Wienand, P.A. Peterson, and H. Saedler (1986) Molecular cloning of the c locus of Zea mays: A locus regulating the anthocyanin pathway. EMBO J. 5:829-833.
24. Pereira, A., Z. Schwarz-Sommer, A. Gierl, I. Bertram, P.A. Peterson, and H. Saedler (1985) Genetic and molecular analysis of the Enhancer (En) transposable element system of Zea mays. EMBO J. 4:17-23.
25. Pereira, A., H. Cuypers, A. Gierl, Z. Schwarz-Sommer, and H. Saedler (1986) Molecular analysis of the En/Spm transposable element system of Zea mays. EMBO J. 5:835-841.
26. Peterson, P. (1953) A mutable pale green locus in maize. Genetics 38:682-683.
27. Peterson, P. (1960) The pale green mutable system in maize. Genetics 45:115.
28. Peterson, P. (1965) A relationship between the Spm and En control systems in maize. Am. Nat. 99:391-398.
29. Peterson, P. (1966) Phase variation of regulatory elements in maize. Genetics 54:249-266.
30. Saedler, H., and P. Nevers (1985) Transposition in plants: A molecular model. EMBO J. 4:585-590.
31. Sanger, F., S. Nicklen, and A.R. Coulson (1977) DNA sequencing with chain-terminating inhibitors. Proc. Natl. Acad. Sci., USA 74:5463-5467.
32. Schiefelbein, J.W., V. Raboy, N.V. Fedoroff, and O.E. Nelson, Jr. (1985) Deletions within a defective Suppressor-mutator element in maize affect the frequency and developmental timing of its excision from the bronze locus. Proc. Natl. Acad. Sci., USA 82:4783-4787.
33. Schiefelbein, J.W., D.B. Furtek, V. Raboy, J.A. Banks, N.V. Fedoroff, and O.E. Nelson (1985) Exploiting transposable elements to study the expression of a maize gene. In Plant Genetics, M. Freeling, ed. Alan R. Liss, Inc., New York, pp. 445-459.
34. Schwarz-Sommer, Z., A. Gierl, R.B. Klösgen, U. Wienand, P.A. Peterson, and H. Saedler (1984) The Spm (En) transposable element controls the excision of a 2-kb DNA insert at the wx^{m-8} allele of Zea mays. EMBO J. 3:1021-1028.
35. Schwarz-Sommer, Z., A. Gierl, H. Cuypers, P.A. Peterson, and H. Saedler (1985) Plant transposable elements generate the DNA sequence diversity needed in evolution. EMBO J. 4:591-597.
36. Schwarz-Sommer, Z., A. Gierl, R. Berndtgen, and H. Saedler (1985) Sequence comparison of "states" of a1-m1 suggests a model of Spm (En) action. EMBO J. 4:2439-2443.
37. Schwarz-Sommer, Z., N. Shepherd, E. Tacke, A. Gierl, W. Rohde, L. Leclercq, M. Mattes, R. Berndtgen, P.A. Peterson, and H. Saedler (1987) Influence of transposable elements on the structure and function of the A1 gene of Zea mays. EMBO J. 6:287-294.

38. Shapiro, J.A., ed. (1983) Mobile Genetic Elements. Academic Press, New York.
39. Tacke, E., Z. Schwarz-Sommer, P.A. Peterson, and H. Saedler (1986) Molecular analysis of states of the A1 locus of Zea mays. Maydica 31:83-91.
40. Tuschall, D.M., and L.C. Hannah (1982) Altered maize endosperm ADP-glucose pyrophosphorylases from revertants of a Shrunken-2-dissociation allele. Genetics 100:105-111.
41. Vieira, J., and J. Messing (1982) The pUC plasmids, and M13mp7-derived system for insertion mutagenesis and sequencing with synthetic universal primers. Gene 19:259-268.
42. Wessler, S.R., G. Baran, M. Varagona, and S.L. Dellaporta (1986) Excision of Ds produces waxy proteins with a range of enzymatic activities. EMBO J. 5:2427-2432.
43. Wessler, S.R., G. Baran, and M. Varagona (1987) The maize transposable element Ds is spliced from RNA. Science 237:916-918.
44. Yanisch-Perron, C., J. Vieira, and J. Messing (1985) Improved M13 phage cloning vectors and host strains: Nucleotide sequences of the M13mp18 and pUC19 vectors. Gene 33:103-119.

MUTAGENESIS USING ROBERTSON'S MUTATOR LINES AND CONSEQUENT INSERTIONS AT THE Adh1 GENE IN MAIZE

Michael Freeling

Department of Genetics
University of California
Berkeley, California 94720

PURPOSE

The Mu transposons in maize and their unique behaviors have been reviewed recently (14), as have results on the structure, function, and regulation of the alcohol dehydrogenase-1 gene (Adh1) in maize (11). My purpose is to review concisely our published and unpublished work on Mu1 and Mu3 insertional mutants at Adh1 and to discuss what I now take to be interesting or useful observations concerning mutagenesis using Mu.

ROBERTSON'S MUTATOR LINES

Donald Robertson, Iowa State University, found that a particular line of maize, when self-pollinated, generated Mendelian recessive mutants at a frequency about 50 times greater than expected spontaneously; many of these mutants were genetically unstable (18). Robertson further showed that, in his particular genetic backgrounds, the ability to generate mutants, the ultimate definition of "mutator activity," was transmitted to a vast majority (approximately 90%) of progeny through either the egg or the pollen. Subsequent work by Robertson showed that mutator activity occurred late in plant development, that known controlling element systems were not responsible, and that Robertson's lines turned "off" following repeated inbreeding (see review, Ref. 14).

We now know that a diverse group of transposons, called Mu transposons, are the active insertional agent causing the mutants in Robertson's Mutator lines. My laboratory's research on a genetically unstable Adh1 mutant (Adh1-S3034) discovered a 1.35-kb insertion within the first intron. The clone of this Adh1 allele led to the first probe for a Mu transposon (see review, Ref. 14). The sequence of this insertion (3), called Mu1, disclosed large (213-bp) inverted repeats flanking an internal domain carrying four small open reading frames (ORFs), two direct repeats, and a high A+T central sequence through which no ORF passed. The overall G+C content of Mu1 is 60%. No mRNA carrying the internal sequences of

Mu1 has been detected (unpubl. data), and Mu1, when added to tomato or tobacco within the borders of a T-DNA, did not transpose during any stage of the plant's life cycle (13).

There is no evidence that Mu1 encodes its own transposase. It is clear, however, that Mu1 elements do transpose and do respond to copy number control. Using probes to the internal sequences of Mu1, it is now clear that most inbred maize lines carry one or two copies of intact Mu1 in a methylated form (13), although some lines carry no Mu1 internal sequences at all. Robertson's Mutator lines carry ten to 50 elements (see review, Ref. 14). In an experiment where some care was taken to homogenize genetic backgrounds, Alleman and Freeling (2) showed that a standard Robertson's Mutator line maintained its Mu1-family (Mu1 + Mu1.7) copy number at 25 through repeated backcrosses to an 1s2p inbred line that carried zero or, perhaps, one copy. Selfed progeny of these lines generated progeny with 50 (double the expected) copies, implying that transposition occurred before fertilization. This implication is very important when designing mutagenesis schemes. The pollen from an active Mutator plant already carry new recessive mutations caused by new Mu transpositions.

There are now five different ways to identify an active Mutator plant. The first and most crucial is the enhanced ability to generate new mutants (18). Unfortunately, progeny tests on individual plants are often rendered ambiguous by small sample sizes. It is now clear from the work of Chandler and Walbot (7) that Mu-off plants that occur at varying frequencies, depending on genetic background, carry Mu1-family elements that are methylated in their terminal inverted repeats. Bennetzen (4) has shown that methylation sometimes occurs internally as well. Sundaresan and Freeling (21, and this Volume) showed that rare Mu-off plants concomitantly gained methyl groups at their terminal Hinf sites and lost an extrachromosomal, supercoiled, circular form of Mu1 family transposons. Thus, presence of these mysterious Mu circles is also a mark of Mutator activity. Several research groups have now shown a positive correlation between various genetically unstable marker genes and Mu activity (5,7,19,21,25). In general, when genetically unstable mutants become stable, Mutator activity by all criteria is "off," but exceptions exist.

Finally, active Mutator plants display "symptoms": slowness to mature, yellowness in the younger leaf, leaf blotching, and frequent yellow, clonal stripes. Any mutagenesis program will require procedures for distinguishing active lines from "off" lines. My laboratory's success in breeding very mutagenic lines was probably due to picking Mutator parents that had Mu symptoms and crossing them onto non-Mutator lines that permit high Mu copy numbers, such as our 1s2p line. It is also interesting that 1s2p with Mutator activity does not methylate Mu elements after five generations of inbreeding (Lillis, this laboratory, unpubl. data), while other lines, like B73 or Funk hybrid G4343, modify Mu elements routinely (unpubl. data). In maintaining active Mutator lines, it is important to cross to an inbred line that will not turn transposition "off."

I suspect that the "symptoms" seen in active Mutator lines, especially apparent in very inbred genetic backgrounds, has something to do with the presence of extrachromosomal Mu circles. I also suspect that Walbot's (25) finding of nonreciprocal male-female transmission of Mu activity, in her lines, involves an off-the-chromosome explanation. However, it is not yet known whether or not Mu circles are involved in the transmission of Mutator activity.

WHERE IS THE AUTONOMOUS Mu TRANSPOSON?

Every well-known transposable element system includes at least one class of transposons that carries its own transposase gene and often other genes affecting the transposition process. If Mu is similar (which is debatable), then other Mu transposons--defined by their inverted repeat sequences and their propensity for creating 9-bp genomic duplications upon insertion--might be the autonomous ones. One of our new Adh1 mutants turned out to carry a new Mu element that is slightly larger and of a different internal structure than Mu1 (9). This element, called Mu3, has been sequenced and compared with Mu1; a companion paper (this Volume) by Karen Oishi and myself summarizes these data. Mu3 and Mu1 share terminal, inverted repeat homologies, but are otherwise completely different. As with Mu1, we have no evidence that the ORFs of Mu3 function, and we do not yet know whether or not Mu3 can transpose in tobacco. We do know that Mu3 elements transpose in active Mutator lines and that Mu3 is methylated as a part of the same global switch that methylates Mu1-family elements.

About when we were at the height of our optimism concerning the possible autonomy of Mu3, we learned that Vicki Chandler and co-workers (pers. comm.; see their report in this Volume) had found a few more Mu elements that, like our Mu3, share the inverted repeats of Mu1 and nothing else. None of Chandler's new Mu transposons are Mu3 (Oishi and Chandler, pers. comm.). Thus, we are left with a vision of the Mu transposons as being composed of several families that can be distinguished using probes internal to the inverted repeats. Perhaps one of them will turn out to encode a transposase. Alternatively, perhaps the transposase is not encoded within a Mu transposon.

Perhaps no single Mu transposon encodes a transposase. One can envision an orgy of extrachromosomal circles somehow piecing together a master element. More realistically, one can imagine a transposon system that parasitizes transposase functions from enzymes expressed naturally during development; enzymes encoded in genes that are not flanked by Mu inverted repeats.

Knowing little is a fine stimulus to the imagination, and is often a sign that careful genetic experiments may be required. Toward that end, I have begun an experiment designed to find rare mutants in genes that modify the pattern of reversion of a Mu-induced bronze-1 mutant recovered by Robertson, bz1-mum9. Venkatesan Sundaresan, when in my laboratory, developed a somatic assay for Mu activity (21, and unpubl. data). Following his lead, I crossed a non-Mutator line which was homozygous for a deletion (of a few map units) of sh and part of bz on chromosome 9S with pollen from an active Mu line of the genotype: deletion/Sh bz1-mum9. This cross, the seed phenotypes, and some of the progeny phenotypes and genotypes are shown in Fig. 1. Half of the progeny kernels are plump (Sh/deletion) and display spots of anthocyanin on a bronze background (bz1-mum9/deletion). Of approximately 30,000 plump seeds examined, about 1% were plump, but stable, bronze; these, when tested for Hinf modification, were usually Mu-off. Several had very few or a great many

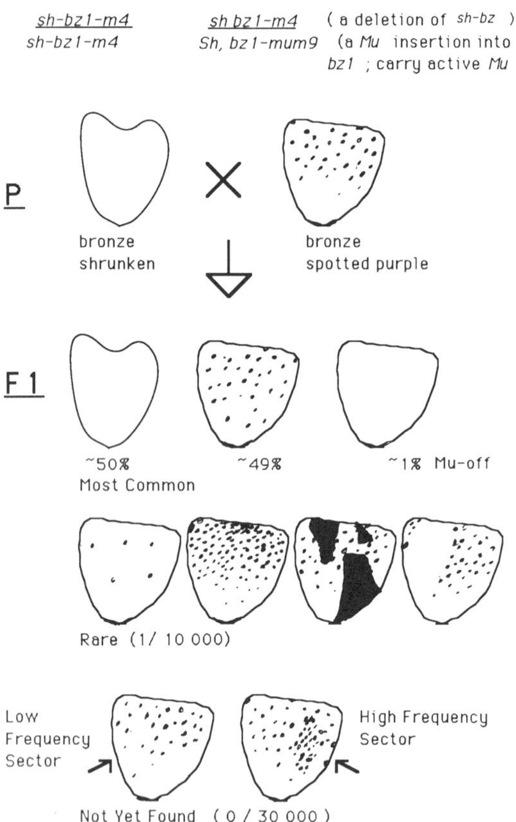

Fig. 1. A genetic strategy for finding mutants that modify the pattern of reversion of a Mu-induced reporter gene. See text for explanation. It is assumed that most of the colored spots are caused by an excision of a Mu transposon from the bz1-mum9 allele. Venkatesan Sundaresan (21 and unpubl. data) developed this reporter gene system. Note especially the absence of "frequency sectors," as diagrammed at the bottom of the figure and as discussed in the text.

spots, or multiple sectors of early reversion events; these have been planted in hopes that some will mark a single gene that is unlinked and acts in trans to the bz1 reporter allele.

What I believe to be the most significant result of these early observations is the phenotype that I did not find. I did not find a spotted seed with a sector of too few or too many spots even though ears were usually segregating kernels with very different spot (reversion) frequencies. The most reasonable hypothesis is that the frequency of reversion, and probably transposition, in the endosperm reflects the status of the endosperm or entire plant, and that status is not easily mutable by altering genes one at a time. Virginia Walbot (25), pursuing germinal transmission studies monitored at a mutable bz2 gene, put forth an explanation for reversion frequency variation and inheritance that involved quantitative thresholds rather than polymorphisms for single genes. My work is too preliminary to test any model, but I do think Walbot's idea is sound.

THE NATURE OF MUTANTS DERIVED FROM ROBERTSON'S Mu LINES

Robertson Mutator lines, when self-pollinated, generate mutants at an exceptionally high frequency. For example, during the summer of 1986, my laboratory (Barbara Lane and co-workers, this laboratory, unpubl. data) planted 20 kernels each of 1,200 self-pollinations. Even after excluding all runts, deformed seedling lethals, all seed traits, male steriles, and brachytics, over 30% of our families were segregating two or more plants with a clear recessive mutant phenotype. When corrections were made for counting the same mutation more than once, our conservative estimate is 10% new recessive mutants per family. Most were white, yellow, green lethals (high chlorophyll fluorescent) or yellow green; but tassel seeds, dwarves with anther ear, and les mutants were also observed. Mutants unlike any recorded previously--a distinctive branched ear phenotype and an attractive "tie-dyed" leaf color pattern phenotype--also appeared among these 1,200 families. Mutants were often genetically unstable.

What is the likelihood that these mutants are really caused by an insertion of a Mu element, and--more to the point--a Mu1-family element? This question is paramount if one hopes to clone these genes using a Mu1 probe. A rigorous answer to this question would require mutant analysis data from mutants induced from each Mutator line in question. Only five mutants have been characterized at the DNA level from my laboratory's lines; all are in Adh1. Four independent Adh1-low mutants turned out to be Mu1 insertions in the first intron, and one organ-specific mutant turned out to be a Mu3 insertion at the TATA box. Thus, for my particular mutator lines, four-fifths of the known mutants were caused by Mu1. The Walbot laboratory (23) found a bz2 mutant in their lines caused by a Mu1-family insertion, a Mu1.7 transposon. Robertson's mutants have been used in a variety of laboratories. An a1 mutant carried Mu1 (17), a sh1 mutant cloned by Ben Burr carried a Mu1 element (Ref. 21), and several bz1 mutants are all Mu1 insertions (Bennetzen, pers. comm.). This Volume also contains reports of other mutants recovered from Robertson's Mu lines; most are Mu1.

All in all, the case is strong that Mu1-family transposons are the major insertional mutagens of Robertson's, Walbot's, and my mutator lines. However, since we now know that there are several Mu transposons in addition to those of the Mu1 family, like Mu3, it is possible that some lines carry disproportionate numbers of particular Mu families. Mutagenesis experiments should probably begin with lines from proved sources, and one should probably check for high copy numbers of Mu transposons for which one has probes.

In our recent review (14), we suggested that new mutants to be used for transposon tagging should be crossed repeatedly to lines known to turn Mu "off." The point was to segregate away all Mu elements unlinked to the mutant gene, eventually leaving only a few or one Mu element genetically linked to the mutant phenotype. Of course, this strategy will not work if the mutant phenotype is dependent on an active Mu insertion.

Mu1 TRANSPOSONS IN THE FIRST INTRON OF Adh1

We now know how one of the Mu1 insertions in the first intron of Adh1 lowers ADH1 mRNA levels (20) in all organs of the corn plant. Run-off transcription studies from nuclei prepared from anaerobic roots showed that polymerase II complexes were present at the control (normal) level in the region of the Adh1 gene 5' to the Mu1 insertion point, but that the polymerase often stalled or dropped off 3' to the insertion (24). Furthermore, it was shown that Mu1 insertion did not alter or add any DNAse I hypersensitive sites around Adh1, nor did the nucleosome phasing pattern change. Apparently, Mu1 constitutes an impediment to polymerase II complex movement along the Adh1 gene.

Since all four Mu1 insertions into at least two positions in intron-1 cause the same 40% message phenotype in all organs, it is unlikely that the exact position of Mu1 insertion within intron-1 is important. The orientation of Mu1 within the intron may be important; the two mutants that have been tested have the same orientation (Adh1-S3034 and Adh1-4477; see review, Ref. 11). We hypothesized that the long inverted repeats of Mu1 might mediate stem-loop structures that would impede polymerase II physically. The methylation state of Mu1 within intron-1 has no detectable effect on the 40%-normal expression of the mutant Adh1 gene (Lillis, this laboratory, unpubl. data), a result that is consistent with the physical impediment model.

All four Mu1 insertions into Adh1 were selected as very rare, allyl alcohol-resistant pollen grains shed from active Mutator plants. Because the mutants were still 40% normal in ADH activity, we could not recover revertants via ability to survive anaerobiosis. However, when pollen from plants that were homozygous for Adh1-S3034, the best studied of these mutants, was reselected with allyl alcohol, over one hundred derivatives with less or no ADH1 activity were recovered (1,20). Those mutants that retained some ADH activity still have Mu1 present, as do many of the nulls. Many other nulls are clear chromosomal aberrations where Mu1 and much adjacent Adh1 DNA is removed or interchanged. One of the nulls that still carried Mu1 was sequenced; there was a small deletion of Adh1 that had as its 5' border the exact beginning of Mu1 (22). I suspect that Mu1 causes nested deletions upon excision.

Mu3 AT THE TATA BOX OF Adh1

After obtaining four of four Mu1-induced Adh1 mutants with the same phenotype and molecular explanation, each recovered via pollen selection, we tried a new mutant selection procedure. Kernels that are homozygous for lesions in any step specifically necessary for glycolysis will die quickly under anaerobic conditions. These mutants cannot start or complete the glycolytic cycle. Che-Hong Chen (8) found that, if the conditions are only partially anaerobic, these "conditional anaerobic lethals" fail to germinate, but remain alive so that development can proceed when air is added. Using this procedure, several conditional anaerobic lethals were selected (8). Among the first confirmed was a new Adh1 mutant; it was recovered from a Mutator line. The mutant immediately captured our interest because it was organ-specific; seed and seedling ADH1 levels were lowered to about 6% normal, but expression in pollen was normal. The mutant turned out to have a Mu3 transposon inserted at the TATA box such that the TATA box was included in the expected 9-bp genomic duplication (9).

Figure 2 shows the cartoon of this 5' Adh1 TATA Mu3 TATA Adh1 situation. Preliminary northern and promoter-usage data are in progress (Barbara Kloeckener-Gruissem, this laboratory); the residual ADH1 message in the seedling is anaerobically inducible, is of normal length, and starts at the expected nucleotide, but there is a trace of a larger ADH1 message that does contain Mu3. ADH1 message in pollen is normal in size and abundance. We have not yet distinguished between models that stress the position of the Mu3 element from those that would have Mu3 sequences acting as enhancers or silencers.

Kernels that are homozygous for this mutant have only 6%-normal ADH levels. This low level is not enough to permit germination under our hypoxic conditions, but revertants germinate readily. We now have many revertants. Table 1 lists some representatives of the four classes identified so far (B. Kloeckener-Gruissem, unpubl. data): (Class A) Complete restoration of ADH activity concomitant with Mu3 excision; (Class B) partial restoration of ADH activity concomitant with Mu3 excision; (Class C) partial restoration of ADH activity without loss of Mu3 sequences plus or minus 50 bp; (Class D) partial restoration of ADH activity in the seed and seedling, but concomitant loss of ADH activity in pollen and excision of Mu3.

In terms of gene regulation, the latter class (Class D; r17 and r19) of mutants are most interesting. First, Mu3 has excised, which simplifies interpretation. If the transposon leaves small deletions near its point of excision, then these deletions must remove pollen activity, while the loss of Mu3 might be sufficient to restore seed and seedling activity. The study of these and similar revertants should shed light on the structure of the Adh1 promoter with respect to organ-specific expression.

The revertants that retained Mu3 sequence, Class C (r16), were entirely unexpected on the basis of our previous work with Mu1 in intron-1.

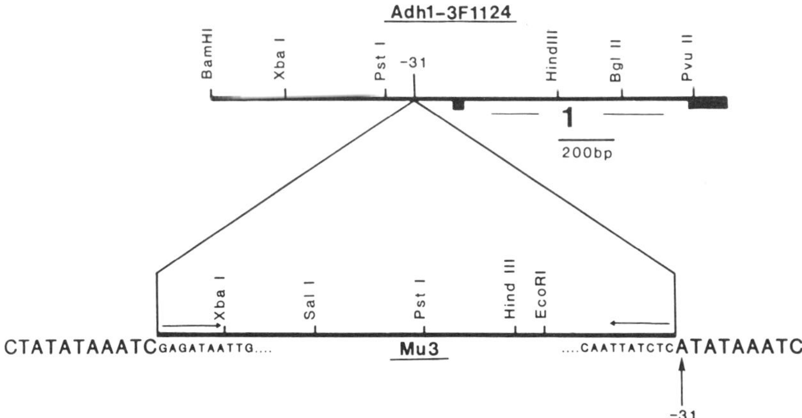

Fig. 2. The structure of an organ-specific Adh1 mutant 3F1124, where Mu3 has inserted such that the TATA box is duplicated. The first two exons of Adh1 are indicated by a bold line; intron-1 is denoted by a "1"; and "-31" indicates the nucleotide that is 31 bp 5' to the first transcribed nucleotide. Diagram is from published data (9).

Tab. 1. ADH1 activity profile for Adh1-3F1124 and its progenitor and revertant alleles.

Genotype	Scutellum	Percent ADH1 activity in: roots	aleurone	pollen
3F (WT)	100	100	normal	normal
3F1124 (mutant)	6.4	6	low	normal
r-16	27	27	normal	normal
r-17	62	68	normal	low
r-18	91	100	normal	normal
r-19	75	100	normal	low
r-21	98	96	normal	normal

On the assumption that Mu1 impedes polymerase II by virtue of its ability to form stem-loop structures during transcription, Mu3 might be expected to act similarly. But, if so, how could a revertant exist that could now read through a Mu element? Clearly, something specific within Mu3 or nearby has changed to permit expression in the seed.

These revertants are providing clues that help us to make more deliberate mutants in vitro for assay in maize cell expression systems.

Mu TRANSPOSONS AND EVOLUTIONARY CREATIVITY

One of the concepts that helps us understand the fundamentally indeterminate nature (10) of the evolutionary process is the ephemeral availability of selectively useful alleles. Goldschmidt (12) was perhaps the first to formulate a compelling argument for the evolutionary importance of what he called "macromutations "--novel regulatory changes in genes that, because of the genes' importance in development (perhaps involved in timing, polarity, or phytomere/segment identity), caused pleiotropic effects on the phenotype. There are now dozens of cases in the literature where transposons have altered or usurped the nonmutant cis-acting regulatory sequence of a gene; those cases involving Mu transposons at Adh1 in maize have already been reviewed. The idea that transposons might have an important role in generating regulatory diversity--in generating "macromutations"--is a notion that can now be found in a few evolution textbooks.

The "genome shock" hypothesis of McClintock (16 and references therein) comprises a framework within which evolutionary progress and transposon-induced macromutations are brought together. McClintock has stressed the body of research in maize indicating that chromosomal breaks, perhaps as a response to biological or environmental stress, can activate cryptic transposons. I think that the beauty of the "genome shock" hypothesis is that it predicts that transposon activation, and the resulting mutational possibilities, might be inducible in organisms that are in a do-or-die situation.

The data supporting the genome shock hypothesis originated in studies of the origin of the autonomous transposon of the Ac-Ds controlling element system in maize (15). However, the Mu transposons seem particularly well suited to be agents of genomic shock. From evidence presented and cited previously in this discussion, Mu transposons are usually present as methylated, intact elements in most inbred maize lines; Mu elements can be active or inactive depending on genetic background; Mu elements can make clear regulatory changes at the Adh1 gene and presumably other genes as well; Mu elements carry diverse internal sequences that might well carry promoters, enhancers, or silencers around the genome. (The relationship between Ds1 and the other Ac-type transposons is similar to this, but Mu elements have several or many different families while Ac has only two.) My laboratory's data at Adh1 and the very high mutation frequencies found in the progeny of self-pollinated Mutator plants suggest that Mu elements are targeted to transcriptional units if not to specific sequences associated with the regulatory components of genes.

All in all, Mu elements comprise a natural mutagenesis system that may persist for more than selfish reasons. Perhaps the survivors of natural and artificial selection in maize survived precisely because they called upon genetic diversity that was transposon-induced.

ACKNOWLEDGEMENTS

I thank all members of my laboratory, past and present, who have contributed to the data reviewed and the notions presented in this discussion. I especially thank Vicki Chandler for sharing her unpublished data with me. Assunta Chytry helped with the graphics and manuscript. Unpublished data are derived from research supported by grants from National Institutes of Health and National Science Foundation.

REFERENCES

1. Alleman, M. (1985) The Mu transposons of maize. Ph.D. dissertation, University of California, Berkeley.
2. Alleman, M., and M. Freeling (1986) The Mu transposable elements of maize: Evidence for transposition and copy number regulation during development. Genetics 112:107-119.
3. Barker, R.F., D.V. Thompson, D.R. Talbot, J. Swanson, and J. Bennetzen (1984) Nucleotide sequence of the maize transposable element Mu1. Nucleic Acids Res. 12:5955-5967.
4. Bennetzen, J.L. (1987) Covalent DNA modification and the regulation of mutator element. Molec. Gen. Genet. 208:454-451.
5. Bennetzen, J.L., R.P. Fracasso, D.W. Morris, D. Robertson, and M.J. Skogen-Hagenson (1986) Concomitant regulation of Mu1 transposition and mutator activity in maize. Molec. Gen. Genet. (in press).
6. Chandler, V.L., C. Rivin, and V. Walbot (1986) Stable, non-Mutator stocks of maize have sequences homologous to the Mu1 transposable element. Genetics 114:1007-1021.
7. Chandler, V., and V. Walbot (1986) DNA modification of a maize transposable element correlates with a loss of activity. Proc. Natl. Acad. Sci., USA 83:1767-1771.

8. Chen, C.-H., M. Freeling, and A. Merckelbach (1986) Enzymatic and morphological consequences of Ds excisions from maize Adh1. Maydica 31:93-108 (the McClintock commemorative issue).
9. Chen, C.-H., K.K. Oishi, B. Kloeckener-Gruissem, and M. Freeling (1987) Organ-specific expression of maize Adh1 is altered after a Mu transposon insertion. Genetics 116:469-477.
10. Dobzahnsky, T. (1966) Determination and indeterminism in biological evolution. In Philosophical Problems in Biology, V.E. Smith, ed. St. John's University Press.
11. Freeling, M., and D.C. Bennett (1985) Maize Adh1. An. Rev. Genet. 19:297-323.
12. Goldschmidt, R.B. (1952) Evolution as viewed by one geneticist. Amer. Sci. 40:84-135.
13. Lillis, M., A. Spielmann, and R.B. Simpson (1985) Transfer of a Robertson's mutator element to tomato and tobacco cells. In Plant Genetics, M. Freeling, ed. Alan R. Liss, New York, pp. 213-224.
14. Lillis, M., and M. Freeling (1986) The Mu transposons of maize. Trends in Genet. 2:183-188.
15. McClintock, B. (1950) The origin and behavior of mutable loci in maize. Proc. Natl. Acad. Sci., USA 36:344-355.
16. McClintock, B. (1984) The significance of responses to genome challenge. Science 226:792-801.
17. O'Reilly, C., N.S. Shepherd, A. Pereira, Zs. Schwarz-Sommer, I. Bertram, D. Robertson, P. Peterson, and H. Saedler (1985) Molecular cloning of the a1 Locus of Zea mays using the transposable elements En and Mu. EMBO J. 4:877-882.
18. Robertson, D. (1978) Characterization of a mutator system in maize. Mutat. Res. 51:21-28.
19. Robertson, D.S., P.S. Stinard, J.G. Wheeler, and D.W. Morris (1985) Genetic and molecular studies on germinal and somatic mutability in Mutator-induced aleurone mutants of maize. In Plant Genetics, M. Freeling, ed. Alan R. Liss, New York, pp. 317-331.
20. Strommer, J.N., S. Hake, J. Bennetzen, W.C. Taylor, and M. Freeling (1982) Regulatory mutants of the maize Adh1 gene induced by DNA insertions. Nature 300:542-544.
21. Sundaresan, V., and M. Freeling (1987) An extrachromosomal form of the Mu transposon in maize. Proc. Natl. Acad. Sci., USA 84:4924-4928.
22. Taylor, L.P., and V. Walbot (1985) A deletion adjacent to the maize transposable element Mu1 accompanies loss of Adh1 expression. EMBO J. 4:869-876.
23. Taylor, L.P., V. Chandler, and V. Walbot (1986) Insertion of 1.4KB and 1.7KB elements into the Bronze1 gene of Zea mays L. Maydica 31:31-45.
24. Vayda, M.E., and M. Freeling (1986) Insertion of the Mu1 transposable element into the first intron of maize Adh1 interferes with transcript elongation but does not disrupt chromatin structure. Plant Molec. Biol. 6:441-454.
25. Walbot, V. (1986) Inheritance of mutatory activity in Zea mays as assayed by somatic instability of bz2-mu1 allele. Genetics 114:1293-1312.

A NEW Mu ELEMENT FROM A ROBERTSON'S MUTATOR LINE

Karen K. Oishi[2] and Michael Freeling[1]

[1]Department of Genetics
University of California
Berkeley, California 94720

[2]Molecular and Cellular Biology Department
and Plant Sciences Department
University of Arizona
Tucson, Arizona 85721

Robertson (8) described a maize line that, when self-pollinated, generated a 50-fold higher frequency of recessive mutants than expected spontaneously. Approximately 40% of these mutations were unstable genetically, a characteristic of transposon-induced mutants. The first insertional mutagen from a Robertson's Mutator line, Mu1, was identified and cloned from an unstable mutant allele of Adh1 (3,9). Nine bp of genomic intron-1 DNA was duplicated upon insertion. A complete sequence analysis (2) of the Mu1 element determined that the 1,367-bp element contained 215-bp terminal repeats and 100-bp internal direct repeats. All Mutator lines were found to contain 10-70 copies of Mu1-homologous sequences (1,3). Zero-2 copies of Mu1 exist in non-Mutator lines (3,6). There are one larger and three smaller Mu1-related elements in Mutator lines (10). The largest-sized Mu1-family element described, Mu2 (also called Mu1.7), has a few hundred additional base pairs of internal DNA as compared to Mu1 (10).

A new Mu element, Mu3, was isolated from an organ-specific mutant of the Adh1 gene. Mu3 was inserted 31 bp 5' from the transcriptional start of Adh1 such that the TATA box was included in the expected 9-bp genomic duplication (6).

Complete sequence analysis indicates that the Mu3 element is different from Mu1 and Mu2 (Mu1.7). At 1,824 bp, Mu3 is the largest Mu element reported to date. The element contains a 200-bp terminal inverted repeat sequence and no direct repeat sequences. One of the terminal inverted repeats of Mu3 has 90% sequence homology to the 215-bp and 89% homology to the 213-bp terminal inverted repeats of Mu1. The other terminal inverted repeat of Mu3 displays less sequence homology: 83% to the 215-bp repeat of Mu1 and 80% to the 213-bp repeat of Mu1. The DNA between the terminal inverted repeats of Mu3 and Mu1 contains no sequence homology.

This total difference in sequence homology is reflected in the relative (G+C) content of Mu1, 60%, versus Mu3, 49%. Even though there is no internal sequence homology between Mu1 and Mu3, the relative distribution of the bases in the two sequences is very similar. There are two high (G+C) domains separated by a more central (A+T)-rich domain. Without knowing the mechanism of transposition for Mu elements, it is impossible to determine whether this shared structure has significance.

The internal sequence of Mu3 was found to contain three open reading frames (ORFs). As shown in Fig. 1, ORF 1 and ORF 2 are overlapping. Only one ORF, ORF 3, on the opposite strand, is bordered by a TATA box and a polyadenylation signal sequence. All three ORFs are hypothetical and have not yet been shown to encode a product.

The Mu3 element is represented in much lower copy number, 3-15, than Mu1-family elements (25-50) in a few Mutator lines examined. As is the case with Mu1, there is sequence conservation among the Mu3 elements within a genome. In contrast to the case with Mu1-family elements, all maize lines and a Tripsacum species were found to contain at least one copy of sequences homologous to the internal domain of Mu3. Whether these endogenous Mu3 sequences are intact Mu3 elements has yet to be determined.

In conclusion, the Mu3 element shares a few molecular characteristics with the Mu1-family elements, but shares sequence homology in the inverted terminal repeats only. This situation is similar to the relationship between Ds1 and the Ds elements that are clearly defective Acs; they only share their terminal inverted repeat sequences (7, and references

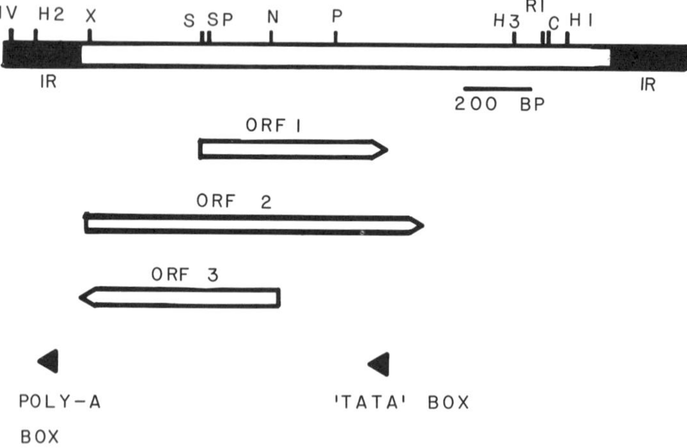

Fig. 1. Open reading frames (ORFs) of Mu3 as determined by sequence analysis. The black boxes represent the terminal repeats (TR) and the open boxes the internal region of Mu3. The arrowed bars show the direction and size of the three ORFs. The closed triangles show the location of the internal TATA box and polyadenylation addition signal. C, ClaI; H1, HaeI; H2, HaeII; H3, HindIII; N, NcoI; S, SalI; Sp, SphI; RI, EcoRI; X, XbaI; IV, EcoIV.

therein). In the case of Mu elements, the first 200 bp of the terminal inverted repeats may be necessary for transposition. As more sequence information becomes available for different Mu elements, perhaps we may be able to pinpoint the target sequence for the Mu transposase.

ACKNOWLEDGEMENTS

This research was supported by a grant from National Institutes of Health to M.F. K.O. was a National Science Foundation postdoctoral fellow.

REFERENCES

1. Alleman, M. (1985) The Mu transposons of maize. Ph.D. dissertation, University of California, Berkeley.
2. Barker, R.F., D.V. Thompson, D.R. Talbot, J. Swanson, and J. Bennetzen (1984) Nucleotide sequence of the maize transposable element Mu1. Nucleic Acids Res. 12:5955-5967.
3. Bennetzen, J.L. (1984) Transposable element Mu1 is found in multiple copies only in Robertson's mutator maize lines. J. Molec. Appl. Genet. 2:519-524.
4. Bennetzen, J.L., J. Swanson, W.C. Taylor, and M. Freeling (1984) DNA insertion in the first intron of maize Adh1 affects message levels: Cloning of projector and mutant Adh1 alleles. Proc. Natl. Acad. Sci., USA 18:4125-4128.
5. Chandler, V.L., C. Rivin, and V. Walbot (1986) Stable, non-Mutator stocks of maize have sequences homologous to the Mu1 transposable element. Genetics 114:1007-1021.
6. Chen, C.-H., K.K. Oishi, B. Kloeckener-Gruissem, and M. Freeling (1987) Organ specific expression of maize Adh1 is altered after a Mu transposon insertion. Genetics 116:469-477.
7. Merckelbach, A., H.-P. Döring, and P. Starlinger (1986) The aberrant Ds element in the Adh1-2F11::Ds2 allele. Maydica 31:109-122.
8. Robertson, D. (1978) Characterization of a mutator system in maize. Mutat. Res. 51:21-28.
9. Strommer, J.N., S. Hake, J. Bennetzen, W.C. Taylor, and M. Freeling (1982) Regulatory mutants of the maize Adh1 gene induced by DNA insertions. Nature 300:542-544.
10. Taylor, L.P., V. Chandler, and V. Walbot (1986) Insertion of 1.4KB and 1.7KB elements into the Bronze1 gene of Zea mays L. Maydica 31:31-45.

ALTERATIONS IN GENE EXPRESSION MEDIATED BY DNA INSERTIONS IN THE *waxy* GENE OF MAIZE

Susan R. Wessler, George Baran,* and
Marguerite Varagona

University of Georgia
Botany Department
Athens, Georgia 30602

INTRODUCTION

Studies on the *waxy* (*wx*) locus have played a central role in maize genetics and molecular biology. The locus encodes a starch granule-bound enzyme responsible for the synthesis of amylose in the endosperm, ovule sac, and pollen grain. The *waxy* mutants are viable and easily distinguished. For this reason, over 50 *wx* mutant alleles have been identified since the turn of the century.

All *wx* mutations map to the short arm of chromosome 9. Phenotypically, the mutant alleles fall into two classes, stable and unstable. To understand the relationship between these two classes of alleles, Nelson (16) constructed a fine-structure map of the locus. This analysis was accomplished by determining whether different combinations of *wx* heteroalleles exhibit recombination. The restoration of *Wx* expression via recombination was scored by staining the pollen grains of these heteroalleles with I/KI, and counting the rare black-staining recombinants that contain amylose.

In addition to mapping most of the stable alleles, Nelson was able to map the *Ds* alleles *wx-m1*, *wx-m6*, and *wxB4* and the *dSpm* allele *wx-m8*. These insertions cannot transpose in the absence of their cognate autonomous element. Nelson was also able to map *wxB3*, an allele harboring an active *Ac* element that does not give rise to germinal revertants. One important conclusion of this study was the finding that the unstable alleles did not cluster in one region of the locus, but were dispersed throughout the fine-structure map. Thus, they behaved like the stable mutations.

---

*Present address: Molecular Genetics Inc., 10320 Bren Road East, Minnetonka, Minnesota 55343.

Recently, the waxy gene products have been characterized (3,23) and the locus has been cloned (23) and sequenced (8). In addition, most of the unstable alleles have been cloned and their respective Ac, Ds, dSpm, or En insertions have been characterized (1,5,17,21,26,27). Finally, most of the stable alleles have been characterized at the level of Southern blots. Over half have been found to harbor either insertions or deletions within the Wx transcription unit (25).

PHENOTYPIC DIVERSITY OF THE UNSTABLE waxy ALLELES

Each of the Ac and Ds alleles of waxy can be distinguished by a unique pattern of expression in endosperm tissue (10-14). Our primary focus is to understand the molecular basis for this phenotypic diversity. An understanding of the subtle ways in which transposable elements alter gene expression helps to elucidate the function of the cryptic elements found in all maize genomes.

The waxy alleles are ideal for an analysis of the mechanisms underlying phenotypic diversity because: (a) unlike genes expressed only in aleurone tissues, the wx phenotype can be analyzed throughout endosperm development; (b) the lesions are simple insertions and relatively small (0.4 kb to 4.5 kb); and (c) the waxy gene products are so abundant [approximately 0.5% of endosperm poly(A)$^+$ RNA (Baran and Wessler, unpubl. data) and 0.5% of endosperm protein (23)] that mutant gene products are readily detectable, even when reduced 100-fold from wild-type levels.

The Influence of the Position of Insertion

The phenotypes characteristic of each Ac and Ds allele described in this paper are displayed diagrammatically in Fig. 1. One of the major factors influencing the phenotype is the position of the insertion within the gene. Each of the mutants described has a simple insertion in either exon sequences [wx-m1 (26), wxB4 (27), wxB3 (Baran and Wessler, unpubl. data), wx-m9 (4,8)] or in 5' flanking sequences [wx-m7 (15)]. The precise site of insertion is displayed in Fig. 2.

Müller-Neumann et al. (15) demonstrated the important contribution of the position of insertion on the mutant phenotype. They showed that the diverse phenotypes displayed by strains harboring the Ac alleles wx-m9 or wx-m7 resulted solely from the position of the element within the gene, since the DNA sequence of the Ac elements is identical. Not yet understood is why an Ac element inserted in 5' flanking sequences (wx-m7) gives rise to very early excision events and many germinal revertants. The Ac element inserted in exon sequences apparently excises later in endosperm development, and produces a lower frequency of germinal revertants.

Inactive derivatives of wx-m7(Ac) and wx-m9(Ac) have been isolated (14,20). In both cases, DNA modification of Ac has been associated with the inactive state (2,20). Inactivation of the Ac element harbored by the wx-m7 allele has an additional effect on Wx gene expression (Fig. 1). McClintock (14) noted that when the mutator component of Ac is inactive, there is a characteristic pattern of Wx expression during endosperm development. This nonclonal pattern is characterized by intermediate Wx expression in the periphery of the endosperm with little or no expression in the region that was the central cell of the embryo sac. Since the Wx

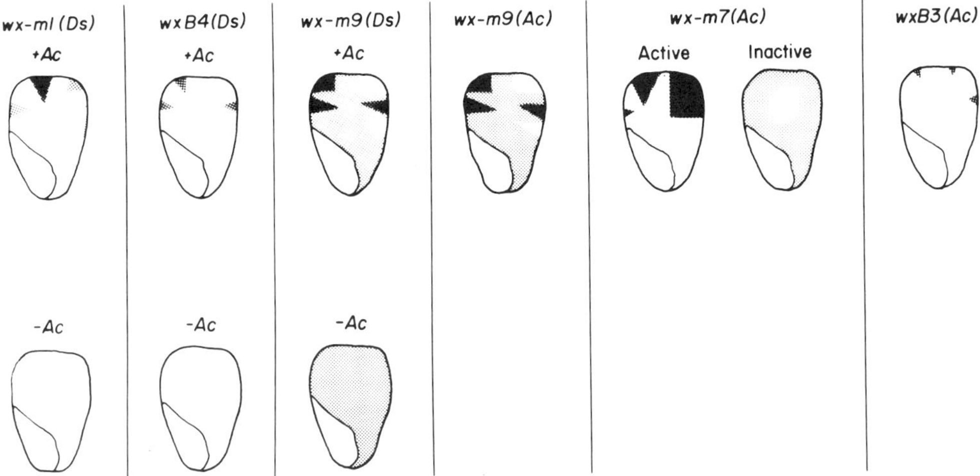

Fig. 1. The phenotypic diversity displayed by Wx alleles harboring Ac or Ds insertions. The representation of an average kernel is presented. Kernels have been symbolically cut longitudinally and stained with I/KI. The region stained in some of the kernels represents the endosperm; the embryo is unstained throughout. In the absence of Ac, the Ds alleles are somatically and germinally stable. In the presence of Ac, excision of Ds results in the distinctive clonal endosperm sectors shown for each allele. The wx-m9(Ds) allele was derived from the wx-m9(Ac) allele (12) and differs from the Ac allele by a small deletion within the transposable elements (4).

alleles are normally expressed uniformly in endosperm tissue, the wx-m7 inactive allele has apparently come under a new program of gene expression, one that senses temporal or positional information.

Although we do not understand the molecular mechanism underlying this pattern of Wx expression in the wx-m7 inactive allele, it is interesting to note that the Ac element is inserted 46 bp upstream of the TATA box. Furthermore, with the recent characterization of the Ac transcription unit (9), we now know that in this allele, Ac and Wx are divergently transcribed and less than 350 bp separates the 5' ends of their respective transcription units. With the Ac and Wx promoters so close together, Wx expression is conceivably controlled by factors that interact with the ends of Ac or with its promoter.

The position of insertion also has a bearing on the quality of somatic excision and germinal derivatives. The phenotype of the wx-ml allele is characterized by sectors with varying amounts of amylose, as indicated by the intensity of staining in Fig. 1. In the presence of Ac, stable germinal derivatives can be isolated that reflect this somatic pattern. McClintock (10) found that the amount of amylose in the endosperm of these new alleles ranged from 0% to 28% of total starch (nonmutant is about 24%).

To understand at what level gene expression was altered in these derivatives, we cloned the wx-ml allele and two germinal derivatives that had 5% (S5) and 9% (S9) amylose in endosperm tissue (26). We found that the wx-m1 allele harbored a 409-bp Ds1-like element in exon 9. In addition, analysis of S5 and S9 revealed that the Ds element had excised and left behind 9 bp (S5) or 6 bp (S9) at the site of insertion, adding three and two amino acids, respectively, to the Wx protein. Since the amount of Wx mRNA and Wx protein was identical to wild-type levels but enzymatic activity was reduced, we concluded that Ds excision had produced new Wx proteins with altered enzymatic properties. This study helped to support the notion that the maize transposable elements may serve a function in natural populations to generate genetic diversity (22).

THE SPLICING OF Ds FROM THE waxy GENE

Recently, we determined that the 4.3-kb Ds element of the wx-m9(Ds) allele (see legend to Fig. 1 for origin of this allele) and the 1.5-kb Ds element of the wxB4 allele (Fig. 2) are spliced from Wx RNA (27). Northern blot analysis of endosperm poly(A)+ transcripts encoded by these alleles is presented in Fig. 3. For both alleles the predominant transcript is wild-type in size despite large Ds elements in exon sequences. DNA sequence analysis of cDNAs revealed that both elements were spliced in a similar manner (Fig. 4). The new 5' splice junction (donor) is in the Ds termini, while the 3' splice junction (acceptor) is in Wx sequences. Although all new donor and acceptor sequences are homologous to pre-mRNA consensus sequences (6), both 3' splice junctions are cryptic splice acceptors that are not utilized in a nonmutant strain.

Alternative splicing is responsible for the production of two distinct mRNAs from the wxB4 allele. The cDNA of mRNA #1 differs from nonmutant Wx cDNA by the addition of 23 bp: 24 bp added from the Ds terminus and 1 bp deleted from Wx sequences (Fig. 4A). The cDNA from mRNA #2 maintains the correct reading frame with the addition of 27 bp to the coding region: 28 bp added from Ds and 1 bp deleted from Wx sequences. Although this transcript is in frame, a Wx protein encoded by this transcript would have an additional nine amino acids. These acids must interfere with enzyme function or protein stability because no amylose is present in wxB4 (-Ac) endosperms and no starch granule-bound Wx enzymatic activity is detected (Varagona and Wessler, unpub. data).

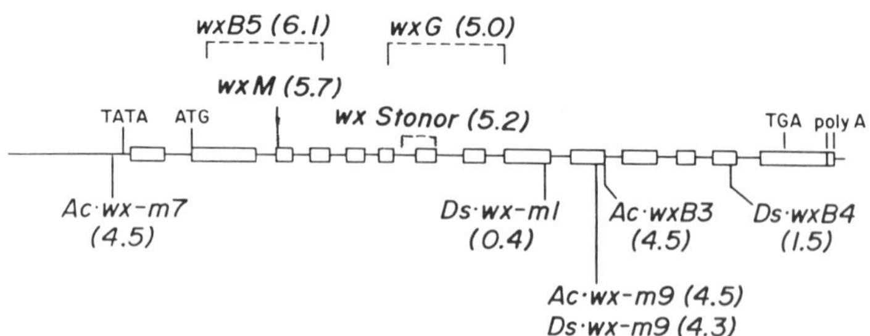

Fig. 2. The location of DNA insertions within the Wx gene. The lengths of each, in kilobase pairs, are shown in parentheses. The unstable Ac and Ds elements are below the gene while the stable insertions are above the gene. Landmarks within the Wx transcription unit are from Klosgen et al. (8). Boxes represent the 14 exons; the connecting lines are introns.

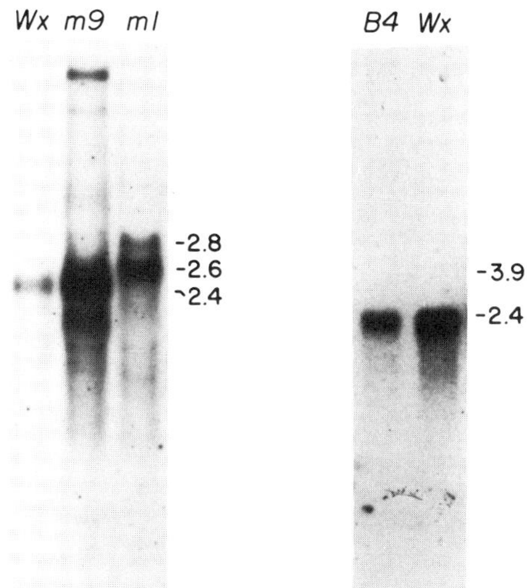

Fig. 3. Northern blot analysis of endosperm poly(A)+ RNA probed with Wx-specific sequences. The amount of RNA loaded was: Wx, 0.5 μg; m9, 5 μg; m1, 5 μg; B4, 3 μg. The molecular weights are in kilobases.

Unlike wxB4, strains harboring the wx-m9(Ds) allele have significant amylose in endosperm tissue, even when Ac is not in the genome (Fig. 1, wx-m9 -Ac). Consistent with this mutant phenotype is the finding that this strain has about 10% of wild-type enzymatic activity (Varagona and Wessler, unpub. data) and produces reduced amounts of a wild-type-sized Wx protein (19). This activity would not be expected to arise from the protein encoded by the mRNA represented by the cDNA in Fig. 4B, since the additional 7 bp (24 bp of Ds added and 17 bp deleted from Wx sequences) does not maintain the correct reading frame. The sequencing of other wx-m9 cDNAs should reveal whether the alternative splicing described for the wxB4 allele produces an in-frame transcript. It is also possible that a different splice acceptor may be utilized (see below).

A Model for the Splicing of Ds from Exons

The 5' splice junctions. Our data have revealed a short sequence adjacent to the Ds terminal inverted repeat that contains at least two, and possibly three, donor splice sites (Fig. 5). These splice sites are positioned so that if each is joined to the same splice acceptor site (as is the case for wxB4 cDNAs I and II), then one of the three resultant transcripts will maintain the correct reading frame, regardless of the site of insertion.

The fact that 15 bp of this 20-bp splice donor sequence is conserved in the highly divergent Ds1-type elements provides additional support for its significance. The Ds elements in the wx-m9 and wxB4 alleles are, like most Ds elements analyzed to date, deletion derivatives of Ac. The Ds1-type elements are very different (24). These elements are short (about 400 bp) and are only homologous with two regions of Ac. One region is the 11-bp inverted repeat termini which are presumably required for

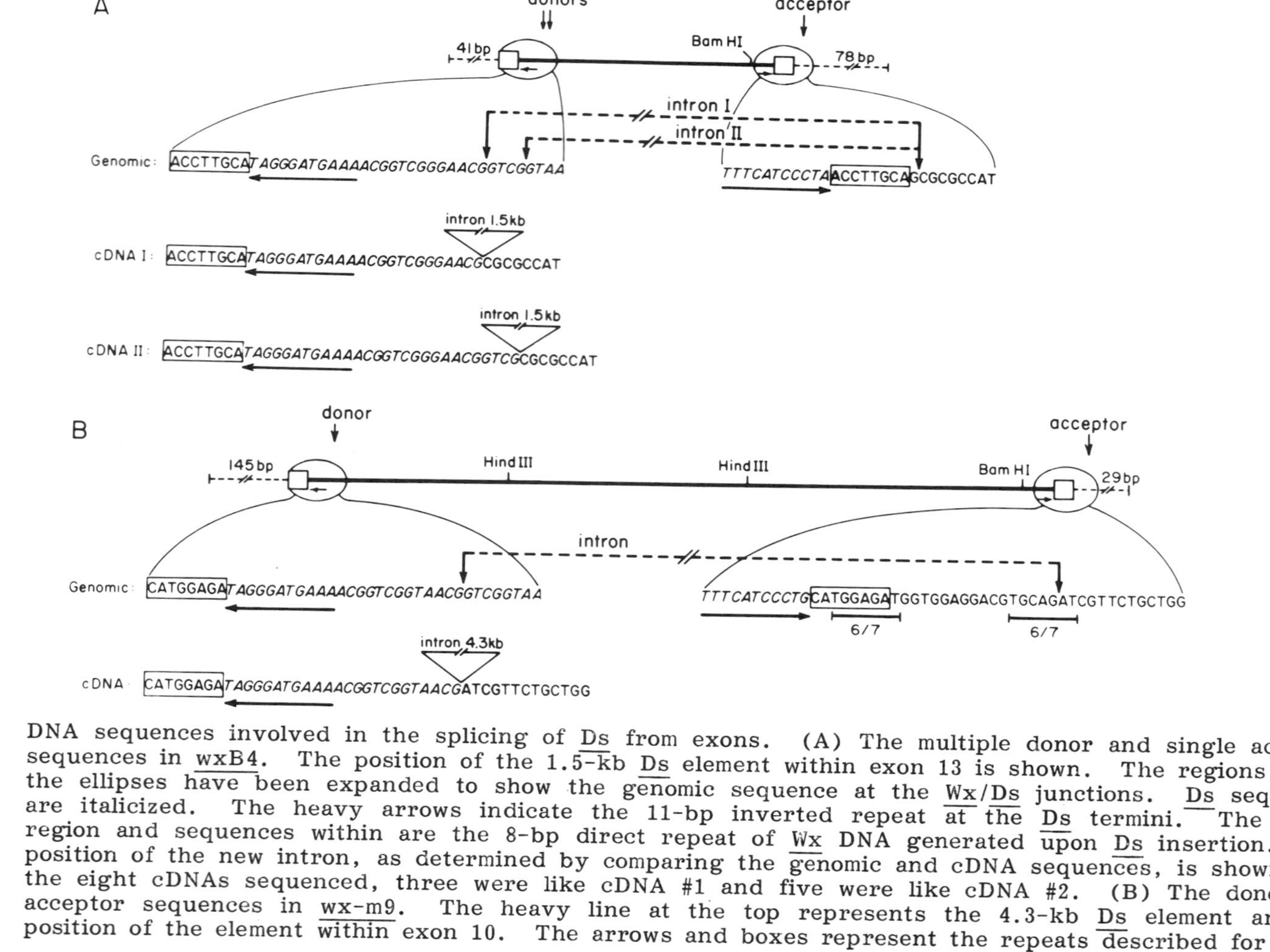

Fig. 4. DNA sequences involved in the splicing of Ds from exons. (A) The multiple donor and single acceptor sequences in wxB4. The position of the 1.5-kb Ds element within exon 13 is shown. The regions within the ellipses have been expanded to show the genomic sequence at the Wx/Ds junctions. Ds sequences are italicized. The heavy arrows indicate the 11-bp inverted repeat at the Ds termini. The boxed region and sequences within are the 8-bp direct repeat of Wx DNA generated upon Ds insertion. The position of the new intron, as determined by comparing the genomic and cDNA sequences, is shown. Of the eight cDNAs sequenced, three were like cDNA #1 and five were like cDNA #2. (B) The donor and acceptor sequences in wx-m9. The heavy line at the top represents the 4.3-kb Ds element and the position of the element within exon 10. The arrows and boxes represent the repeats described for A.

trans-activation by Ac. The other region is the sequence containing the splice donor sites just within one of the inverted termini.

The splicing of Ds1 elements from exon sequences has been investigated for two Ds1-containing alleles: Adh-Fm335 (20) and wx-m1 (Baran and Wessler, unpub. data). The Ds1 element in the Adh-Fm335 allele is inserted in untranslated exon sequences. Northern blot and S1 analysis suggests that it is spliced from Adh mRNA (20). The Ds1 element of wx-m1 is also inserted in exon sequences [(26), Fig. 2]. Northern blot analysis of Wx transcripts from a strain harboring this allele in the absence of Ac reveals multiple RNAs, the most abundant being about 2.6 kb in size (Fig. 3). Only the larger 2.8-kb transcript hybridizes with a Ds1 specific probe (Baran and Wessler, unpubl. data), suggesting that part of the element is spliced from the smaller transcript. For both wx-m1 and Adh-Fm335, the putative splice donor region adjacent to the Ds terminus is in the correct orientation to permit splicing. To determine if these donor splice sites are in fact utilized in vivo, wx-m1 cDNAs are currently being isolated and sequenced.

The 3' splice junction. The fact that the donor splice sites in the Ds elements of wxB4 and wx-m9(Ds) are joined to Wx sequences means that an acceptor site is not encountered in either the 4.3-kb wx-m9 Ds or the 1.5-kb wxB4 Ds. For both alleles, a splice acceptor is encountered in Wx sequences adjacent to the inverted repeat of the Ds element. We do not believe this to be coincidental, and propose that the Ac transposase may have some sequence preference for splice acceptor sites. The target site for the wxB4 allele contains the splice acceptor site (boxed regions, Fig. 4A and 4B). For wx-m9, the acceptor site is encountered 17 bp downstream from the 8-bp target sequence (Fig. 4B). However, a comparison of the actual acceptor site, utilized in vivo with the Ds target site of wx m9, reveals an almost perfect direct repeat (underlined DNA in Fig. 4B; 6/7-bp direct repeat). With only one wx-m9 cDNA examined, it is possible that others utilize this as an acceptor site.

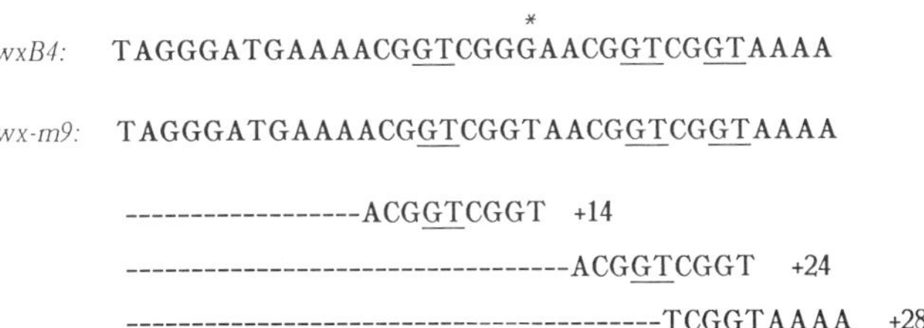

Fig. 5. Homologies with the donor splice consensus sequence in the Ds termini. The +24 and +28 bp sites are utilized in vivo; the donor at +14 is predicted. The GT residues at the 5' splice junction are underlined throughout.

In most introns, the acceptor sequence described above is found downstream of a polypyrimidine stretch (6). For the wxB4 acceptor shown in Fig. 4A, this stretch is furnished by the Ds terminus, which contains pyrimidine residues at nine of 11 positions. Thus, insertion of Ds at a splice acceptor core sequence will position the polypyrimidine-rich Ds terminus just upstream from this core sequence, and result in a complete splice acceptor site.

Taken together, we believe that the splicing mechanism described above has evolved to permit some gene expression, despite insertion of Ds into exon sequences. This is accomplished in two ways. First, the three donor sites of Ds are coupled with a single acceptor site, thereby ensuring a correct reading frame. Second, the intron created by insertion contains virtually all of the element and the direct repeat generated upon insertion. With the 5' splice junction adjacent to the inverted repeat and the 3' splice junction within or near the host target sequence, a minimum amount of RNA is added to or deleted from the transcription unit. The chances of producing a functional gene product are maximized by maintaining the correct reading frame and by minimizing the addition and deletion of RNA within the coding region.

One argument against this model is that most Ds alleles are like wxB4. That is, they have a null phenotype in the absence of Ac. Our data suggest that gene function is not restored in wxB4 because splicing does not generate a normal transcript. Part of the Ds terminus persists in all cDNAs examined. However, in support of the notion that splicing has evolved to permit gene expression, it could be argued that all Ds alleles were selected for their mutant phenotypes. Ds alleles might never be detected if splicing restored gene function. Accordingly, the Ds alleles available for analysis may represent a biased sample and may not reflect the behavior of the 25 to 100 copies of cryptic Ds sequences estimated to be in all maize genomes (4,5,24).

THE STABLE MUTANT ALLELES OF THE waxy LOCUS

Of 17 stable wx mutant alleles of spontaneous origin, we have determined that seven are associated with DNA insertions and five with deletions (25). The Wx insertion mutations have been crossed with strains harboring the autonomous elements of four maize controlling element families: Ac-Ds, Spm, Bg, and Uq. None of these crosses resulted in an unstable phenotype. This does not, of course, rule out the possibility that these elements are defective in their ability to transpose, or that they are the nonautonomous members of a different two-element system.

Four of the mutants have significant gene expression, despite the presence of large DNA insertions (5.0 kb to 6.1 kb) within the Wx transcription unit. Stonor, M, B5, and G have between 5% and 15% of wild-type Wx enzymatic activity. The 5.7-kb wxM insertion has been localized to exon 3 (Fig. 2). The insertions in wxStonor (5.2 kb), wxG (5.0 kb), and wxB5 (6.1 kb) have been delimited to the regions displayed in Fig. 2 (Varagona and Wessler, unpubl. data). These alleles are currently being cloned to determine both the structure and location of the insertion.

To determine the molecular basis for the Wx expression exhibited by these mutant alleles, we have examined endosperm poly(A)+ RNA. Northern blot analysis of Wx homologous sequences is presented in Fig. 6. For

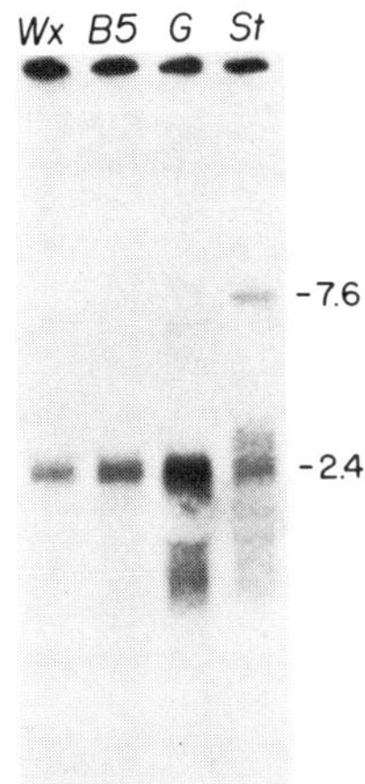

Fig. 6. Northern blot analysis of endosperm poly(A)+ RNA probed with Wx-specific sequences. The amount of RNA loaded was: Wx, 0.5 µg; B5, 5.0 µg; G, 5.0 µg; Stonor, 5.0 µg. The molecular weights are in kilobases.

the mutants analyzed, the predominant transcript is wild-type in size. Larger transcripts, corresponding in size to pre-RNA readthrough transcripts, are seen in some cases. The length of these larger RNAs is the sum of the Wx transcript plus the DNA insertion. From these data, we conclude that the Stonor, B5 and G inserts are either spliced from exons or spliced with introns. The net result, in either case, is the removal of the insert and the generation of a functional Wx protein. We are isolating cDNAs for each mutant allele to determine the precise nature of the mutant transcripts.

CONCLUSION

A variety of molecular mechanisms have been shown to underlie the phenotypic diversity of the waxy mutant alleles containing DNA insertions. For the wx-m1 allele, Ds excision results in the addition of amino acids to the Wx protein. These extra amino acids produce new Wx proteins with altered enzymatic properties. For the inactive derivatives of wx-m7 and wx-m9, a correlation has been made between the loss of Ac activity and the modification of Ac sequences. In addition, the novel pattern of Wx expression in endosperms harboring the inactive Ac-wx-m7 allele may be due to the close proximity of the Ac and Wx promoters. Finally, examination of the wxB4 and Ds-wx-m9 alleles has revealed a mechanism whereby Ds elements are efficiently spliced from exon sequences.

The ability to be spliced from exons may be a feature common to many different maize elements. Recently, Kim et al. (7) reported that the splicing of a dSpm element from the bz locus is responsible for the Bz enzymatic activity observed in strains harboring the bz-m13 Cs9 allele. In addition, our analysis of four leaky Wx alleles (M, Stonor, G, and B5) containing large DNA insertions suggests that splicing may also provide the molecular basis for the Wx expression observed in strains harboring these alleles.

ACKNOWLEDGEMENTS

We would like to thank Beth Johnston and Marion Spell for excellent technical assistance. This work has been supported by NIH grant GM32528 to S.R.W. and an NIH postdoctoral fellowship to G.B.

REFERENCES

1. Behrens, U., N. Fedoroff, A. Laird, M. Muller-Neumann, P. Starlinger, and J. Yoder (1984) Cloning of the Zea mays controlling element Ac from the wx-m7 allele. Molec. Gen. Genet. 194:346-347.
2. Chomet, P.S., S.R. Wessler, and S.L. Dellaporta (1987) Inactivation of the maize transposable element Activator (Ac) is associated with its DNA modification. EMBO J. 6:295-302.
3. Echt, C.S., and D. Schwartz (1981) Evidence for the inclusion of controlling elements within the structural gene of the waxy locus in maize. Genetics 99:275-284.
4. Fedoroff, N., S. Wessler, and M. Shure (1983) Isolation of the transposable maize controlling elements Ac and Ds. Cell 35:235-242.
5. Freeling, M. (1984) Plant transposable elements and insertion sequences. Ann. Rev. Plant Physiol. 35:277-298.
6. Green, M.R. (1986) Pre-mRNA splicing. Ann. Rev. Genet. 20:671-708.
7. Kim, H.-Y., J.W. Schiefelbein, V. Raboy, D.B. Furtek, and O.E. Nelson (1987) RNA splicing permits expression of a maize gene with a defective Suppressor-mutator transposable element insertion in an exon. Proc. Natl. Acad. Sci., USA (in press).
8. Klösgen, R.B., A. Gierl, Z. Schwarz-Sommer, and H. Saedler (1986) Molecular analysis of the waxy locus of Zea mays. Molec. Gen. Genet. 203:237-244.
9. Kunze, R., U. Stochaj, J. Laufs, and P. Starlinger (1987) Transcription of transposable element Activator (Ac) of Zea mays L. EMBO J. 6:1555-1563.
10. McClintock, B. (1948) Mutable loci in maize. Carnegie Institution Washington Yearbook 47:155-169.
11. McClintock, B. (1961) Further studies of the suppressor-mutator system of control of gene action in maize. Carnegie Institution Washington Yearbook 60:469-476.
12. McClintock, B. (1963) Further studies of gene-control systems in maize. Carnegie Institution Washington Yearbook 62:486-493.
13. McClintock, B. (1964) Aspects of gene regulation in maize. Carnegie Institution Washington Yearbook 63:592-602.
14. McClintock, B. (1967) Development of the maize endosperm as revealed by clones. In The Clonal Basis of Development, S. Subtelny and I.M. Sussex, eds. Academic Press, New York, pp. 217-237.
15. Müller-Neumann, M., J.I. Yoder, and P. Starlinger (1984) The DNA sequence of the transposable element Ac of Zea mays. Molec. Gen. Genet. 198:19-24.
16. Nelson, O.E. (1968) The waxy locus in maize. II. The location of the controlling element alleles. Genetics 99:275-284.

17. Pereira, A., Zs. Schwarz-Sommer, A. Gierl, I. Bertram, P. Peterson, and H. Saedler (1985) Genetic and molecular analysis of the Enhancer (En) transposable element system of Zea mays. EMBO J. 4:579-583.
18. Peacock, W.J., E.S. Dennis, W.L. Gerlach, M.M. Sachs, and D. Schwartz (1984) Insertion and excision of Ds controlling elements in maize. Cold Spring Harbor Symposium on Quantitative Biology 19:347-354.
19. Schwartz, D., and E. Echt (1982) The effect of Ac dosage on the production of multiple forms of Wx protein by the wx-m9 controlling element mutation in maize. Molec. Gen. Genet. 187:410-413.
20. Schwartz, D., and E. Dennis (1986) Transposase activity of the Ac controlling element in maize is regulated by its degree of methylation. Molec. Gen. Genet. 205:476-482.
21. Schwarz-Sommer, Zs., A. Gierl, R.B. Klösgen, U. Wienand, P. Peterson, and H. Saedler (1984) The Spm (En) transposable element controls the excision of a 2 kb DNA insert at the wx-m8 locus of Zea mays. EMBO J. 3:1021-1028.
22. Schwarz-Sommer, Zs., A. Gierl, H. Cuypers, P.A. Peterson, and H. Saedler (1985) Plant transposable elements generate the DNA sequence diversity needed in evolution. EMBO J. 4:591-597.
23. Shure, M., S. Wessler, and N. Fedoroff (1983) Molecular identification and isolation of the waxy locus in maize. Cell 35:225-233.
24. Sutton, W.D., W.L. Gerlach, D. Schwartz, and W.J. Peacock (1984) Molecular analysis of Ds controlling element mutations at the Adh1 locus of maize. Science 223:1265-1268.
25. Wessler, S.R., and M. Varagona (1985) Molecular basis of mutations at the waxy locus of maize: Correlation with the fine structure genetic map. Proc. Natl. Acad. Sci., USA 82:4177-4181.
26. Wessler, S.R., G. Baran, M. Varagona, and S.L. Dellaporta (1986) Excision of Ds produces waxy proteins with a range of enzymatic activities. EMBO J. 5:2427-2432.
27 Wessler, S.R., G. Baran, and M. Varagona (1987) The maize transposable element Ds is spliced from RNA. Science 237:916-918.

DISCOVERY OF Ac ACTIVITY AMONG PROGENY OF TISSUE CULTURE-DERIVED MAIZE PLANTS

R.L. Phillips and V.M. Peschke

Department of Agronomy and Plant Genetics, and
Plant Molecular Genetics Institute
University of Minnesota
St. Paul, Minnesota

ABSTRACT

A high frequency of structurally altered chromosomes in maize plants regenerated from tissue culture led us to predict that newly activated transposable elements could be detected in regenerated plants. Test-crosses of 1,200 progeny from 301 regenerated maize plants resulted in 56 positive tests for the Activator (Ac) transposable element. Further testing of these progenies confirmed that ten regenerated plants from two independent embryo cell lines contained an active Ac element; indications that an eleventh regenerated plant from a third cell line may contain Ac have not yet been confirmed. Thus, Ac activity was observed in 2 to 3% of the embryo cell lines and regenerated plants tested. Ac activity has not been found in plants from noncultured control kernels, and only 30% of the plants regenerated from the positive cell lines had Ac activity, demonstrating that no active Ac elements were present in the explant sources. Nine regenerated plants derived from a single embryo cell line had Ac activity; this indicates that activation occurred during tissue culture rather than during plant regeneration. DNA analysis of the Ac-containing plants and noncultured control plants is now in progress.

Other researchers at this university are currently investigating possible mechanisms for the chromosomal aberrations observed in tissue culture-derived plants; these breakage events may be responsible for the activation of transposable elements in tissue culture. Recovery of transposable element activity in regenerated plants indicates that some tissue culture-derived genetic variability may be the result of insertion and/or excision of transposable elements.

INTRODUCTION

The plant tissue culture process generates genetic variation among cultured cells and regenerated plants (25,36). First- and/or second-generation progeny of regenerants often segregate for genetic and cytogenetic variants not present in the donor tissue. Chromosome breakage, as well as single-gene recessive mutations, occur at relatively high frequencies. The research reported here is an attempt to understand why the frequency of chromosome breakage and mutation are both elevated as the result of tissue culture. We are interested in finding the common link causing the increased frequency while recognizing that the two events, breakage and mutation, are not obviously correlated; that is, plants carrying mutations do not always possess detectable chromosomal aberrations, and vice versa. The results reported here lead us to believe that events that cause a delay in the duplication of late-replicating DNA (present in heterochromatic blocks) result in chromosome bridges during mitotic anaphases and, thereby, generate broken chromosome(s) and subsequent chromosomal rearrangements. These breakage events are hypothesized to have an effect on transposable elements, resulting in their release or activation. These elements can then transpose into a gene, giving a noticeable phenotypic change. Transposable elements may be the common link between the occurrence of chromosomal aberrations and mutations. DNA methylation changes may underlie the activation of transposable elements in plant tissue cultures.

An understanding of the mechanism underlying tissue culture-induced variation may lead to methods for controlling the occurrence of cytological and genetic alterations. Variation might then be enhanced for various breeding purposes and avoided for others.

CHROMOSOME BREAKAGE

Although nearly all common types of chromosomal alterations have been reported in plant tissue cultures or among regenerated plants, our studies indicate that aberrations resulting from chromosome breakage are much more common than chromosome number changes (aneuploidy and polyploidy). Our earlier work with oats (Avena sativa L.) made this point evident (33). Heterozygosity for near-telocentric chromosomes occurred at a frequency of 20% among regenerated plants of two different oat varieties, Tippecanoe and Lodi. Plants carrying such chromosomes seldom occur among seed-grown plants. Root-tip analysis of about 6,000 seedlings of the variety "Garry" revealed no plants with telocentric chromosomes (34). Thus, the occurrence of such broken chromosomes is rare in oats, and must reflect an unusual cellular event occurring during tissue culture. We hypothesized that oats might have pericentromeric heterochromatin that replicates improperly in the tissue-culture environment. Specifically, we thought that the late-replicating DNA of these regions might be delayed in replication, leading to anaphase bridges and chromosome breakage. The existence of late-replicating centromeric heterochromatin in oats has recently been documented (24).

The cytological data on maize plants regenerated from tissue culture indicate that chromosome breakage is the predominant cytological change and that heterochromatic blocks (knobs) again are involved (2,6,28,32,39). Summarizing these studies, only 46 polyploid and 16 aneuploid plants were found among 1,036 regenerants analyzed at meiosis (Tab. 1); 37 of the polyploid plants were derived from one embryo-derived callus line, thus inflating the apparent polyploid frequency. Over 200 of the plants carried either a translocation (107 plants) or a deletion or duplication (108 plants). Only two plants were heterozygous for a paracentric inversion.

Tab. 1. Chromosomal variation among plants regenerated from maize tissue cultures.

Chromosomal alterations	Aberration frequency	No. (%)
Polyploidy	46	4.4 *
Aneuploidy	16	1.5
Structural alterations		
Interchanges	107	10.3
Inversions	2	0.2
Deletions/Duplications	108	10.4
Total plants scored	1,036	

*75% from one embryo-derived callus line.

Most of the break positions occurred in a chromosome arm with a knob and were located between the knob and the centromere (6,28). Models, diagramming how the delayed duplication of the late-replicating DNA can lead to such aberrations, are presented in Lee and Phillips (28) and Johnson et al. (24).

INDUCED MUTATIONS

Early reports on lines derived from maize tissue cultures indicated a high frequency of mutations. Edallo et al. (16) reported a frequency of 1.2 variants per R_0 plant (initial regenerant) for inbred W64A and 0.8 variants for inbred S65, based on segregations in the second selfed generation (R_2). Table 2 shows that qualitative genetic variation is common among progeny of regenerants representing several genotypic backgrounds. The variants reported in Tab. 2 each segregate as if controlled by a single, recessive gene. We have observed over 50 different phenotypes segregating in the progeny of various regenerants (38) with up to five new mutations occurring in one regenerant (28). Few of the mutants appear to be unstable. Armstrong (1) discovered a mutant in which the seedlings had green stripes on an otherwise albino background.

Many of the R_0 plants are sectored for the new mutation, based on the segregation of the mutant phenotype in the R_2 but not in the R_1 generation. This is interpreted as being the result of genetic nonconcordance between the tassel and the ear of regenerated plants. About 50% of the R_0 plants segregating for mutants were sectored (Tab. 3).

In addition to qualitative genetic changes, lines derived from tissue culture also have altered quantitative traits. Beckert et al. (3) reported small but significant variation for a number of agronomic traits among

Tab. 2. Frequency of variants per regenerated plant.

Genotype	(Reference)	Variants/R_0
W64A	(16)	1.2
S65	(16)	0.8
(Oh43 ms isolines/A188) F_2	(28)	
4-month cultures		0.52
8-month cultures		1.32
A188 x W22 R-nj R-nj	(32)	0.3
A188	(40)	0.15
PG6	(40)	0.20
B73	(43)	0.35
A188 and A188/B73	(2)*	0.41

*Armstrong (1) found that the frequency of genetic variants was nearly the same, regardless of whether the regenerants arose from Type I (organogenic) or Type II (embryogenic) cultures.

progeny of regenerated plants from several inbreds. Earle and Gracen (15) also reported agronomic trait variation as the result of maize tissue culture. Significant variation was detected for vigor, flowering date, and plant height in progeny of inbred W1B2BN regenerants. Lee et al. (26) evaluated a total of 305 tissue-culture derived lines and 48 control lines in the S_2 or in testcrosses. The top-ranked line, based on an index for yield and moisture in five of six trials, was derived from tissue culture. In general, plant height was reduced and maturity was earlier in the tissue culture-derived lines. The occasional dramatic changes in maturity while maintaining high yield may be of particular interest in plant breeding.

Tab. 3. Genetic nonconcordance in regenerated plants, based on mutant segregation in the R_2 but not in the R_1 generation.

Reference	Percent	Total number of plants segregating
Benzion (6)	64	45
Lee and Phillips (27)	53	81
Armstrong (1)	33	34

TRANSPOSABLE ELEMENTS

The activation of transposable elements, as a possible cause of the variation among tissue culture-derived materials, has been suggested by several workers (10,25,31). The rationale that led us to test (37) for the presence of an active transposable element related to the observation of chromosome breakage as the principal cytogenetic alteration resulting from tissue culture.

Transposable elements in maize occur in genetic materials that undergo the chromosome type of break-fusion-bridge cycle (14,29,30). Other methods that result in broken chromosomes, such as X-rays and UV light, were also shown to generate active transposable elements (7,35). Based on the previous research, the relatively high frequency of chromosome breakage occurring in tissue culture would be expected to generate active transposable elements. For this reason, we decided to test for the presence of the transposable elements among regenerated materials. Our first tests were for the transposable element Activator (Ac) because of the availability of an efficient tester strain.

Genetic Stocks

The genetic materials tested were those previously analyzed in other studies for segregation of new variants and cytological alterations (2,23,27,28). A total of 301 regenerated (R_0) plants from 94 immature embryo-derived cell lines of three inbred backgrounds (A188, B73, and Oh43) were tested (Tab. 4). The R_0 plants had been regenerated 4 to 22 months after culture initiation. The material from the Lee and Phillips study (27,28) represented 161 R_0 plants, regenerated from organogenic (Type I) cultures, initiated from 24 immature embryos that were produced by sib pollinations of F_2 plants of Oh43 x A188. The Armstrong and Phillips material (2) included 90 R_0 plants from 41 embryo cell lines of A188 x B73 crosses; regenerated plants were from both organogenic (Type I) and embryogenic (Type II) cultures. The Hartloff material (23) was from organogenic cultures of A188, and represented 50 R_0 plants from 29 embryo cell lines. The Ac tester stock contained Ds about 33 map units proximal to C in chromosome 9. The Ds was the double-Ds type that causes chromosome breakage in the presence of Ac. The tissue culture-derived genetic material all contained c. An active Ac was detected by a colored/colorless aleurone pattern, indicative of the break-fusion-bridge cycle. All independent Ac's detected were later double-checked by crossing with c-m2, which has a different form of Ds inserted in the C locus. An active Ac is detected by causing the appearance of small, colored spots of aleurone tissue on an otherwise colorless background.

Occurrence of Ac

Most R_0 plants were tested in the R_1 generation (about five R_1 progeny per R_0 plant) or in the R_2 generation (second selfed generation) in a few cases. Among over 1,200 testcrosses, 56 indicated the presence of a single active Ac element. These 56 traced back to 11 of the 301 R_0 plants and three of the 94 embryo cell lines. Thus, the frequency of Ac activity was about 3%. The Ac-containing materials were from the Lee and Phillips study (27,28), involving the inbreds Oh43 and A188.

The putative Ac in one embryo cell line was confirmed in eight of the nine subsequent tests with the c-m2 stock; in two of these, reduced

Tab. 4. Tests for Ac transposable element.

Embryo cell lines		R_0 Tests		R_0 with Ac*
94	⟶	301	⟶	11 (3 embryo cell lines)

Frequency

Ac among total R_0 tested, 11/301 = 3.7%

Ac among total embryo cell lines tested, 3/94 = 3.2%

*Sib-pollinated ears of (Oh43 x A188) F_2.

transmission of Ac through the pollen is indicated. Two out of three testcrosses involving the Ac from a second embryo cell line indicated close linkage of Ac to the C locus. Ac activity in the third embryo cell line has not yet been confirmed; further tests are in progress.

Evidence that an active Ac was not present in the explant tissue comes from various sources. First, from the three cell lines which produced regenerated plants with an active Ac, only about 30% of the regenerated plants had an active Ac based on testcrosses to the Ds stock. Second, 30 plants were grown from mature seed borne on the same ears from which embryos for culture initiation were extracted (only half the ear was brought into the lab for embryo extraction); all gave negative results in the testcrosses (Tab. 5). Thus, active Ac's were not present in the donor plants, nor was Ac uniformly present in the Ac-containing embryo cell lines. A tissue culture origin is suggested. In addition, 22 testcrosses of noncultured kernels of the parent of the rows that gave Ac-positive embryo cell lines all indicated the absence of Ac, as did 45 testcrosses of the two inbreds Oh43 and A188. Tests of related F_2 plants of Oh43 ms13 x A188 were also negative for Ac activity.

This discovery of active transposable elements supports earlier reports, based on more limited tests, of tissue culture activation of Ac and Spm (17,18). Induced Ac function has also been reported in maize endosperm tissue cultures (13). Endosperm culture of c-m2 resulted in a high frequency of colored sectors that were not present in cultures of non-Ds containing c cultures. Southern analyses examining the PvuII sites within Ac indicate hypomethylation of the resident Ac copies in the cultures which are methylated in field-grown material (Culley, pers. comm.). An unstable flower color allele has been observed in tissue culture-derived alfalfa (Medicago sativa), and was also suggested to be a result of transposable element activation (21,22).

Southern Analyses

We anticipated the possibility that inactive forms of Ac might be resident in Oh43 and/or A188, based on the report of four to eight copies

Tab. 5. Control materials and inbreds tested for Ac activity.

Seed source	Number of tests for Ac
Controls for cell lines 4-41 and 1-42	
From same ear as embryo 4-41	6
From same ear as embryo 1-42	10
Parent of rows 23341 and 23342*	22
Other (Oh43 ms13 x A188) F_2 plants	46
Controls for cell line 8-27	
From same ear as embryo 8-27	14
Other (Oh43 ms1 x A188) F_2 plants	3
Inbreds	
A188	41
Oh43	4
B73	4

*Embryos 4-41 and 1-42 were from sib-crosses within rows 23341 and 23342, respectively.

of an Ac-like sequence in several maize inbreds which did not contain genetically active Ac elements (19). Using an Ac-specific probe in Southern blots of DNA digests, we have observed multiple Ac-like sequences in both of these inbreds.

IMPLICATIONS

Several testable ideas come from the hypothesis that the underlying cause of tissue culture-induced variation is a problem in the timely replication of late-replicating DNA, and that the subsequent chromosome breakage event(s) activates transposable elements that, in turn, cause mutations. Implied is the possibility that the degree of variation might be line-specific, based at least in part on differences in the amount of heterochromatin and the presence of silent forms of transposable elements. Although mutations probably occur by various mechanisms, the hypothesis predicts that many of the induced mutations would be the result of transposable element insertions. Because there is no a priori reason to believe that only Ac can be activated, the element involved might be a member of any of the known systems, or even others not yet defined. Several tissue culture-induced mutations need to be analyzed to determine how frequently they are the result of insertional events. To date, the only tissue culture-derived mutant tested was an alcohol dehydrogenase mutant from maize tissue culture that resulted from a single base change (8).

Methylation of DNA appears to be involved in transposable element activity. Methylation studies on Ac's that cycle between inactivity and activity have shown that active Ac's are unmethylated at one or more sites which are methylated in inactive Ac's (12,41). Loss of Mutator activity of the unstable bz2-mul allele upon inbreeding has also been shown to correlate with DNA methylation (4,5,11,42). Because A188 also possesses Mu elements (Morris and Stadler, pers. comm.), one might expect to detect Mu activity among regenerants or their progenies. Such Mu activity could lead to unexpectedly high mutation frequencies in later generations. Evidence has been reported for DNA methylation changes as the result of tissue culture. DNA from 120 regenerated A188 plants that came from immature embryo-initiated embryogenic cultures was often altered in methylation, based on HpaII and MspI restriction patterns (Ref. 9, and Brown and Lorz, pers. comm.). DNA from most plants had limited digestion with HpaII and more with MspI. Some plants had the reverse of this response, while others did not digest well with either enzyme, indicating very high methylation. One plant had identical HpaII and MspI restriction patterns with extensive cleavage, indicating little methylation. Brown and Lorz also studied the methylation of the promoter sequences of Sh and Adh and found that they were always hypomethylated. This effect carried over through at least four selfings. Plants from a single embryo cell line were often quite different in methylation.

The activation of silent forms of transposable elements, and perhaps even the modification of quantitative traits in progeny of regenerants, could be the result of methylation effects in culture. How chromosome breakage might relate to methylation is not known. However, breakage as a consequence of a delay in the duplication of late-replicating DNA could be related to culture-induced perturbations in the cell cycle which might alter the normal methylation of DNA. Methylation following DNA repair can lead to hypomethylation or hypermethylation, based on studies of UV-treated prokaryotic cells (20). Perhaps transposable elements are activated by hypomethylation as the result of disturbances in the normal methylation processes. Such disturbances might be a result of the chromosome breakage events, since DNA replication and repair are tightly coupled to methylation. This process could be the phenomenon described by McClintock, known as "genome shock."

Further studies of DNA methylation in tissue cultures and regenerated progenies should be informative. We know that DNA is methylated upon replication and repair, transposable element activity can be related to methylation, and gene activity can be affected by methylation. The various genetic and cytogenetic effects of the tissue culture process, including chromosome breakage, single-gene mutations, transposable element activation, and quantitative trait alteration, could therefore all relate directly or indirectly to disturbances in the normal methylation of DNA.

REFERENCES

1. Armstrong, C.L. (1986) Genetic and cytogenetic stability of maize tissue cultures: A comparative study of organogenic and embryogenic cultures. Ph.D. Thesis, University of Minnesota, St. Paul.

2. Armstrong, C.L., and R.L. Phillips (1988) Genetic and cytogenetic variation in plants regenerated from organogenic and friable, embryogenic tissue cultures of maize. Crop Sci. (in press).
3. Beckert, M., M. Pollacsek, and M. Gaenen (1983) Etude de la variabilite genetique obtenue chez le maiz apres callogenese et regeneration de plantes in vitro. Agronomie 3:9-18.
4. Bennetzen, J.L. (1985) The regulation of Mutator function and Mu1 transposition. In Plant Genetics, M. Freeling, ed. Liss, New York, pp. 343-353.
5. Bennetzen, J.L. (1987) Covalent DNA modification and the regulation of Mutator element transposition in maize. Molec. Gen. Genet. 208:45-51.
6. Benzion, G., and R.L. Phillips (1988) Cytogenetic stability of maize tissue cultures: A cell line pedigree analysis. Genome (in press).
7. Bianchi, A., F. Salamini, and R. Parlavecchio (1969) On the origin of controlling elements in maize. Genet. Agrar. 22:335-344.
8. Brettell, R.I.S., E.S. Dennis, W.R. Scowcroft, and W.J. Peacock (1986) Molecular analysis of a somaclonal mutant of maize alcohol dehydrogenase. Molec. Gen. Genet. 202:235-239.
9. Brown, P.T.H., and H. Lorz (1986) Molecular changes and possible origins of somaclonal variation. In Somaclonal Variation and Crop Improvement, J. Semal, ed. Martinus Nijhoff Publ., Dordrecht, The Netherlands, pp. 148-159.
10. Burr, B., and F. Burr (1981) Transposable elements and genetic instabilities in crop plants. Stadler Symp. 13:115-128.
11. Chandler, V.L., and V. Walbot (1986) DNA modification of a maize transposable element correlates with loss of activity. Proc. Natl. Acad. Sci., USA 83:1767-1771.
12. Chomet, P.S., S. Wessler, and S.L. Dellaporta (1987) Inactivation of the maize transposable element Activator (Ac) is associated with its DNA modification. EMBO J. 6:295-302.
13. Culley, D.E. (1986) Evidence for the activation of a cryptic transposable element Ac in maize endosperm cultures. In Proceedings of VI International Congress on Plant Tissue and Cell Culture, Minneapolis (abstract), p. 263.
14. Doerschug, E.B. (1973) Studies of Dotted, a regulatory element in maize. I. Induction of Dotted by chromosome breaks. II. Phase variation of Dotted. Theor. Appl. Genet. 43:182-189.
15. Earle, E.D., and V.E. Gracen (1985) Somaclonal variation in progeny of plants from corn tissue cultures. In Tissue Culture in Forestry and Agriculture, R. Henke, K. Hughes, Milton J. Constantin, and A. Hollaender, eds. Plenum Press, New York, pp. 139-152.
16. Edallo, S., C. Zucchinali, M. Perenzin, and F. Salamini (1981) Chromosomal variation and frequency of spontaneous mutation associated with in vitro culture and plant regeneration in maize. Maydica 26:39-56.
17. Evola, S.V., F.A. Burr, and B. Burr (1984) The nature of tissue culture induced mutations in maize. Eleventh Annual Aharon Katzir--Katchalsky Conf., Jerusalem (abstract).
18. Evola, S.V., A. Tuttle, F. Burr, and B. Burr (1985) Tissue culture associated variability in maize: Molecular and genetic studies. First International Congress on Plant Molecular Biology, Savannah, Georgia (abstract), p. 10.
19. Fedoroff, N., S. Wessler, and M. Shure (1983) Isolation of the transposable maize controlling elements Ac and Ds. Cell 35:235-242.
20. Grafstrom, R.H., D.L. Hamilton, and R. Yuan (1984) DNA methylation: DNA replication and repair. In DNA Methylation:

Biochemistry and Biological Significance, A. Razin, H. Cedar, and A.D. Riggs, eds. Springer Verlag, New York, pp. 111-126.

21. Groose, R.W., and E.T. Bingham (1984) Variation in plants regenerated from tissue culture of tetraploid alfalfa heterozygous for several traits. Crop Sci. 24:655-658.
22. Groose, R.W., and E.T. Bingham (1986) An unstable anthocyanin mutation recovered from tissue culture of alfalfa (Medicago sativa). 1. High frequency of reversion upon reculture. 2. Stable nonrevertants derived from reculture. Plant Cell Rep. 5:104-110.
23. Hartloff, H.J. (1984) Altered expression of male fertility and disease resistance in maize plants regenerated from tissue culture. M.S. Thesis, University of Minnesota, St. Paul.
24. Johnson, S.S., R.L. Phillips, and H.W. Rines (1987) Possible role of heterochromatin in chromosome breakage induced by tissue culture in oats (Avena sativa L.). Genome 29:439-446.
25. Larkin, P.J., and W.R. Scowcroft (1981) Somaclonal variation--A novel source of variability from cell cultures for plant improvement. Theor. Appl. Genet. 60:197-214.
26. Lee, M., J.L. Geadelmann, and R.L. Phillips (1988) Agronomic evaluation of inbred lines derived from tissue cultures of maize. Theor. Appl. Genet. (in press).
27. Lee, M., and R.L. Phillips (1987) Genetic variants in progeny of regenerated maize plants. Genome 29:834-838.
28. Lee, M., and R.L. Phillips (1987) Genomic rearrangements in maize induced by tissue culture. Genome 29:122-128.
29. McClintock, B. (1939) The behavior in successive nuclear divisions of a chromosome broken at meiosis. Proc. Natl. Acad. Sci., USA 25:405-416.
30. McClintock, B. (1950) The origin and behavior of mutable loci in maize. Proc. Natl. Acad. Sci., USA 36:344-355.
31. McClintock, B. (1984) The significance of responses of the genome to challenge. Science 226:792-801.
32. McCoy, T.J., and R.L. Phillips (1982) Chromosome stability in maize (Zea mays) tissue cultures and sectoring in some regenerated plants. Can. J. Genet. Cytol. 24:559-565.
33. McCoy, T.J., R.L. Phillips, and H.W. Rines (1982) Cytogenetic analysis of plants regenerated from oat (Avena sativa) tissue cultures: High frequency of partial chromosome loss. Can. J. Genet. Cytol. 24:37-50.
34. McGinnis, R.C. (1962) Aneuploids in common oats, Avena sativa. Can. J. Genet. Cytol. 4:296-301.
35. Neuffer, M.G. (1966) Stability of the suppressor element in two mutation systems of the A-1 locus in maize. Genetics 53:541-549.
36. Orton, T.J. (1983) Somaclonal variation: theoretical and practical considerations. In Gene Manipulations in Plant Improvement, J.P. Gustafson, ed. Plenum Press, New York, pp. 427-468.
37. Peschke, V.M., R.L. Phillips, and B.G. Gengenbach (1987) Discovery of transposable element activity among progeny of tissue culture-derived maize plants. Science 238:804-807.
38. Phillips, R.L., D.A. Somers, and K.A. Hibberd (1988) Cell/tissue culture and in vitro manipulation. In Corn and Corn Improvement, G.F. Sprague, ed. Amer. Soc. Agron., Madison (in press).
39. Rhodes, C.A., R.L. Phillips, and C.E. Green (1986) Cytogenetic stability of aneuploid maize tissue cultures. Can. J. Genet. Cytol. 28:374-384.

40. Rice, T.B. (1982) Tissue culture induced genetic variation in regenerated maize inbreds. In Proceedings 37th Annual Corn and Sorghum Research Conference, Chicago, Illinois, pp. 148-162.
41. Schwartz, D., and E. Dennis (1986) Transposase activity of the *Ac* controlling element in maize is regulated by its degree of methylation. Molec. Gen. Genet. 205:476-482.
42. Walbot, V., V. Chandler, and L. Taylor (1985) Alterations in the *Mutator* transposable element family of *Zea mays*. In Plant Genetics, M. Freeling, ed. Liss, New York, pp. 333-342.
43. Woodman, J.C., and D.A. Kramer (1986) The recovery of somaclonal variants from tissue cultures of B73, an elite inbred line of maize. VI International Congress on Plant Tissue and Cell Culture, Minneapolis, Minnesota (abstract), p. 215.

ACTIVATION OF SILENT TRANSPOSABLE ELEMENTS

Benjamin Burr and Frances A. Burr

Biology Department
Brookhaven National Laboratory
Upton, New York 11973

ABSTRACT

It is well known among maize geneticists that agents that cause chromosome breakage can activate quiescent transposable elements. However, other than temporarily relieving position effect, it is difficult to understand how these events can lead directly to activation. One possibility is that chromosome breakage can initiate a process in the cell resulting in a higher rate of spontaneous mutation. Such a system could be analogous to the SOS response of *Escherichia coli* in which an error-prone repair system is induced. Chemical mutagens that cause little chromosome breakage but add bulky adducts to the DNA can induce the SOS response. In seed homozygous for *a1-m2(8004)*, *wx-m8*, no active *Spm*, that had been treated with ethyl methanesulfonate, we observed activation of *Spm* at the rate of 1.1×10^{-4}. The spontaneous rate of activation in this material was 1.2×10^{-5}. Most of the activation events occurred as single kernels. This result contrasts with sectors covering at least one-eighth of the ear that would have been expected if activation had occurred as a direct result of mutagenesis in the mature kernel. The late timing of these events suggests that the activation, in most instances, may not be the direct result of chemical mutagenesis.

INTRODUCTION

It is a central belief among maize geneticists that silent endogenous copies of most, if not all, transposable element families are present in the normal maize genome and that a sufficient shock can activate them (22). As each element has been cloned, multiple sequences hybridizing to at least a portion of the element have been observed on whole genome Southern blots. Even *Mu1*, once thought to be strain-specific, has been shown to be present in *non*-Mutator lines (7).

From stocks exposed to the breakage-fusion-bridge cycle early in development, McClintock (18, 19) obtained evidence for the activation of

Ac, Spm, and Dt. When the chromosomes of plants that had begun development with the breakage-fusion-bridge cycle on 9S were examined at meiosis, it was observed that other chromosomes of the complement had been altered by this process (20). Thus, elements on chromosomes other than 9S were activated. McClintock's experimental induction of Dt activity was repeated by Doerschug (13). Both McClintock and Doerschug used a terminally duplicated chromosome 9 to produce dicentric chromatids after crossover between the inverted repeats to induce the breakage-fusion-bridge cycle. Of the two germinally inherited events Doerschug recovered, one was on 9S, although not in the same location as the original Dt described by Rhoades (32), and the second was eventually shown to reside on chromosome 4.

Further proof that chromosome breakage could cause the activation of endogenous elements came from the work of Neuffer (26), who irradiated pollen with X-rays and ultraviolet light (UV) to induce both Dt and Spm. Although Neuffer noted that UV was less efficient than X-rays in causing chromosome breakage, he pointed out that UV induced more activation. Bianchi, Salamini, and Parlavecchio (4) subjected pollen to X-ray treatment, and after pollination selected progeny that underwent the breakage-fusion-bridge cycle. Among these they obtained evidence for the activation of both Ac and Spm.

It is now general knowledge that chromosome breaks induce transposable activity in maize, but this does not explain how the elements are actually activated. Three ways in which chromosome breaks might activate endogenous elements come to mind.

Position Effects

McClintock (18), noting that Dt1 mapped to the heterochromatic knob of 9S, was impressed by the possibility of the heterochromatic origin of these elements. Genes that are variably suppressed by adjacent heterochromatin can give rise to a variegated phenotype, at least superficially extending the analogy between position effect and transposable elements. The simplest mechanism by which chromosome breaks could lead to the activation of quiescent elements is to assume that a break can temporarily relieve an element imbedded in heterochromatin from control by a position effect. Position effects are, of course, well known in *Drosophila* (36) and *Oenothera* (6); however, it has been difficult to find evidence for them in maize.

Stadler (37) realized that X-ray-induced chromosomal rearrangements could give rise to apparent mutations if gene action were suppressed by position effect. He found no correlation in the occurrence of mutations and translocations and no mutations that mapped to translocation breakpoints. Roberts (33) examined thirteen reciprocal translocations in the inbred C20 that among them involved all but chromosomes 7 and 9. He measured six quantitative traits in a well-replicated experiment. No obvious phenotypic differences between homozygous or heterozygous translocations and the inbred parent were observed, although some statistically significant small differences were found. He concluded that these small differences could not be interpreted as position effects, since the occurrence of mutations from the X-ray treatment used to generate the rearrangements could not be ruled out.

Methylation Control

The major thinking for the past two years in the maize community has been centered on the relation of methylation and transposable element inactivity (2,8,9). Virginia Walbot (this Volume) suggests that rapid replication following chromosome breakage leads to undermethylation, which in turn may lead to element activation. There is much that is attractive about the hypothesis that undermethylation will lead to element activation. One of the most provocative experiments in this area is that of Jaenisch and co-workers (14), who injected mice with the drug 5-azacytidine, known to lead to undermethylated DNA, and observed activation of silent retroviral genomes. It seems likely that upon incorporation into DNA, 5-azacytidine and analogous drugs, that cannot be methylated at position 5 of cytosine, trap methylating enzymes in nonproductive complexes and lead to subsequent undermethylation of newly replicated DNA (34). The protein-DNA complexes most likely interfere with DNA replication (3), leading to cell death, increased mutation rate (1,5,41), accelerated recombinational activity (18,41), decreased chromosome condensation, and the appearance of chromosome breaks (38). Because of its manifold effects, results obtained with 5-azacytidine and its analogs should be interpreted with caution.

Recent work on sexual imprinting (31,35) indicates that methylation patterns are stably inherited and can be reprogrammed at meiosis. The parallels with these observations and those of McClintock (21) on presetting and erasure are striking.

Point Mutations

A final possibility is that chromosome breaks can lead to the production of point mutations that are ultimately the agent responsible for transposable element activation. In *E. coli*, events that lead to the disruption of DNA replication and the exposure of single-stranded DNA induce the SOS response. The error-prone repair system induced is responsible not only for mutations at the site of damage but also for mutations in undamaged DNA (40). For instance, in an extension of Weigle mutagenesis, bacterial cells were irradiated with UV light to induce the response, then infected with bacteriophage. Mutations appeared in the bacteriophage DNA that was never exposed to UV light (29). We are suggesting that chromosome breaks induced by a variety of events might lead to the induction of a system analogous to the SOS repair system in *E. coli* and that, once induced, this system can further induce heritable changes through the production of point mutations.

Inducible systems of DNA repair analogous to the SOS response of *E. coli* have been demonstrated in other bacteria and in the fungi (40). A variety of DNA-damaging agents induce the expression of integrated SV40 proviruses in mammalian cells (15). These results are similar to the induction of bacteriophage lambda by the RecA protein after exposure to agents eliciting the SOS response (14).

SOME RESULTS AND DISCUSSION

It may be some time before the third hypothesis stated above can be entirely tested. However, it is possible to test parts of the hypothesis.

The first question we can ask is, "Do agents that primarily cause point mutations activate transposable element activity?" The least clastogenic mutagens are base analogs, but these need to be introduced during DNA replication and so far have not been useful mutagens in maize. On the other hand, ethyl methanesulfonate (EMS) is a potent mutagen in maize (27). EMS lists in the middle among alkylating agents ranked for their ability to induce chromosome breaks (39). However, these events are very rare (0.01 to 0.02) compared to point mutations in Drosophila (25). In barley, EMS induces high frequencies of chromosomal aberrations only when applied at concentrations greater than LD_{50} or after prolonged storage of treated kernels (28).

We have asked whether EMS treatment can activate cryptic Spm elements in the maize genome. To do this we used a dSpm reporter allele, a1-m2, that has a unique response to trans-acting signals from an active Spm. The Sp function activates rather than suppresses the gene (20). Thus we looked for red events on a background on white kernels. The allele also shows a dosage response to Spm, giving more intense color and more frequent transpositions from the A1 locus with increasing copies of active Spm in the genome (22). The cross that we used to look for induction was the following:

a1-m2(8004) Sh2, wx-m8, no Spm X a1 sh2, wx-m8, no Spm

Female plants were either untreated controls or were treated as mature seeds with 34 mM EMS for 4 hr prior to planting. All kernels with a red aleurone or sectors indicative of the activation of Spm were planted in the greenhouse and crossed reciprocally with a1-m1(8004) Sh2, wx-m8, no Spm to see if germinal activation events occurred. Although we observed a number of apparent somatic activation events, we found the following rates of induction of germinal events:

EMS treated: $2 / 1.7 \times 10^4 \rightarrow 1.1 \times 10^{-4}$

Control: $1 / 0.85 \times 10^5 \rightarrow 1.2 \times 10^{-5}$

In all cases, both Sp and M functions were activated, as judged by pigmentation, due to the activation of a1-m2 and transposition events from wx-m8. M.G. Neuffer (pers. comm.) has conducted similar experiments and has observed similar rates of induction. We plan to clone the activated Spms, using the strategy of first identifying a band hybridizing to an Spm probe in DNA of a segregating population cut with methyl-sensitive enzymes on whole-genome Southern blots (12). We will use fragments adjacent to the Spm termini as probes to see if an inactive element is present at the same location in parental seed. If so, the inactive elements will be cloned and compared with the activated copies. Techniques are now available that allow one to look for the presence of point mutations (24) which would permit us to see if a mutational event had activated the element.

The next question we would like to address is the possibility that chromosome breaks, or other DNA-damaging agents, can induce an error-prone DNA repair system in higher plants. There is much less information available on this subject, and the following remarks are therefore speculative. EMS, in adding an ethyl group to guanine residues, causes the replication machinery to pause and induces both an adaptive response (40) and the SOS response in E. coli (30). If a system analogous to the SOS response is induced by DNA damage in plants, we then might expect to see a delayed response in the occurrence of mutations. Experiments conducted by Coe and Neuffer (11) indicated that in the meristem of the mature maize seed there are 2-4 cells that will give rise to each ear. A mutation fixed in one DNA strand would therefore be expected to give rise to a sector comprising at least one-eighth of the ear. One can calculate that the half-life of EMS in water is 48 hr and that it is particularly reactive with protein sulfhydrals (28). It is therefore likely that the period of EMS mutagenesis is of short duration.

If chemical mutagenic treatment induces an ongoing error-prone repair system that is capable of causing further mutations, then these might occur several cell divisions after the treatment period and result in very small sectors on the ear. In fact, such sectors are seen. Chourey and Schwartz (10) treated mature seed with EMS and then pollinated resultant plants with pollen carrying an sh1 reference allele. Of the 16 new mutations they found, two occurred as sectors that comprised about one-fiftieth of the ear. In our experiment, both the spontaneous mutation and one of the EMS mutations occurred as a single seed on otherwise nonmutant ears. The second EMS-induced Spm event was found in about one-fourth of the kernels scattered over the ear. There were four additional apparent activation events in the EMS-treated plants that could not be confirmed by germinal inheritance. Three of these were single-kernel events, and the fourth occupied a sector of three kernels on a small ear with 55 kernels.

Clearly, more direct experiments will have to be conducted to see whether an error-prone repair system can be induced in plants, but the work cited here provides an indication that such a system may be operating.

ACKNOWLEDGEMENT

Research was supported by grant GM31093 from the National Institutes of Health and by the Office of Basic Energy Science of the U.S. Department of Energy.

REFERENCES

1. Amacher, D.E., and G.N. Turner (1987) The mutagenicity of 5-azacytidine and other inhibitors of replicative DNA synthesis in the L5178Y mouse lymphoma cell. Mutat. Res. 176:123-132.
2. Bennetzen, J.L. (1987) Covalent modification and the regulation of Mutator element transposition in maize. Molec. Gen. Genet. 208:45-51.
3. Bhagwat, A.S., and R.J. Roberts (1987) Genetic analysis of the 5-azacytidine sensitivity of Escherichia coli K-12. J. Bact. 169:1537-1546.
4. Bianchi, A., F. Salamini, and R. Parlavecchio (1969) On the origin of controlling elements in maize. Genetica Agraria 22:335-344.

5. Call, K.M., J.C. Jensen, H.L. Liber, and W.G. Thilly (1986) Studies of mutagenicity and clastogenicity of 5-azacytidine in human lymphoblasts and Salmonella typhimurium. Mutat. Res. 160:249-258.
6. Catcheside, D.G. (1947) The P-locus position effect in Oenothera. J. Genet. 48:31-42.
7. Chandler, V., C. Rivin, and V. Walbot (1986) Stable non-Mutator stocks of maize have sequences homologous to the Mu1 transposable element. Genetics 114:1007-1021.
8. Chandler, V.L., and V. Walbot (1986) DNA modification of a maize transposable element correlates with loss of activity. Proc. Natl. Acad. Sci., USA 83:1767-1771.
9. Chomet, P.S., S. Wessler, and S.L. Dellaporta (1987) Inactivation of the maize transposable element Activator (Ac) is associated with its DNA modification. EMBO J. 6:295-302.
10. Chourey, P.S., and D. Schwartz (1971) Ethylmethanesulfonate-induced mutations of the Sh_1 protein in maize. Mutat. Res. 12:151-157.
11. Coe, E.H., and M.G. Neuffer (1978) Embryo cells and their destinies in the corn plant. In The Clonal Basis of Development, S. Subtelny and I. Sussex, eds. Academic Press, New York, pp. 113-129.
12. Cone, K.C., F.A. Burr, and B. Burr (1986) Molecular analysis of the maize anthocyanin regulatory locus C1. Proc. Natl. Acad. Sci., USA 83:9631-9635.
13. Doerschug, E.B. (1973) Studies of Dotted, a regulatory element in maize. Theor. Appl. Genet. 43:182-189.
14. Elespuru, R.K. (1984) Induction of bacteriophage lambda by DNA-interacting chemicals. Chem. Mutagens 9:213-231.
15. Elespuru, R.K. (1987) Inducible responses to DNA damage in bacteria and mammalian cells. Env. Molec. Mutagen. 10:97-116.
16. Jaenisch, R., A. Schnieke, and K. Harbers (1985) Treatment of mice with 5-azacytidine efficiently activates silent retroviral genomes in different tissues. Proc. Natl. Acad. Sci., USA 82:1451-1455.
17. Katz, A.J. (1985) Genotoxicity of 5-azacytidine in somatic cells of Drosophila melanogaster. Mutat. Res. 143:195-199.
18. McClintock, B. (1950) The origin and behavior of mutable loci in maize. Proc. Natl. Acad. Sci., USA 36:344-355.
19. McClintock, B. (1951) Mutable loci in maize. Carnegie Institution Washington Yearbook 50:174-181.
20. McClintock, B. (1965) The control of gene action in maize. Brookhaven Symposia on Quantitative Biology 18:162-182.
21. McClintock, B. (1967) Genetic systems regulating gene expression during development. Develop. Biol. Suppl. 1:84-112.
22. McClintock, B. (1968) The states of a gene locus in maize. Carnegie Institution Washington Yearbook 66:20-28.
23. McClintock, B. (1984) The significance of responses of the genome to challenge. Science 226:792-801.
24. Meyers, R.M., Z. Larin, and T. Maniatis (1985) Detection of single base substitutions by ribonuclease cleavage at mismatches in RNA:DNA duplexes. Science 230:1242-1246.
25. Muñoz, E.R., and B.M. Barnett (1987) Comparative study of the clastogenic efficiency of ethyl methanesulfonate and diethyl sulfate in Drosophila melanogaster mature sperm. Mutat. Res. 178:217-223.
26. Neuffer, M.G. (1966) Stability of the suppressor element in two mutator systems at the A1 locus in maize. Genetics 53:541-549.
27. Neuffer, M.G., and E.H. Coe (1977) Paraffin oil technique for treating corn pollen with chemical mutagens. Maydica 22:21-28.

28. Osterman-Golkar, S., L. Eherenberg, and C.A. Wachtmeister (1970) Reaction kinetics and biological action in barley of mono-functional methanesulfonic esters. Rad. Bot. 10:303-327.
29. Quillardet, P., and R. Devoret (1982) Damaged-site independent mutagenesis of phage produced by inducible error-prone repair. Biochimie 64:789-796.
30. Quillardet, P., O. Huisman, R. D'ari, and M. Hofnung (1982) SOS chromotest, a direct assay of induction of an SOS function in Escherichia coli K-12 to measure genotoxicity. Proc. Natl. Acad. Sci., USA 79:5971-5975.
31. Reik, W., A. Collick, M.L. Norris, S.C. Barton, and M.A. Surani (1987) Genomic imprinting determines methylation of parental alleles in transgenic mice. Nature 328:248-251.
32. Rhoades, M.M. (1945) The genetic control of mutability in maize. Proc. Natl. Acad. Sci., USA 31:91-95.
33. Roberts, L.M. (1942) The effects of translocation on growth in Zea mays. Genetics 27:584-603.
34. Santi, D.V., C.E. Garrett, and P.J. Barr (1983) On the mechanism of inhibition of DNA-cytosine methyltransferases by cytosine analogues. Cell 33:9-10.
35. Sapienza, C., A.C. Peterson, J. Rossant, and R. Balling (1987) Degree of methylation of transgenes is dependent on gamete of origin. Nature 328:251-254.
36. Spofford, J.B. (1976) Position-effect variegation in Drosophila. In The Genetics and Biology of Drosophila, M. Ashburner and E. Novitski, eds. Academic Press, New York, 1c:955-1018.
37. Stadler, L.J. (1941) The comparison of ultraviolet and X-ray effects on mutation. Cold Spring Harbor Symposia on Quantitative Biology 9:168-178.
38. Viegas-Pequignot, E., and B. Dutrillaux (1976) Segmentation of human chromosomes induced by 5-azacytidine. Human Genet. 34:247-254.
39. Vogel, E., and A.T. Natarajan (1979) The relation between reaction kinetics and mutagenic action of mono-functional alkylating agents in higher eukaryotic systems. I. Recessive lethal mutations and translocations in Drosophila. Mutat. Res. 62:51 100.
40. Walker, G.C. (1984) Mutagenesis and inducible responses to deoxyribonucleic acid damage in Escherichia coli. Microbiol. Rev. 48:60-93.
41. Zimmermann, F.K., and I. Scheel (1984) Genetic effects of 5-azacytidine in Saccharomyces cerevisiae. Mutat. Res. 139:21-24.

ACTIVATION OF A MUTABLE ALLELE IN ALFALFA TISSUE CULTURE

E.T. Bingham, R.W. Groose, and I.M. Ray

Agronomy Department
University of Wisconsin
Madison, Wisconsin 53706

ABSTRACT

A white-flowered mutant (WFM) was regenerated from tissue culture of a purple-flowered alfalfa donor carrying one dominant allele (C2) for anthocyanin synthesis. Chromosome counts confirmed that WFM was not due to loss of the chromosome carrying C2. The dominant allele had mutated. This mutation has occurred only once among about 2,500 plants regenerated from the original donor. When WFM was recultured, many regenerated plants (about 20%) were purple-flowered. Genetic transmission of the unstable recessive through the zygote (single cell) to progeny that reverted proved that revertants of WFM were not due to culturing chimeral tissue. Hence, the dominant allele (C2) in the donor mutated to an unstable recessive (mutable), c2-m4, which is carried by WFM, and transmitted to its progeny. The c2-m4 allele has now been transmitted through three sexual generations, and in single dosage behaves as it did in WFM.

Allele c2-m4 is unstable in vitro and frequently reverts to the functional state. Reversion occurs early in culture and may be the result of a genome shock associated with callus formation. Callus growth studies indicated that the high frequency of revertant regenerated plants was not due to more rapid growth of revertant callus. Nonrevertant regenerated plants are alike phenotypically, but differ in reversion frequency upon reculture. Allele c2-m4 is relatively stable in planta, and revertant sectors in flowers and revertant gametes are fewer than 1/1,000. The data are discussed in terms of a transposable element which is especially active in vitro.

INTRODUCTION

Regeneration of plants from tissue and cell cultures often results in recovery of variants of the donor plants, termed somaclonal variants (12). Somaclonal variation in plants has been reviewed and discussed in terms of the chromosomal and genetic mechanisms of variation in plants in general (12,15,17) and in alfalfa (*Medicago sativa* L. 2n=4x=32) (3,16).

This report will focus on the origin, phenotypic behavior, and genetic transmission in alfalfa of a white-flowered mutant (WFM) that arose as a somaclonal variant of a purple-flowered tetraploid plant that possessed only one functional allele for pigment production (9). WFM and its progeny are stable white-flowered mutants in planta, but they are unstable in vitro, as evidenced by a high frequency of revertant purple-flowered plants regenerated from tissue culture of WFM (10). Genetic analysis established that a functional allele, C2, of a locus required for anthocyanin pigmentation mutated to an unstable recessive (mutable) allele, c2-m4, which is carried by WFM. The mutable behavior of WFM is unique among several other white-flowered stocks that have been stable during cycles of culture and regeneration, including another white-flowered mutant that arose in a cell selection experiment. The mutable behavior of WFM and its derivatives is most simply explained by the action of a transposable element that is especially active in vitro.

McClintock (13,14) proposed that genomic stress can trigger mechanisms, including transposable element systems, capable of reorganizing the genome. Tissue culture may produce such a genomic stress or shock. Several authors have suggested that the action of transposable elements in vitro may be one of the mechanisms producing somaclonal variation (3,5-7,10,12). Recently, transposable element activity has been found among the progeny of tissue culture-derived maize plants (18). Other chapters in this Volume also report current information about activation of transposable elements in plants. In this chapter, we review published information on the mutable nature of the c2-m4 allele discovered in WFM, and report new information about (i) the behavior of the mutable allele in new genetic backgrounds, (ii) the stable dominant progenitor allele, and (iii) the characteristics of other stable and mutable alleles in alfalfa that have been studied both in vitro and in planta.

DESCRIPTION OF PLANT MATERIALS

Following the discovery that WFM (Fig. 1 and 2) carried a mutable allele of the C2 locus, the parents and grandparents of WFM (Fig. 1 and Tab. 1) were genetically analyzed to characterize the progenitor C2 allele that mutated to c2-m4, and to identify the ancestral stock harboring the transposable element. Another interesting mutable system, involving a purple to white mutant (PWM), was discovered in a sibling of the purple-flowered donor 11 that gave rise to WFM. PWM is helping provide a complete understanding of the origin of mutable allele c2-m4.

The progenitor C2 allele of c2-m4 came from 6-4 ms (Fig. 1 and Tab. 1), a purple-flowered male-sterile with the capacity to regenerate after tissue culture (9). Plant 6-4 ms was a progeny from a cross of two plants from the cultivar "Saranac." The seed parent behaved as a cytoplasmic male-sterile, and the pollen parent carried nuclear genes for maintaining the cyto-sterile condition. Plant ccmf (Tab. 1) was a stable white-flowered, multifoliate segregate descending from a cross of cultivars "Vernal" and "Multileaf." Plant 09 (Tab. 1) was a colchicine double diploid from "CADL" 76-20 (2) that was heterozygous at the C2 locus and carried genes for plant regeneration (4,20). Other stocks described in Tab. 1 were used in experiments discussed in the following sections. The stocks will be discussed in terms of respective experiments.

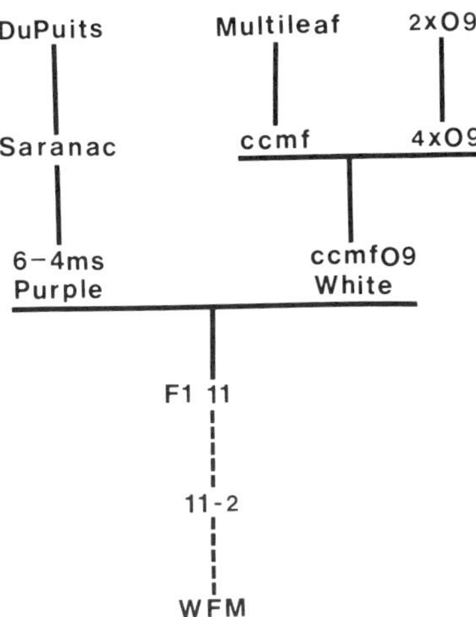

Fig. 1. Pedigree of the purple-flowered tissue donor that gave rise to white-flowered mutant WFM. Stocks are described in Tab. 1 and the text.

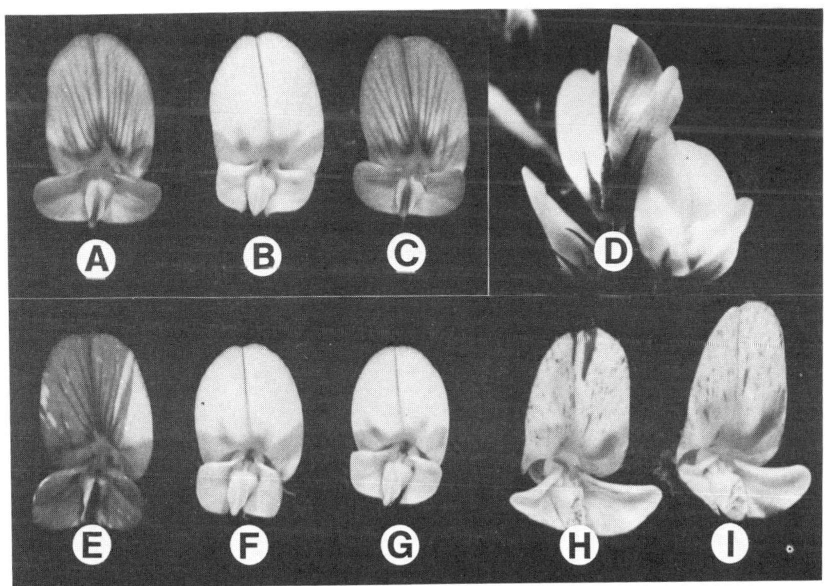

Fig. 2. Flower colors of mutable alfalfa plants. A. Purple-flowered donor 11. B. White-flowered mutant (WFM). C. Purple revertant of WFM. D. Rare somatic sector in WFM. E. A test cross progeny of F_1-14 with white sectors. F. White regenerate of E. G. White regenerate of F. H. Mutable c2-m2 with frequent somatic sectors. I. Regenerate of H that is unaffected by in vitro culture and regeneration.

Tab. 1. Description of tetraploid alfalfa stocks in the pedigree of mutable WFM (see Fig. 1) in tissue culture experiments and in test crosses.

Stock	Genotype at C2 locus	Description
6-4ms	CCCC	Cytoplasmic male-sterile plant from cultivar "Saranac."
09	CCcc	Cultivated alfalfa at the diploid level (CADL) (2), which was chromosomally doubled using colchicine.
ccmf	cccc	White-flowered, multifoliolate plant from cultivar "Multileaf."
WFM	c2-m4ccc	White-flowered mutant regenerated from tissue culture.
HG2	CCCC	Purple-flowered regenerator (3,20).
HGW	cccc	White-flowered regenerator stock produced by sib-mating white F_2 progeny HG2 x white genetic stock.
R64c	cccc	Advanced generation of a somaclonal variant of HG2 that contained a mutation to recessive c2.
T-1c	cccc	White-flowered regenerator stock from cultivar Saranac.
W4 and W5	cccc	White-flowered, nonregenerator stocks used as test-cross parents in genetic analysis.
M2-6	c2-m2ccc	Mutable regenerator stock selected from the F_2 generation of HG2 x mutable c2-m2 stock.

IDENTIFICATION OF MUTABLE ALLELE c2-m4

Tissue Culture and Regeneration

Tissue culture procedures, callusing medium (SHII), and regeneration medium (BOi2Y) have been described (4,9,20). In most experiments, callus was initiated from excised ovaries or petiole segments because they are convenient explants that produce callus readily. Following about four weeks of callus growth from explants, calli were transferred to regeneration medium for four to five weeks. Regenerating calli were then subdivided and again subcultured on regeneration medium for another four to five weeks.

Reversion was based on the presence (revertant) or absence (nonrevertant) of anthocyanin pigmentation in the shoots of regenerated plantlets. Only plantlets with at least two true leaves were scored for reversion. It was possible to classify thousands of regenerates from hundreds of cultures for reversion at the plantlet stage, although it should be noted that a small percentage of revertants do not express pigmentation at this stage. Several samples of plantlets regenerated from WFM have been grown to maturity and scored for reversion on the basis of flower pigmentation. Between 5% and 10% of revertants usually are misclassified as nonrevertant at the plantlet stage. On the other hand, all regenerates classified as revertant at the plantlet stage exhibited full pigmentation when in flower. The small underestimation of reversion frequencies in experiments where reversion was scored at the plantlet stage does not alter the conclusions that may be drawn from these experiments.

Discovery of the White-Flowered Mutant WFM

WFM (Fig. 2) occurred among the first 45 regenerates from petiole-derived calli of purple donor 11-2 (Fig. 1) in the second cycle of an experiment designed to monitor loss of genetically marked chromosomes (9). Donor 11-2 has been recultured from ovary and petiole explants periodically for the past five years, and no additional white-flowered mutants have been found among about 2,500 regenerated plantlets. This result is surprising because, based on aneuploidy due to chromosome loss in other experiments (9), eight white regenerates would be expected among 2,500 regenerates on the basis of chromosome loss alone. We have no explanation for the fortuitous discovery of WFM among the first small group of plants regenerated from donor 11-2.

Initially, it was assumed that WFM had lost the chromosome carrying the functional allele C2 for anthocyanin production. However, root-tip chromosome counts revealed that WFM had the same chromosome number as the donor. Both donor 11-2 and WFM are aneuploid 2n-1=31. Genetic tests established that donor 11-2 was simplex (C2c2c2c2) at the locus, and that aneuploidy did not involve homologous chromosomes carrying this locus (8,9). Donor 11-2 segregated 1 purple: 1 white in a test cross (simplex segregations from a tetraploid are the same as those from a heterozygous diploid). WFM in a test cross produced all white progeny, indicating that the functional C2 allele was not present in the germline (cell layer 2) of WFM. Hence, WFM was due to a mutation, rather than due to chromosome loss.

In Vitro Reversion of Mutable Allele c2-m4

WFM has been cloned from shoot cuttings and recultured in several experiments using various explants, including ovaries and segments of petioles, leaflets, petals, sepals, raceme rachises, and internodes. The reversion frequency was about the same, regardless of the source of explants (8). In original experiments (8,10,11), and in ongoing studies, 23% of 1,624 plants regenerated from WFM have been revertant, with the revertant percentage ranging from 16% to 43% over experiments (Tab. 2).

Revertants tended to occur in clusters in callus cultures rather than independently, and a X^2 test showed lack of homogeneity for frequency of reversion per culture (10). When several revertant plantlets were regenerated from a single culture, perhaps they had a common origin in a sector of callus derived from a progenitor cell that reverted early in culture. Nonetheless, reversion was relatively frequent whether based on plants or cultures.

Alternatives to the proposition that there was a high frequency of reversion in the in vitro culture of WFM included: (i) that reversion is rare

Tab. 2. Instability of mutable allele c2-m4 in vitro as measured by reversion to C2 among plants regenerated from WFM and its sexual derivatives.

Stock	Number: c2-m4 alleles	Number: genotypes tested	Number: regenerated plants	Average reversion frequency	Range
					over explants
WFM	1	1	1,624	23%	16%-43%
WHGW-3	1	1	314	21%	13%-27%
WHGW-4	1	1	95	4%	4%-5%
					over genotypes
WHGW-3 x HGW	0	7	316	0%	-
	1	5	176	23%	2%-40%
WFM x HG2 (White F_2)	1	2	70	14%	14%-15%
WHGW-3 S_1	0	5	175	0%	-
	1	4	196	19%	7%-43%
	2	1	63	67%	64%-70%

but that revertant cells proliferate faster and/or regenerate better than nonrevertant cells; and (ii) that WFM is a periclinal chimera with mutant and nonmutant tissue layers. Both alternatives were rejected on the basis of the following experiments.

A time course experiment indicated that revertant cells (from revertant donors) have no growth advantage over nonrevertant cells (from nonrevertant donors) (10). Also, it was shown that revertant and nonrevertant cells have an equal potential to regenerate from culture, and to grow equally well as plants (8,10). Finally, genetic transmission of the mutable condition through the zygote (single cell) to progeny that reverted proved that a chimera was not involved. Genetic transmission performed two functions: (i) it provided individuals of zygotic origin, and (ii) it transferred the mutable allele to new genetic backgrounds.

In Vitro Behavior of Mutable Allele c2-m4 in New Genetic Backgrounds

Genetic transmission and segregation of the mutable allele must be based on plants regenerated from tissue cultures. This is labor intensive, and is further complicated by the fact that genes controlling regeneration also are segregating. Hence, only a portion of test-cross and self progeny will regenerate, and segregation data are limited to regenerators. Numbers of regenerates that can be scored for reversion are small for some progeny.

WFM was crossed to a white-flowered tester (HGW) which was bred to regenerate (Tab. 1). All progeny of the cross are white, and about one-fourth of them regenerate. Two of five progeny that regenerated also reverted (Tab. 2). Progeny WHGW-3 reverted at about the same rate as WFM (21%), while WHGW-4 reverted at a much lower rate (5%) (10). No explanation is offered for the low reversion frequency of WHGW-4; however, apparent changes in mutability of c2-m4 will be discussed later. A cross of WFM x HG2 (a purple-flowered regenerator described in Tab. 1) produced all-purple F_1 progeny, which were sibmated to produce an F_2 generation segregating tetrasomically for about 35 purple: 1 white progeny. White-flowered progeny also were segregating for regeneration ability. Two of four white-flowered progeny regenerated, and both reverted (14% and 15%, respectively). The important point is that the mutable allele was transmitted through two sexual generations, and retained its ability to revert.

WHGW-3 was the best regenerator among the progeny of WFM, and was selfed and crossed with the white-flowered tester HGW (Tab. 2). The frequency of good regenerators was low, and of 81 test-cross progeny challenged to regenerate, only 12 regenerated samples large enough to analyze (19). Of these, five reverted and seven did not (goodness of fit to expected 1:1 ratio, X^2, 1 d.f., $.50<P<.75$).

In the S_1 progeny test, if WHGW-3 were simplex for the mutable allele (c2-m4c2c2c2), the genotypic array of the S_1 progeny would be 1 duplex (c2-m4c2-m4c2c2):2 simplex (c2-m4c2c2c2):1 nulliplex (c2c2c2c2). Thus, the expected phenotypic array would be 3 mutable:1 nonmutable. Of ten S_1 progeny that could be classified, five reverted and five did not (goodness of fit to expected 3:1 ratio, X^2, 1 d.f., $.05<P<.10$). Plants regenerated from one S_1 donor averaged 67% revertant over two experiments. This may have been due to duplex dosage of the mutable allele, which is expected in one-fourth of the S_1 progeny.

Further research is in progress on effects of dosage on stability in vitro and in planta. Thus far, c2-m4 has behaved as if it is controlled by an autonomous element.

Stable Nonrevertants Derived from Reculture

Early experiments where WFM was recultured suggested that nonrevertant plants regenerated from WFM differed in their ability to revert upon reculture (8). Subsequent experiments using WFM and WHGW-3 (11) indicated that there were three classes of nonrevertants: (Class 1) nonrevertants that revert in vitro at a high frequency (10-30%) typical of WFM; (Class 2) nonrevertants that revert upon reculture, but at lower frequencies (1-9%) than WFM; and (Class 3) nonrevertants that do not revert upon repeated reculture, i.e., stable nonrevertants. Furthermore, samples of nonrevertants that were regenerated from donors of all three classes of nonrevertants of WFM and WHGW-3 were tested for reversion in vitro in another cycle of tissue culture (11). The results indicated that: (i) reculture of Class 1 nonrevertants may produce nonrevertants of all three classes; (ii) reculture of Class 2 nonrevertants produces Class 2 and Class 3 nonrevertants; and (iii) reculture of Class 3 nonrevertants produces only Class 3 nonrevertants.

Most of the nonrevertants of both WFM and WHGW-3 are stabilized Class 3 nonrevertants (50%), the next most common is Class 2 (31%), and

Tab. 3. Stability in planta of WFM and its sexual derivatives carrying a single c2-m4 allele.

Stock	Number			
	Revertant flowers	Flowers scored	Revertant gametes	Progeny scored
WFM	29	42,000	1	1,273
WHGW-3	4	40,000	0	610
WHGW-3S_1 A	5	10,000	0	228
C	0	10,000	0	165
E	3	6,000	0	165
S	13	10,000	0	205

the smallest is Class 1 (19%) (11). Nonrevertants of WFM were arbitrarily ordered into three classes; however, it may be that no two nonrevertants are alike. More extensive tests might differentiate more classes of nonrevertants. The basis for the reduced reversion potential of Class 2 and Class 3 nonrevertants remains to be determined. The stabilization of these nonrevertants could be the result of genetic changes, either at the c2-m4 allele or elsewhere in the genome. Full knowledge of the mechanisms of mutability and stabilization will require molecular genetic analysis of c2-m4 in all classes of nonrevertants and revertants.

STABILITY OF MUTABLE ALLELE c2-m4 IN PLANTA

Reversion occurs in planta, but at a low frequency (Tab. 3; Fig. 2). A direct comparison of in vitro and in planta reversion (e.g., per cell division) has not been attempted due to the differences in the respective tissues. Nonetheless, a single dose of mutable allele c2-m4 in several different genotypes (Tab. 1) is relatively stable in planta (revertant sectors <<1%) compared with the overall in vitro reversion rate (about 20%) (10,19).

Vegetative propagules of WFM and WHGW-3 have been observed as flowering plants for about five years, and at least five-fold more flowers have been observed than reported in Tab. 3 (10,19). These observations, along with the data in Tab. 3, indicate that WHGW-3 is more stable in planta than WFM, in which it is less difficult to find a revertant flower or revertant branch. In addition to the data in Tab. 3, comparable stability of white-flowered plants carrying c2-m4 has been observed while developing isogenic progenitor, mutant, and revertant stocks of c2-m4.

In segregating S_1 and S_2 generations where two to four doses of c2-m4 are expected in some individual plants, in vitro reversion has increased to as high as 70%, and in planta reversion also has increased. The preliminary indication is that each allele of c2-m4 maintains its original capacity to revert, and reversion increases with increased dosage of the mutable allele.

REVERTANT ALLELES ARE STABLE

Revertant alleles are stable, both in vitro and in planta. Several hundred revertant regenerates have been examined over the past five years, and all showed stable anthocyanin pigmentation. Reculture of over 100 revertants of WFM, WHGW-3, and S_1 progeny of revertant WHGW-3 has resulted in over 1,000 pigmented regenerates, at least 200 of which have been grown to mature purple-flowered plants. As discussed earlier for donor 11-2, at least a few white-flowered regenerates would have been expected, based on the known aneuploidy among regenerated plants of alfalfa. The stability of the chromosome carrying the functional allele is some evidence that it is not one of the chromosomes commonly lost in aneuploid variants.

White sectors were observed at low frequency on a purple-flowered S_1 plant of a revertant of WHGW-3 (19). Although this plant was not a good regenerator, a total of 82 plants were regenerated from it in two experiments, and in all cases the regenerates were purple. The mechanism of sectoring in the clone is not known, although some possible mechanisms for similar sectoring in other stocks will be discussed in a later section. In any case, data thus far support the conclusion that revertant alleles are stable.

Stability of revertant alleles in planta also has been observed in several hundred sexual progeny of revertant plants in the process of producing isogenic genetic stocks with progenitor, mutant, and revertant alleles. Phenotypes of stocks carrying progenitor and revertant alleles are essentially indistinguishable. A small percentage of regenerates of WFM and its sexual derivatives over many experiments are weakly pigmented or "partial revertants" (0-3%) (8,19). These partial revertants are variable in their expressivity, but basically stable. Partial revertants add to the list of derivatives of mutable allele c2-m4.

It was noted several years ago that homozygous recessive c2 plants sometimes have a faint tinge of pigment, especially under conditions of low light intensity and cool temperatures (1). Partial revertants exhibit relatively more pigmentation under these conditions. Partial revertants have been tested for in vitro reversion (8), but regenerates resembled the partial revertant donors. Furthermore, partial revertants have not reverted to full pigmentation in planta or produced revertant progeny in test crosses (8). Partial revertants may carry "pale" alleles derived from c2-m4 that are incapable of further reversion to wild-type.

STABLE WHITE-FLOWERED ALFALFA STOCKS

Several white-flowered alfalfa stocks that are unrelated to WFM have been stable in vitro as well as in planta. Their stability is evidence that the high frequency of reversion in WFM and its descendants carrying the c2-m4 alleles is unique among white-flowered alfalfa plants. White-flowered stocks HGW, R64c, and T-1c (Tab. 1) have been stable in tissue culture and regeneration experiments for several years, both as primary regenerates and when regenerated plants were recultured. The mutation from C2 to stable recessive c2 in R64c occurred during a cell selection experiment that involved chemical mutagenesis (3); however, all of the other stable white-flowered stocks were isolated from populations (1). White-flowered stocks W4 and W5 do not regenerate, but some white-flowered segregates of crosses involving W4 and W5 have been stable when used in tissue culture and regenerated. Stocks W4 and W5 have been used as test-cross parents in genetic analyses of mutable alleles.

GENETIC ANALYSIS OF RELATIVES OF WFM

Genetic analysis of parent 6-4 ms and its progeny has revealed that two dominant C2 alleles lost function to produce WFM. One dominant allele mutated to a stable recessive (or a deletion) and the other dominant became mutable allele c2-m4. The evidence is as follows.

Parent 6-4 ms Is Homozygous Dominant

The F_1 progeny of a cross between 6-4 ms and stable white tester (W4) are all duplex (CCcc), as indicated by test crosses of 50 F_1 progeny (Tab. 4). The test-cross progeny segregated homogeneously near 5:1, as expected for progeny of duplex parents, and the combined results (Tab. 4) were a good fit to 5:1 (X^2 1 d.f., $.20<P<.30$). Thus, 6-4 ms is homozygous dominant at the C2 locus and the F_1 progeny of 6-4 ms x white are duplex (CCcc). These included F_1-11, from which 11-2 and WFM were regenerated.

It should be noted that a tetraploid such as 6-4 ms could possess as many as four different C alleles, e.g., CiCjCkCl. The F_1 progeny would each possess two alleles in any one of six combinations: ij, ik, il, jk, jl, and kl. The fact that F_1 progeny may possess different combinations of C alleles may be helpful in explaining abnormal ratios. Certain alleles may be stable and others may be inherently unstable due to a transposable element located at or near the allele.

Aberrant Ratios Among Progeny of 6-4 ms x ccmf09

The F_1 progeny of a cross between 6-4 ms and ccmf09 (Tab. 1 and Fig. 1) were not all duplex as expected. Only 13 of 32 F_1 progeny of 6-4 ms x ccmf09 segregated near 5:1, expected for duplex in the second test cross (Group I, Tab. 4). Among the remaining F_1 progeny, four were near 2:1 (Group II) and exhibited a deficiency in purple test-cross progeny that was highly significant (X^2 1 d.f., $P<.01$), as were all remaining F_1 progeny. Progeny F_1-11 and F_1-14 gave rise to mutables (discussed below) and segregated near 1:1, expected for simplex (Cccc), as did nine others (Group III). This condition indicated that one C2 allele already had lost function or was deleted. Group IV was especially interesting because it was deficient even under a simplex segregation, and indicated that the second C allele was sometimes lost in the germline, during gametogenesis, or failed to be transmitted. It would not be surprising if the different groups represented different combinations of C2 alleles, but this is unknown at this time. Thus, fewer than half of the F_1 progeny of 6-4 ms x ccmf09 were duplex as expected. Test-cross segregations of the remaining F_1 progeny were significant departures from those expected. Clearly, there was much mutant activity at or near the C2 locus in offspring of the cross 6-4 ms x ccmf09.

Tab. 4. Test cross segregations of F_1 progeny of 6-4 ms x white-flowered stocks.

Entry	Number of F_1 progeny test crossed	Test-cross progeny Purple	White	Expected ratio Purple:White	Actual ratio Purple:White
(6-4ms x W4) x W5	50	1,172	256	5:1	4.6:1
(6-4ms x ccmf09) x W5					
Group I-near expected	13	202	45	5:1	4.5:1
Group II-abnormal	4	111	48	5:1	2.3:1
Group III-abnormal	9	122	110	5:1	1.1:1
Group IV-abnormal	4	40	76	5:1	0.5:1
F_1-11-2[a] -abnormal	1	49	53	5:1	0.9:1
F_1-14 -abnormal	1	62[b]	50	5:1	1.3:1

[a]Tissue donor of WFM.

[b]Included 12 purple-mutable test cross progeny (Fig. 2 and text).

Although two dominant alleles lost function to produce WFM, several lines of evidence indicate that only one of them is mutable. Segregations when the mutable allele was transferred to new backgrounds, as discussed earlier, were consistent with a model based on a single allele. Next, test crosses of 54 revertants all segregated 1 purple:1 white, indicating that all of the revertants were simplex (10). If WFM carried two mutable alleles, both capable of reversion in culture, then at least a few duplex revertants with test-cross ratios of 5 purple:1 white would be expected.

Finally, a rigorous test involved the culture and regeneration of white-flowered S_1 progeny of revertants of mutable WHGW-3. If any revertants carried a mutable allele in addition to the dominant revertant, most of the white S_1 progeny of the revertants also would carry it and show reversion in culture. A group of 21 white S_1 progeny of four revertants were all stable in culture. Thus, all tests support the existence of a single mutable allele, c2-m4.

Purple-to-White Mutable Relative of WFM

A striking purple-to-white mutable (PWM) condition was discovered among the test-cross progeny of F_1-14 (Fig. 2; Tab. 4). Flowers are basic purple with many random white sectors. A large plant may exhibit large white sectors including white side branches. It is significant that F_1-14 was the only F_1 progeny that segregated for this condition, and this mutable condition (about 10% of total progeny) did not fit a genetic model. Genetic transmission and tissue culture behavior both indicate that this mutable condition reflects the loss of function of one dominant allele to become a stable recessive or deletion. Test-cross progeny of over 25 PWM plants varied from about 90% white to all-white; i.e., neither dominant purple nor the PWM is transmitted at an expected frequency. In addition, the white progeny have been stable in tissue culture and regeneration (no reversion among over 500 regenerated plants from ten white progeny; E.T. Bingham, unpubl. data). When PWM plants are cultured and regenerated, about 10% of the regenerates are PWM like the donor and the remainder are white. Moreover, when white regenerates are recultured, no reversion has been observed. Therefore, the PWM phenomenon results in stable white c2 alleles, and it appears to behave the same way in planta and in vitro.

The PWM phenomenon provides an explanation for the loss of function of one of the C2 alleles in several of the F_1 progeny of 6-4 ms x ccmf09, including F_1-11, which gave rise to WFM. The origin of WFM involved two events: (i) one dominant C2 allele mutated to a stable recessive (or was lost by deletion), either in planta or in vitro, in the first pass through culture and regeneration; and (ii) the other dominant C2 allele mutated to become mutable allele c2-m4.

BEHAVIOR OF UNRELATED MUTABLE ALLELE c2-m2 IN TISSUE CULTURE

Mutable allele c2-m2 was discovered in a field population in 1968 (21), and recently was bred into regenerable stock M2-6 in order to test its behavior in culture (Tab. 1 and Fig. 2). In contrast to c2-m4 that is stable in planta, c2-m2 exhibits frequent random reversions to pigment formation that result in pigmented streaks and sectors in white flowers. At this time, the c2-m2 allele has been passed through two cycles of tissue culture and regeneration in only one genotype (M2-6). Among 30 plants regenerated from M2-6, 28 resembled the mutable donor and only two were purple revertants. This is about the same frequency of reversion as observed in the germline of M2-6 (21). Six mutable and one purple revertant regenerates were recultured and regenerated, with essentially the same result as in the first culture cycle.

Mostly, mutables were regenerated from mutables, with an occasional purple revertant; only purple plants were regenerated from the purple revertant from the first culture cycle. Hence, the mutability of c2-m2 does not appear affected by the tissue culture process. Developmental timing of expression may be involved. Mutable allele c2-m4, discovered in WFM, reverts early in development, as evidenced by frequent revertant regenerates and rare, but large, revertant sectors on plants and in flowers. Mutable allele c2-m2 reverts randomly, but most often later in development, and appears unaffected by tissue culture.

ACKNOWLEDGEMENTS

Research was supported by the College of Agricultural and Life Sciences, University of Wisconsin, Madison, Wisconsin 53706, and by a grant from the National Science Foundation (PMC-8409115).

REFERENCES

1. Bingham, E.T. (1973) Interaction of two basic color factor genes in alfalfa. Crop Sci. 13:393-394.
2. Bingham, E.T., and T.J. McCoy (1979) Cultivated alfalfa at the diploid level: Origin, reproductive stability, and yield of seed and forage. Crop Sci. 19:97-100.

3. Bingham, E.T., and T.J. McCoy (1986) Somaclonal variation in alfalfa. Plant Breed. Rev. 4:123-152.
4. Bingham, E.T., L.V. Hurley, D.M. Kaatz, and J.W. Saunders (1975) Breeding alfalfa which regenerates from callus tissue in culture. Crop Sci. 15:719-721.
5. Burr, B., and F. Burr (1981) Transposable elements and genetic instabilities of crop plants. Stadler Genet. Symp. 13:115-128.
6. Chaleff, R.S. (1983) Isolation of agronomically useful mutants from plant cell cultures. Science 214:676-682.
7. Freeling, M. (1984) Plant transposable elements and insertion sequences. An. Rev. Plant Physiol. 35:277-298.
8. Groose, R.W. (1985) An unstable anthocyanin mutation in alfalfa (Medicago sativa L.) which reverts at high frequency in tissue culture. Ph.D. thesis, University of Wisconsin, Madison, Wisconsin 53706.
9. Groose, R.W., and E.T. Bingham (1984) Variation in plants regenerated from tissue culture of tetraploid alfalfa heterozygous for several traits. Crop. Sci. 24:655-658.
10. Groose, R.W., and E.T. Bingham (1986) An unstable anthocyanin mutation recovered from tissue culture of alfalfa (Medicago sativa). 1. High frequency of reversion upon reculture. Plant Cell Rep. 5:104-107.
11. Groose, R.W., and E.T. Bingham (1986) An unstable anthocyanin mutation recovered from tissue culture of alfalfa (Medicago sativa). 2. Stable nonrevertants derived from reculture. Plant Cell Rep. 5:108-110.
12. Larkin, P.J., and W.R. Scowcroft (1981) Somaclonal variation--A novel source of variability from cell cultures for plant improvement. Theor. Appl. Genet. 60:197-214.
13. McClintock, B. (1978) Mechanisms that rapidly reorganize the genome. Stadler Genet. Symp. 10:25-48.
14. McClintock, B. (1984) The significance of responses of the genome to challenge. Science 226:792-801.
15. Meins, F., Jr. (1983) Heritable variation in plant cell culture. An. Rev. Plant Physiol. 34:327-346.
16. Mroginski, L.A., and K.K. Kartha (1984) Tissue culture of legumes for crop improvement. Plant Breed. Rev. 2:215-249.
17. Orton, T.J. (1980) Somaclonal variation: Theoretical and practical considerations. In Gene Manipulations in Plant Improvement, 16th Stadler Genetics Symposium, J.P. Gustafson, ed. Plenum Press, New York, pp. 427-468.
18. Peschke, V.M., R.L. Phillips, and B.G. Gengenbach (1987) Discovery of transposable element activity among progeny of tissue culture derived maize plants. Science 238:804-807.
19. Ray, I.M. (1987) Genetic transmission and behavior of an unstable basic color factor allele, c2-m4, which reverts at high frequencies in tissue culture of alfalfa (Medicago sativa L.). M.S. thesis, University of Wisconsin, Madison, Wisconsin 53706.
20. Reisch, B., and E.T. Bingham (1980) The genetic control of bud formation from callus cultures of diploid alfalfa. Plant Sci. Lett. 20:71-77.
21. Talbert, L.E., and E.T. Bingham (1986) Somatic analysis of a mutable trait for anthocyanin pigmentation in alfalfa. Crop Sci. 26:823-827.

STRUCTURE AND DNA MODIFICATION OF ENDOGENOUS Mu ELEMENTS

Vicki L. Chandler, Luther E. Talbert, Laura Mann, and Catherine Faber

Institute of Molecular Biology
University of Oregon
Eugene, Oregon 97403

SUMMARY

The high mutation rate of Mutator stocks of maize is the result of transposition of Mu elements. Most standard stocks of maize contain sequences homologous to Mu elements, yet they show no genetic activity characteristic of Mu transposition. We have determined the sequence, structure, and genomic organization of several Mu-homologous elements from non-Mutator lines. Our results suggest that these elements transposed into their current genomic positions. We discuss the relationship of these endogenous Mu elements to active elements and the role DNA modification may play in their apparent inactivity.

INTRODUCTION

About ten years ago, Robertson (18) described a Mutator stock of maize which had a high mutation rate, producing new mutant alleles at frequencies ranging from 10^{-3} to 10^{-5}. This mutagenic activity is labile and occasionally is lost when Mutator plants are outcrossed or inbred (20). For example, when a Mutator stock is crossed to a standard stock of maize with a normal mutation rate, approximately 10% of the F_1 progeny lose the high mutation rate while 90% retain it (18). Approximately one-third of the mutations which occur in Mutator stocks are unstable, and several related tranposable elements have been isolated from Mutator plants by their insertion into well-characterized genes. A 1.4-kbp insertion, Mu1, has been cloned from Mutator-induced mutations at Adh1 (2), A1 (16), and Bz2 (15). Approximately 10-60 Mu1-like elements are found in Mutator plants, and most mutant alleles that have been examined with molecular probes contain Mu1 insertions (11). A related element, Mu1.7, which differs from Mu1 primarily by an extra 380-bp segment (25), transposed into the Bz1 locus (24).

Other Mu elements have been studied that share terminal inverted repeats with Mu1 and Mu1.7, but have unrelated internal sequences (7; this ms.). Two elements with this structure have transposed in Mutator stocks. A 1.8-kbp element, termed Mu3, was isolated as a result of its insertion into the Adh1 gene (8; Freeling et al., this Volume), and a 1.4-kbp element was cloned from a Mutator-induced waxy allele (Wessler et al., this Volume). The latter element is the size of Mu1, but shares no internal homology. Mu elements do not interact genetically or share sequence homology with several other characterized maize transposable elements such as Ac/Ds and Spm (19). It is not known which Mu elements encode factors required for their transposition, or if other factors are necessary. One level of regulation appears to be DNA modification, as there is a correlation between modification of certain methylation-sensitive restriction sites within Mu elements and loss of Mutator activity in previously active lines (3,6).

Standard non-Mutator stocks of maize whose pedigrees show no Mutator ancestors contain sequences with homology to Mu elements (7). There is no genetic evidence of a high mutation rate in non-Mutator stocks, suggesting that the elements are not transposing at detectable frequencies. In this paper we describe the structure of several Mu-homologous elements from non-Mutator lines, and speculate on their relationship to active Mu elements and the role DNA modification may play in their apparent inactivity.

RESULTS

The lack of activity of Mu elements in non-Mutator stocks could be the result of nucleotide changes that have rendered the elements defective. Alternatively, the elements may be intact but their low numbers, the absence of regulatory factors, or DNA modification may prevent their transposition at a detectable frequency. To investigate these possibilities, we have cloned and characterized several types of Mu-homologous elements from non-Mutator lines. Three distinct types of Mu-homologous sequences have been identified (7). First, intact Mu1- or Mu1.7-like elements have been found in low numbers in some non-Mutator stocks, but not in others. Second, all maize stocks examined contain a unique sequence with homology to internal portions of Mu1 and Mu1.7 elements. This sequence is not associated with Mu termini and does not have the structure of a transposable element (23). Third, all stocks examined have approximately 40-50 copies of sequences homologous to the Mu terminal repeats that are not associated with Mu1 or Mu1.7 internal sequences.

An Intact Mu1 Element May Be Inactive Due to DNA Modification

An intact Mu1 element, termed Mu1.4-B37, was cloned from the non-Mutator inbred line B37 (7). The sequence of Mu1.4-B37 is identical to that of Mu1 (Chandler, Talbert, and Raymond, Genetics in press). Mu1.4-B37 is not found in the same genomic location in most other maize lines, and is flanked by 9-bp direct repeats. These facts suggest that Mu1.4-B37 transposed into its current genomic location in B37. However, Mu1.4-B37 does not appear to be an active transposable element. A single homozygous copy of this element is found in both B37 and W23, and these lines have normal mutation rates. Southern blot analysis suggests that DNA modification of Mu1.4-B37 might contribute to its inactivity (7).

As shown in Fig. 1 (lanes a and b), the HinfI sites in the Mu1.4-B37 termini are modified in W23 DNA such that digestion with HinfI yields fragments larger than those predicted from the cloned sequences (diagrammed in the map in Fig. 1). Complete digestion is expected to yield 1.3-kbp and 0.95-kbp fragments homologous to the Mu1 and flanking probes, respectively. Instead, both the Mu and flanking probes hybridize strongly to the same 2.55-kbp fragment and, to a lesser extent, to a 2.25-kbp fragment. This demonstrates that the HinfI sites in the Mu element are not cleaved, resulting in fusion of the Mu and flanking sequences. The 2.55-kbp fragment corresponds to inhibition at both of the HinfI sites in the Mu1.4-B37 element, and the 2.25-kbp fragment represents cleavage of the left site and inhibition of the right site as drawn on the map. The 2.25-kbp fragment is less than one copy per genome, suggesting that in most cells neither site within the Mu element is cleaved. Both the Mu and right flanking probes hybridize to a few additional fragments. The 1.65- and 1.85-kbp HinfI fragments homologous to the Mu1 probe (Fig. 1, lane a) result from other Mu-homologous sequences in the genome (Talbert and Chandler, Molec. Biol. and Evo., in press). The right flanking DNA probe also hybridizes to one additional fragment unlinked to Mu1.4-B37 (lane b).

The correlation between DNA modification and lack of activity in Mutator stocks suggests that a prerequisite for activation of Mu elements may be the loss of DNA modification. To test whether modification of the HinfI sites in Mu1.4-B37 is stable in the presence of Mutator activity, the Mu1.4-B37 element was introduced into an active Mutator background by genetic crosses. Individual W23 plants containing a single homozygous Mu1.4-B37 element were crossed to several active Mutator plants homozygous for the bz2mu1 allele (6). F_1 progeny were self-pollinated to generate F_2 individuals, and Mutator activity assessed by following the ability of the Mu1 element at bz2mu1 to promote somatic reversion (6,26). DNA was isolated from F_1 and F_2 individuals, and accessibility of the Mu1.4-B37 HinfI sites was examined using Southern blot analysis.

As expected, all the Mu1 and Mu1.7 elements are digestible with HinfI in DNA from active Mutator plants (Fig. 1, lanes c and d). This is also true for the Mu1.4-B37 element, as evidenced by the loss of the 2.55-kbp and 2.25-kbp fusion fragments and the appearance of the 0.95-kbp right flanking fragment (Fig. 1, lanes e and f). This demonstrates that the modification state of the Mu1.4-B37 element can change, depending on the genetic background of the stock. Furthermore, in F_2 Mutator stocks that lost activity, Mu1.4-B37 again becomes modified (Fig. 1, lanes j-l), as do all the Mu elements in the plant (Fig. 1, lanes g-i). Only the 2.25-kbp or the 2.55-kbp fragments are observed, showing that HinfI sites relatively close to the element are cleavable. The amount of the 2.55-kbp and 2.25-kbp fusion fragments varies among different F_2 individuals, demonstrating that the modification of the left Mu1.4-B37 HinfI site does not always occur.

We hypothesize that the DNA modification of Mu1.4-B37 contributes to its inactivity, since several lines of evidence suggest that modified Mu elements are nonfunctional. First, a correlation between an increased level of DNA modification of Mu elements and loss of Mutator activity as measured by somatic reversion has been reported (6,26). Second, Bennetzen (3) showed that deoxycytosine modification of the internal EcoRII site in Mu elements correlates with loss of mutagenic activity in highly inbred Mutator stocks (21). Third, analysis of the transmission of Mu elements to progeny in active Mutator stocks versus stocks that have lost activity suggests that modified elements are not transposing.

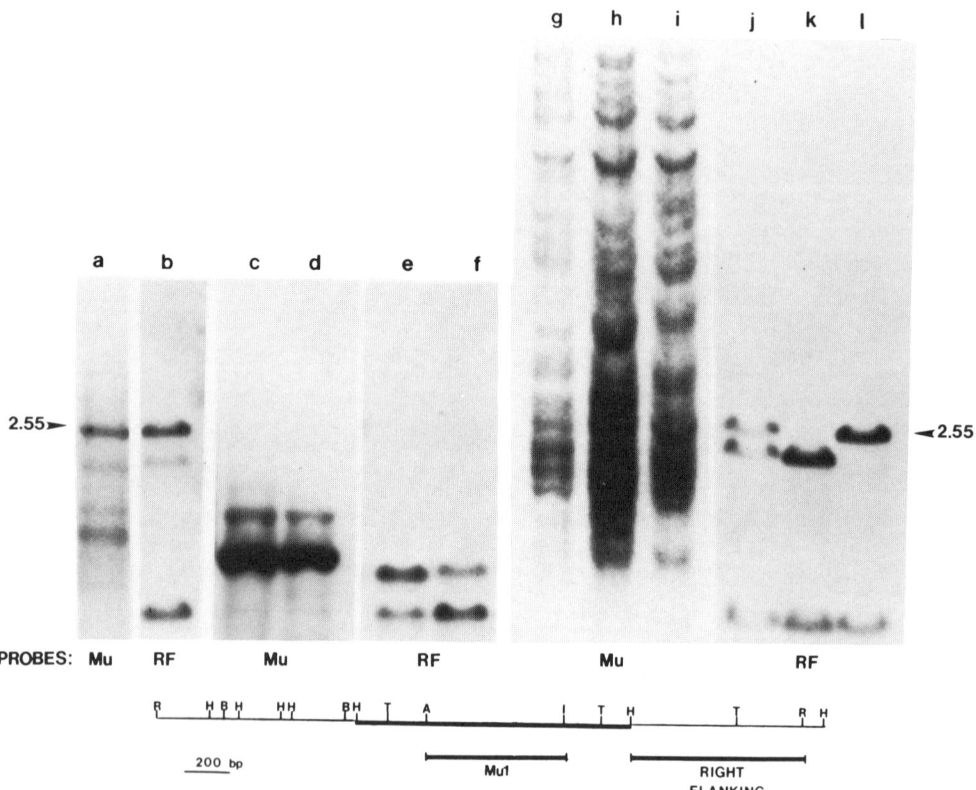

Fig. 1. Modification of Mu1.4-B37. DNA samples were prepared from leaves or immature, unfertilized ears, digested with HinfI, and analyzed on Southern blots. The Mu1 and Mu1.7 elements in each plant were monitored using the internal Mu1 probe (Mu), while the individual Mu1.4-B37 element was monitored using the right flanking probe (RF). Both probes are indicated in the restriction map of the cloned Mu1.4-B37 element and its flanking DNA. Lanes a and b are DNA samples from the non-Mutator W23 parent hybridized with the Mu probe (a) or the right flanking probe (b). Lanes c-f are DNA samples from two progeny of a cross between the non-Mutator W23 parent and an active Mutator plant hybridized with the Mu probe (c,d) or the right flanking probe (e,f). These individuals segregated the bz2mu1 allele normally. Lanes g-l are DNA samples from three F_2 progeny that failed to segregate the bz2mu1 allele normally. Lanes g-i are hybridized with the Mu probe, lanes j-l with the right flanking probe. Restriction enzyme sites on the map are: R, EcoRI; H, HinfI; B, BamHI; T, Tth111-I; A, AvaI; and I, BstNI. The HinfI sites in the left flanking DNA have not been accurately mapped with respect to each other, so that the sites indicated are approximate.

One characteristic of active Mutator stocks is maintenance of a high number of Mu1-like elements (1,4,27). Southern blot analysis of parents and progeny suggests that this is the result of frequent transposition combined with transmission of parental elements to the progeny. In contrast, when Mutator stocks containing modified elements are crossed to non-Mutator stocks, the number of Mu elements decreases substantially in each generation.

An example of this is shown in Fig. 2, Panel A. The non-Mutator W23/K55 hybrid (Panel A, lane b) was crossed to a Mutator stock (Panel A, lane c), and the resulting F_1 progeny subsequently backcrossed to the W23/K55 hybrid. DNA samples were prepared from parental plants, F_1 progeny, and backcross progeny, digested with an enzyme that does not cleave within Mu1, and Southern blots prepared. Several new restriction fragments were observed in most of the F_1 progeny examined. One example is shown in Panel A, lane d. Several parental fragments were transmitted to the individual shown, and examination of a larger number of progeny demonstrated that all the parental fragments were transmitted (data not shown). This suggests that the new fragments represent transposition events rather than rearrangement of existing Mu-homologous fragments.

The Mutator parental DNA contains unmodified Mu1- and Mu1.7-like elements, as shown in Panel B, lane c. The HinfI sites in the Mu1-like elements were modified in DNA from the F_1 plant as shown in Panel B, lane b. When this F_1 plant was backcrossed to the W23/K55 hybrid, no new restriction fragments were observed in the 15 progeny examined, and the number of Mu1-like elements decreased. One example is shown in Panel A, lane e. We have examined a total of approximately 200 individuals from five different Mutator lineages with similar results. When we observe progeny with a significant decrease in the number of elements relative to their parents, this has correlated with parental DNA containing a majority of elements with modified HinfI sites. This demonstrates that plants with modified elements are not able to maintain the high copy number of Mu1-like elements associated with Mutator activity, suggesting that modified elements are inactive.

The few copies of intact Mu1-like elements found in some non-Mutator lines may be relics of a previously active system. The elements, once modified, would quickly become diluted in number in subsequent crosses. Breeders may select for the absence of Mutator activity in the construction of stable inbred lines so that only inactive elements are found in these stocks.

All Maize Lines Contain Multiple Elements Homologous to Mu Termini

Two elements homologous to the Mu termini, but not homologous to internal regions of Mu1 and Mu1.7, were cloned from the inbred line B37. Their structures as determined by DNA sequencing are shown in Fig. 3. Both elements, designated Mu4 and Mu5, have Mu terminal inverted repeats that are approximately 90% homologous to those of Mu1 (Chandler and Talbert, ms. in prep.). Mu4 is ∿2.0 kbp in length and is flanked by 9-bp direct repeats, while Mu5 is ∿1.4 kbp and is flanked by 9-bp direct repeats. The internal portions of Mu4 and Mu5 bear no sequence homology to Mu1, Mu1.7, or Mu3. An unusual structural characteristic of Mu4 and Mu5 is that the terminal inverted repeats extend internally from the Mu termini approximately 290 and 150 bp, respectively. However, aside from the Mu termini, Mu4 and Mu5 are not homologous to each other.

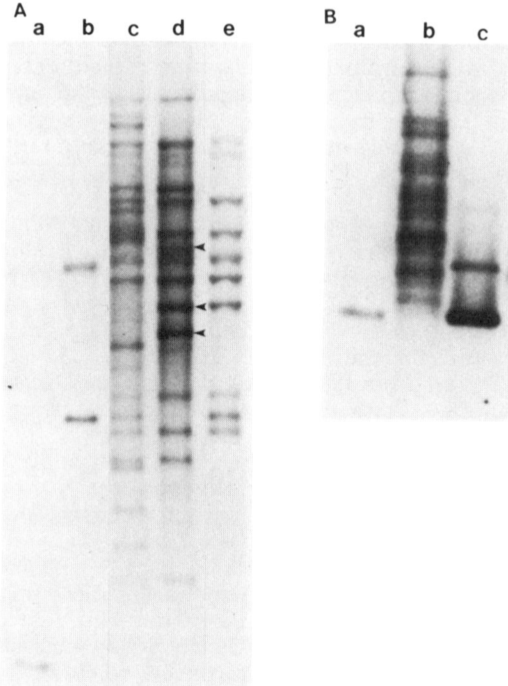

Fig. 2. The number of modified Mu elements decreases each generation. DNA samples were prepared from unfertilized ears, digested with the indicated restriction enzymes, and analyzed on Southern blots. The blots were hybridized with the internal Mu1 probe diagrammed in Fig. 1. Panel A: EcoRI/HindIII double digests of the following DNAs: lane a, the cloned 1.4-kb Mu1 element at approximately one copy per diploid genome; lane b, the W23/K55 hybrid parent; lane c, the Mutator parent; lane d, a single F_1 progeny; lane e, a single backcross progeny of the F_1 in lane d crossed to the W23/K55 hybrid in lane b. The arrows indicate new restriction fragments. Panel B: HinfI digests on the following DNAs: lane a, the cloned 1.4-kb Mu1 element at approximately one copy per diploid genome; lane b, the same F_1 individual shown in Panel A, lane d; lane c, the Mutator parent shown in Panel A, lane c.

Only a subset of the 40-50 copies of Mu termini found in all maize stocks (7) is represented by Mu4 and Mu5 elements. Internal probes for Mu4 and Mu5 hybridize to 2-4 and 8-12 fragments, respectively, in the Mutator and non-Mutator lines (Fig. 4), suggesting that additional related elements remain to be characterized. In contrast to Mu1 elements, there is not an obvious difference in the number of Mu4 and Mu5 elements in these four Mutator and non-Mutator stocks. With both Mu4 and Mu5, one or a few of the fragments homologous to the internal probes are not homologous to the Mu terminal probe (data not shown). This indicates that the internal regions of Mu4 and Mu5 may exist independent of Mu termini.

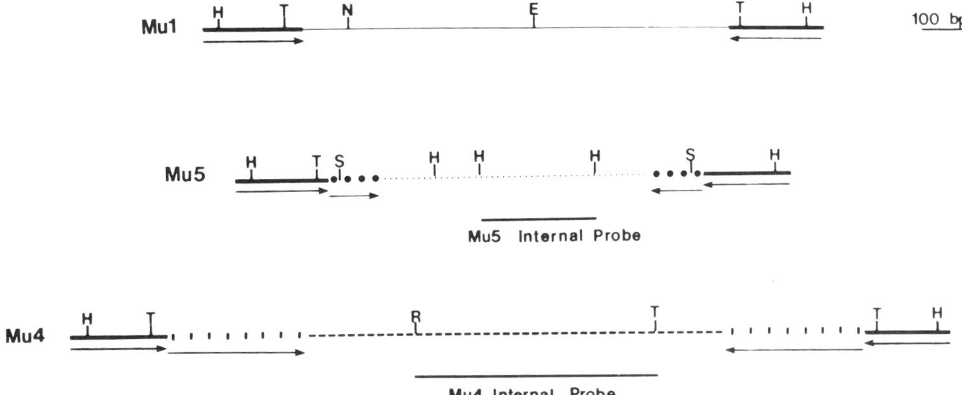

Fig. 3. Comparison of Mu4 and Mu5 with the transposable element Mu1. The Mu terminal inverted repeats which are the only regions of homology are indicated by thickened bars and arrows. The extensions of the inverted repeats for Mu5 are indicated by dots, and for Mu4 by dashes. The restriction sites represented are: H, HinfI; T, Tth111-I; N, NotI; E, BstEII; S, SalI; and R, EcoRI. There are numerous HinfI sites in Mu4 which are not shown.

Sequences homologous to the internal portions of each different Mu element may exist separately from the Mu termini. The sequence homologous to the internal region of Mu1 and Mu1.7 elements has been cloned and sequenced (Talbert and Chandler, Molec. Biol. and Evo., in press). This sequence, termed MRS-A (for Mu Related Sequence), is closely related to the Mu1.7 element and is found in all maize lines and related Zea species we have examined. MRS-A is not a repeated sequence in the genome and does not have the structure of a transposable element in that it lacks Mu termini. The homology between MRS-A and the Mu1 and Mu1.7 elements ceases 1 bp from the inverted repeats. MRS A also has a deletion of ~400 bp relative to the internal region of Mu1 and Mu1.7, yet the remaining 893 bp is 97% conserved. The fact that there are a variety of Mu elements with different internal sequences, and that these internal sequences also exist separately from Mu termini, suggests that the termini may have once been able to transpose independently.

Our sequencing data suggest that Mu4 and Mu5 represent previously uncharacterized transposable elements in the Robertson's Mutator family. We would like to know if these elements are capable of transposing in a Mutator background. We have surveyed a number of Mutator (five) and non-Mutator (ten) lineages looking for an increased number of Mu4 and Mu5 elements. In one Mutator family, shown in Fig. 5, we observed an increased number of Mu4 elements in several individuals. The parental fragments are observed in all plants, suggesting that the new fragments are not the result of recombination or other rearrangements. The new fragments are transmitted to progeny; thus the simplest explanation is that they represent new transposition events. Cloning experiments are in progress to determine if this hypothesis is correct.

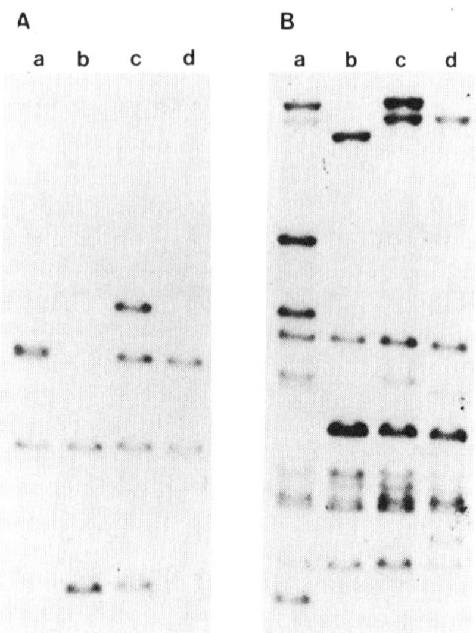

Fig. 4. Number of Mu4 and Mu5 elements in maize lines. A. EcoRI/-HindIII double digests of DNA from two non-Mutator (lanes a, b) and two Mutator (lanes c, d) maize plants, blotted onto Genetran and hybridized to the Mu4 internal probe (as shown in Fig. 3). B. Same blot hybridized to the Mu5 internal probe (as shown in Fig. 3).

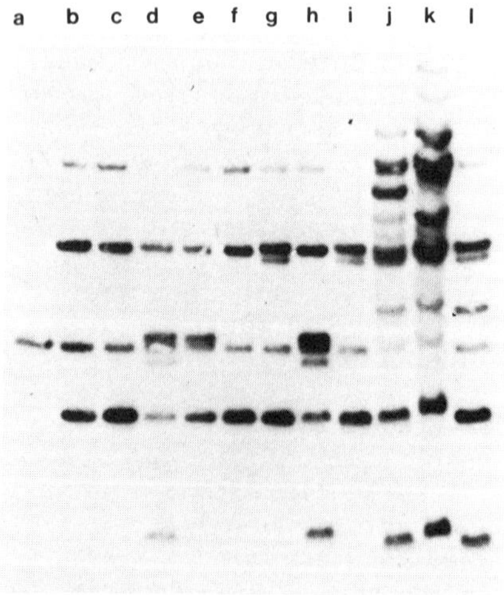

Fig. 5. Increased number of Mu4 elements in some Mutator plants. DNA samples were prepared from unfertilized ears, digested with HindIII and BclI, electrophoresed and blotted onto Genetran. The blot was hybridized with the internal Mu4 probe diagrammed in Fig. 3. Lane a, the cloned Mu4 element at approximately one copy per haploid genome; lane b, the non-Mutator W23 parent; lanes c-l, individual sibling progeny from a single Mutator plant backcrossed to the W23 stock.

DISCUSSION

Our results demonstrate that an intact Mu1 element, termed Mu1.4-B37, normally exists in a modified state in the inbred lines W23 and B37. The strong correlation between DNA modification of Mu elements and the loss of activity in Mutator stocks suggests that activation of Mu1.4-B37 may require the loss of its DNA modification. We show here that this occurs when Mu1.4-B37 is crossed into an active Mutator stock. Additionally, in Mutator stocks that subsequently lose activity, Mu1.4-B37 again becomes modified, as do all the Mu elements in the stock. Thus, the modification state of Mu1.4-B37 reflects the state of all the Mu elements in a Mutator line, demonstrating that Mu1.4-B37 is capable of responding to regulatory factors in a manner similar to other Mu elements.

DNA modification may be a direct mechanism for inactivating Mu elements; alternatively, inactive elements may become more easily modified. Whether modification is the cause or consequence of inactivation, modified elements appear to be nonfunctional, and DNA modification may play a role in maintaining their inactive state. The modification of Mu elements has a practical application in transposon tagging experiments, as it provides a means of diluting the number of Mu elements to identify the element segregating with a particular mutant phenotype. However, DNA modification is not the only mechanism regulating Mu element transposition, as some Mutator losses do not correlate with DNA modification. Bennetzen (3) examined plants that lost the high mutation rate after outcrossing and found that they did not contain modified elements. Only a subset of Mu elements may encode trans-acting factors, and if these elements are lost by segregation, plants may contain only inactive elements. In addition, if the number of active Mu elements become too low, the mutation rate may be decreased to background levels.

Mutator plants differ from non-Mutator plants by the presence of a high mutation rate, which correlates with an increased number of Mu1-like elements. Mu1 elements have been found inserted into most mutant alleles isolated in Mutator lines, suggesting that Mu1-like elements are the most active elements in Mutator stocks. However, it is clear that Mu1 and the closely related Mu1.7 elements are not the only Mu elements capable of transposition in Mutator stocks. Many questions remain, such as how many Mu elements exist, what element or elements encode the transposase activity, how these different elements interact, and what factors influence which subset of elements will transpose in particular plants. An analogy might be drawn with the Ac/Ds transposable element family (reviewed in Ref. 10). Ds elements are defined by their ability to transpose in the presence of an active Ac. This transposition is dependent upon a transposase encoded by Ac. All characterized Ac and Ds elements share an 11-bp terminal inverted repeat sequence. However, the internal regions of Ds elements are quite variable. Some Ds elements appear to be deleted versions of Ac, while others share little homology with Ac except for the 11-bp terminal inverted repeats. However, all of these elements can respond to the Ac transposase. If a similar situation exists for the Mutator family, we might expect Mu4 and Mu5 to respond to the same transposase factors as do Mu1, Mu1.7, and Mu3. Preliminary results presented here suggest that Mu4 may be capable of transposition in some Mutator stocks.

Modification of Mu elements, whether carried out by enzymes encoded by the elements themselves or by other genes in the genome, provides a heritable but potentially reversible mechanism for regulating transposable element activity. Mu elements are not the only transposable elements whose activities may be regulated by DNA modification. The activity of the prokaryotic transposable element IS10 is regulated by adenosine methylation (17), and transposition of Ac and Spm elements in maize is correlated with hypomethylation at certain restriction sites (9,22; Fedoroff et al., this Volume; Schiefelbein et al., this Volume). DNA modification of transposable elements may serve as a general cellular mechanism for regulating a potentially deleterious activity, and active elements may have evolved strategies for escaping the modification systems.

One source of genomic instability may be the activation of previously silent transposable element systems. Environmental and genetic stresses have been correlated with increased mutability in plants and have been hypothesized to activate transposable elements (5,13). Treatments that evoke DNA repair pathways may create hemimodified elements, which in the next round of replication give rise to hypomodified elements. Thus, cryptic elements may become transiently demethylated, which might lead to their activation. It is interesting to note that the Ac transposable element was first detected after chromosome breakage (14), an event that might be expected to stimulate repair synthesis. In addition, a cryptic Ac has been shown to be activated by passage through tissue culture (Phillips and Peschke, this Volume), and a Mu element by gamma irradiation (Walbot et al., this Volume).

We have documented the existence of multiple types of Mu elements in non-Mutator maize stocks. Three cloned examples (Mu1.4-B37, Mu4, and Mu5) are flanked by apparent target site duplications indicative of prior transposition. We are interested to know if these endogenous Mu elements might give rise to an active Mutator system. Activation might occur simply by the genetic construction of a stock which assembled all the necessary components for Mu transposition into a single genome. An alternative, but not mutually exclusive, model is that all the components may exist in certain stocks and activation might occur as a result of a genomic shock which releases cryptic elements, possibly by demethylating them. Once activated, the elements might increase in number up to the level observed for Mu1 in Mutator stocks.

REFERENCES

1. Alleman, M., and M. Freeling (1986) The Mu transposable element of maize: Evidence for transposition and copy number regulation during development. Genetics 112:107-119.
2. Bennetzen, J.L., J. Swanson, W.C. Taylor, and M. Freeling (1984) DNA insertion in the first intron of maize Adh1 affects message levels: Cloning of progenitor and mutant Adh1 alleles. Proc. Natl. Acad. Sci., USA 81:4125-4128.
3. Bennetzen, J. (1987) Covalent DNA modification and the regulation of Mutator element transposition in maize. Molec. Gen. Genet. 208:45-51.
4. Bennetzen, J.L., R.P. Fracasso, D.W. Morris, D.S. Robertson, and M.J. Skogen-Hagenson (1987) Concomitant regulation of Mu1 transposition and Mutator activity in maize. Molec. Gen. Genet. 208:57-62.
5. Burr, B., and F. Burr (1981) Transposable elements and genetic instabilities in crop plants. Stadler Genet. Symp. 13:115-128.

6. Chandler, V., and V. Walbot (1986) DNA modification of a maize transposable element correlates with loss of activity. Proc. Natl. Acad. Sci., USA 83:1767-1771.
7. Chandler, V., C. Rivin, and V. Walbot (1986) Stable non-Mutator stocks of maize have sequences homologous to the Mu1 transposable element. Genetics 114:1007-1021.
8. Chen, C.-H., K. Oishi, B. Kloeckener-Gruissem, and M. Freeling (1987) Organ-specific expression of maize Adh1 is altered after a Mu transposon insertion. Genetics 116:469-477.
9. Chomet, P.S., S. Wessler, and S. Dellaporta (1987) Inactivation of the maize transposable element Activator (Ac) is associated with its DNA modification. EMBO J. 6:295-302.
10. Döring, H.-P., and P. Starlinger (1984) Barbara McClintock's controlling elements: Now at the DNA level. Cell 39:253-259.
11. Lillis, M., and M. Freeling (1986) Mu transposons in maize. Trends in Genet. 2:183-187.
12. McClintock, B. (1951) Chromosome organization and genic expression. Cold Spring Harbor Symposia on Quantitative Biology 16:13-47.
13. McClintock, B. (1978) Mechanisms that rapidly reorganize the genome. Stadler Genet. Symp. 10:25-48.
14. McClintock, B. (1984) The significance of responses of the genome to challenge. Science 226:792-801.
15. McLaughlin, M., and V. Walbot (1987) Cloning of a mutable bz2 allele of maize by transposon tagging and differential hybridization. Genetics 117:771-776.
16. O'Reilly, C., N.S. Shepherd, A. Pereira, Zs. Schwarz-Sommer, I. Bertram, D.S. Robertson, P.A. Peterson, and H. Saedler (1985) Molecular cloning of the a1 locus of Zea mays using the transposable elements En and Mu1. EMBO J. 4:877-882.
17. Roberts, D., B.C. Hoopes, W.R. McClure, and N. Kleckner (1985) IS10 transposition is regulated by DNA adenosine methylation. Cell 43:117-130.
18. Robertson, D.S. (1978) Characterization of a Mutator system in maize. Mutat. Res. 51:21-28.
19. Robertson, D.S., and P.N. Mascia (1981) Tests of 4 controlling-element systems for mutator activity and their interaction with Mu mutator. Mutat. Res. 84:283-289.
20. Robertson, D.S. (1983) A possible dose dependent inactivation of Mutator (Mu) in maize. Molec. Gen. Genet. 191:86-90.
21. Robertson, D.S. (1986) Genetic studies on the loss of Mu mutator activity in maize. Genetics 113:765-773.
22. Schwartz, D., and E. Dennis (1986) Transposase activity of the Ac controlling element in maize is regulated by its degree of methylation. Molec. Gen. Genet. 205:476-482.
23. Talbert, L.E., D. Turks, C. Faber, L. Mann, K. Sylvester, F. Raymond, and V.L. Chandler (1987) Molecular genetic analysis of sequences homologous to Mutator transposable elements in non-Mutator maize stocks. In Plant Molecular Biology, D. von Wettstein and N.-H. Chua, eds. Plenum Press, New York, pp. 181-189.
24. Taylor, L., V. Chandler, and V. Walbot (1986) Insertion of 1.4 kb and 1.7 kb Mu elements into the bronze-1 gene of Zea mays L. Maydica 31:31-45.
25. Taylor, L., and V. Walbot (1987) Isolation and characterization of a 1.7 kb transposable element from a Mutator line of maize. Genetics 117:297-307.

26. Walbot, V. (1986) Inheritance of mutator activity in _Zea_ _mays_ as assayed by somatic instability of the _bz2-mu1_ allele. _Genetics_ 114:1293-1312.
27. Walbot, V., and C. Warren (1988) Regulation of _Mu_ element copy number in maize lines with an active or inactive _Mutator_ transposable element system. _Molec. Gen. Genet._ 211:27-34.

COMPARISON OF METHYLATION OF THE MALE- AND FEMALE-DERIVED *wx-m9Ds-cy* ALLELE IN ENDOSPERM AND SPOROPHYTE

Drew Schwartz

Department of Biology
Indiana University
Bloomington, Indiana 47405

Differential cytosine methylation in the gametes and maintenance methylation are proposed to account for the differential functioning of male- and female-derived genes, such as *R*, in the endosperm. Experiments with the *wx-m9Ds-cy* transposable element mutant, with which it is possible to quantitate cytosine methylation of a particular *PvuII* site, provided evidence for this model.

The functioning of the *R* anthocyanin gene in the aleurone layer of maize endosperm differs strikingly in reciprocal crosses with *r/r* lines. A mottled phenotype, indicative of reduced functioning, results when *R* is introduced by the male gamete, whereas the aleurone is self colored when introduced by the female. The female parent provides two sets of chromosomes to the triploid endosperm, and backcrossed heterozygous endosperms are either *R/R/r* or *R/r/r* depending on the direction of the cross. By the use of the A-B chromosome translocations which undergo nondisjunction at the second microspore division, Roman (H.L. Roman, pers. comm.) was able to manipulate the dosages of the *R* allele. He showed that the phenotypic differences in reciprocal crosses depended only on the direction of the cross and not on gene dosage. This was later confirmed by Kermicle (7). The *R/R/r/r* aleurone is mottled when the *R* allele is paternally introduced, but this genotype is self-colored in the reciprocal cross. This differential functioning is even more extreme with an *R* allele altered by paramutation (2). Furthermore, the differential functioning of male- and female-derived *R* genes persists even in homozygous *R/R/R* endosperm where both are present in each cell (8). This *R* imprinting is limited to the endosperm and is not observed in scutella or other sporophytic tissues (3,8). The functioning capacity of the *R* gene in male and female gametes must differ significantly, and the difference is maintained by some steady-state mechanism for many cell generations during endosperm development.

The phenomenon of maintenance methylation in the replication of ^{m}CG and ^{m}CXG sequences in DNA (1,11) is an obvious candidate to account for this imprinting. According to this scheme, methylation-sensitive genes, including R, are differentially methylated in pollen and embryo sac, and this difference is maintained throughout endosperm but not sporophyte development. Gene inactivation by cytosine methylation has been clearly demonstrated (see Ref. 12). This hypothesis has been subjected to experimental tests in maize, utilizing the wx-m9Ds-cy allele (9), with which one can quantitate the methylation of the cytosines in a particular PvuII site in the transposable Ds element.

The wx-m9Ds-cy allele of the waxy gene is a derivative of wx-m9Ac in which the transposase-inactive Ds element is highly but not completely methylated. The internal cytosines in all of the CCGG sites in the Ds are methylated and not cut by the HpaII restriction enzyme, but methylation of the CAGCTG PvuII sites is variable. There are two PvuII sites in Ac at positions 718 and 3242 (6). As far as can be detected by Southern blot analysis (10), the 3243 site is always methylated and not restricted by the PvuII enzyme, whereas the 718 site is variably methylated, and two PvuII restriction fragments of 4.1 and 6.2 kb are observed in blots hybridized with the internal HindIII probe of Ac. The PvuII sites in the flanking wx sequences are not methylated, and the large 6.2-kb fragment consisting of the entire 4.5-kb Ds plus some of the wx flanking sequences is produced when the 718 site is methylated and not cut. The shorter 4.1-kb fragment is formed when the cytosines in that site are not methylated and the PvuII enzyme makes an internal cut in the Ds element. Thus, the degree of methylation of the cytosines in the 718 PvuII site can be determined by comparison of the relative intensities of the 4.1- and 6.2-kb PvuII bands.

Reciprocal backcrosses to inbred wx/wx tester lines were made using the same wx-m9Ds-cy plants as male and female parent. One-half of each ear was harvested 16 days following pollination for DNA extraction from the endosperm. The remaining portion of each ear was allowed to develop to maturity for DNA extraction from germinating seedlings.

For DNA extraction, the 16-day immature kernels were shaved off the freshly harvested ears into mortars and the endosperm liquefied by gentle pressure with a pestle. Only light pressure was applied so as not to disrupt the cells of the maternal pericarp tissue and thereby contaminate the endosperm extract with sporophyte DNA. The DNA was purified as previously described (5) except that the CsCl centrifugation step was omitted.

The distribution of PvuII restriction fragments was determined using Southern blot transfer and hybridization. Ten micrograms of DNA were cut with the enzyme, transferred to nitrocellulose, and hybridized with the internal HindIII probe of Ac radioactively labeled by nick translation using $\alpha^{32}P$ nucleotides.

Analysis of the restriction patterns of endosperm DNA extracted from the reciprocal crosses clearly indicates a striking difference in the methylation of the Ds element of wx-m9Ds-cy in the endosperm. The cytosines in the 718 PvuII site of male-derived wx-m9Ds-cy DNA appear to be completely methylated, and only the 6.2-kb fragment, from cuts outside the Ds element, is produced. On the other hand, the same 718 site in female-derived wx-m9Ds-cy is variably methylated, and both the 4.1- and 6.2-kb fragments are formed (Fig. 1A). This difference is not found in the embryo, and the DNA extracted from germinating seedlings arising from the reciprocal crosses shows the presence of both the 4.1- and 6.2-kb bands with the same relative intensity (Fig. 1B).

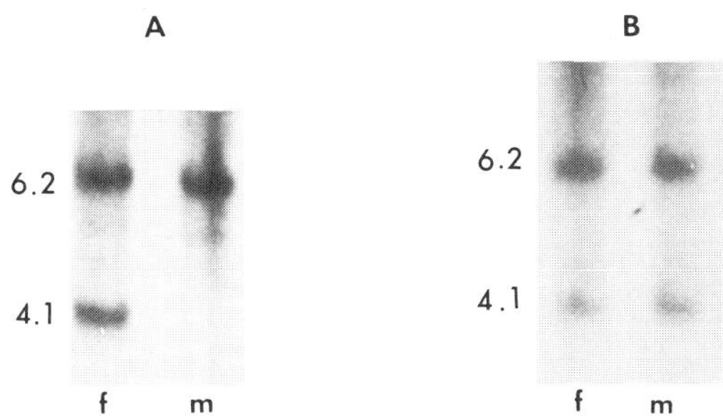

Fig. 1. Comparison of methylation of the 718 PvuII site of the Ds element in the wx-m9Ds-cy gene introduced via the male (m) or female (f). The DNA was extracted from endosperm (A) and seedlings (B). The Southern blots were probed with the internal HindIII fragment of Ac (5). The 6.2-kb PvuII restriction fragment is produced when the 718 site is methylated and not cut. The 4.1-kb fragment is formed when the 718 PvuII site is not methylated and cut by the restriction enzyme.

The molecular basis for the partial methylation of the 718 PvuII site is not known. Sufficient DNA cannot be extracted from single immature endosperm for Southern blot analysis. The presence of both the 4.1- and 6.2-kb bands may result from variation between endosperms in the methylation of the 718 PvuII site, or each endosperm may contain a mixture of cells with the DNA methylated or unmethylated at that site. Both bands are observed in Southern blots of DNA extracted from leaves of individual seedlings (9). However, recent studies suggest that de novo methylation and demethylation are ongoing processes in the sporophyte and that this accounts for the partial methylation and the presence of both the 4.1- and 6.2-kb bands. This conclusion is based on the following observations. There is a striking difference in the digestion of maize DNA by the HpaII and Msp1 restriction enzymes. The DNA is highly digested by Msp1 but only slightly by HpaII, as judged by ethidium bromide staining of enzyme-treated DNA following gel electrophoresis. Both enzymes have a CCGG target but HpaII does not cut when the internal C is methylated, while Msp1 does not cut when the external C is methylated. This indicates that, in contrast to the internal C of CCGG, the external C is not maintenance methylated during DNA replication. The CCGG sites at the BamHI end of the mutant Ac element in wx-m9Ds-cy leaf DNA are not cut by HpaII or Msp1. Since the external C of CCGG is not maintenance methylated, it must be methylated de novo in a continuing process.

The results suggest that only maintenance methylation of the Ac element occurs during endosperm development, since de novo methylation would erase any difference in methylation introduced by the male and female gametes; and, in fact, in endosperm DNA the CCGG sites at the BamHI end of the mutant Ac are cut by Msp1 but not by HpaII.

These studies with the wx-m9Ds-cy allele provide a model to account for R imprinting. According to this scheme, the cytosines in the DNA of the R allele are not methylated in the polar nuclei of the embryo sac and remain unmethylated in the endosperm following fertilization. On the other hand, some of the cytosines in critical regions of the R gene are methylated in the sperm of the pollen grains and persist in this condition throughout endosperm development by maintenance methylation, resulting in reduced gene function. Paramutation may render the R gene more sensitive such that more of the cytosines are methylated, with a further reduction in gene function.

Increased methylation of particular transposons in pollen may be responsible for the altered organ-specific expression of the Mu-suppressed Adh1-3F1124 allele (4). If the differential functioning of the Adh gene in pollen and sporophyte does in fact reflect a difference in methylation of Mu, Adh1-3F1124 functioning in backcross endosperm should vary with the direction of the cross.

REFERENCES

1. Bird, A.P. (1978) Use of restriction enzymes to study eukaryotic DNA methylation. II. The symmetry of methylated sites supports semi-conservative copying of methylation pattern. J. Mol. Biol. 118:49.
2. Brink, R.A. (1956) A genetic change associated with the R locus in maize which is directed and potentially reversible. Genetics 41:872.
3. Brink, R.A., J.L. Kermicle, and N.K. Ziebur (1970) Depression in the female gametophyte in relation to paramutant R expression in maize endosperm, embryos and seedlings. Genetics 66:87.
4. Chen, C., K.K. Oishi, B. Kloeckener-Gruissem, and M. Freeling (1987) Organ-specific expression of maize Adh1 is altered after a Mu transposon insertion. Genetics 116:469.
5. Dennis, E.S., M.M. Sachs, W.G. Gerlach, E.J. Finnegan, and W.J. Peacock (1985) Molecular analysis of the alcohol dehydrogenase 2 (Adhz) gene of maize. Nucl. Acids Res. 13:727.
6. Fedoroff, N., S. Wessler, and M. Shure (1983) Isolation of the transposable maize controlling elements Ac and Ds. Cell 35:235.
7. Kermicle, J.L. (1970) Dependence of the R-mottled aleurone phenotype in maize on mode of sexual transmission. Genetics 66:69.
8. Kermicle, J.L. (1978) Imprinting of gene action in maize endosperm. In Maize Breeding and Genetics, D.B. Walden, ed. John Wiley and Sons, New York, p. 357.
9. Schwartz, D., and E. Dennis (1986) Transposase activity of the Ac controlling element in maize is regulated by its degree of methylation. Mol. Gen. Genet. 205:476.
10. Southern, E.M. (1975) Detection of specific sequence among DNA fragments separated by gel electrophoresis. J. Mol. Biol. 98:503.
11. Wigler, M., D. Levy, and M. Perucho (1981) The somatic replication of DNA methylation. Cell 24:33.
12. Yisraeli, J., and M. Szyf (1984) Gene methylation patterns and expression. In DNA Methylation: Biochemistry and Biological Significance, A. Razin, H. Cedar, and A.D. Rigg, eds. Springer-Verlag, New York, p. 353.

CONTENTS: POSTER ABSTRACTS

TRANSPOSITION OF Mu1 IN TRANSGENIC FLAX

Phil Armitage, Nazir Basiran, Rod Scott, and John Draper

Botany Department
Leicester University
University Road, Leicester, United Kingdom

The transposable element Mu1 has been stably introduced into flax. In some constructs, Mu1 is inserted into the untranslated leader sequence of a nos/npt-II gene; transposition from this site thus restoring the kanamycin resistance (kan^r) phenotype. Using the latter type of construct, it has been found that kan^r transformants can be recovered under conditions allowing dedifferentiation of the transformed flax cells. The frequency is around 30% of that achieved using a transformation vector carrying a normal constituitively expressed nos/npt-II gene. Assays reveal that all of the kan^r transformants contain high levels of NPT-II activity. In some lines NPT-II activities are also found with an abberant molecular weight. Southern blotting analysis of Mu1 transposition is complex, possibly due to the cell lines not being clonal. In some cell lineages the nos/npt-II is no longer linked to Mu1 and the transposable element has been found at one or several new locations in the genome. Strongly hybridizing bands homologous to Mu1 in some transformed tissues suggests that some new Mu1 insertions may be repressed for further transposition. It is clear that if transposition can be induced and then repressed, controlled random transposition could be used as a valuable tool for gene cloning in the future.

TAGGING THE cms S RESTORER GENES IN MAIZE

Prasanna Athma, J.R. Laughnan, and Susan Gabay-Laughnan

University of Illinois
Urbana, Illinois 61801

We are following two routes to tag the cms-S fertility restorer genes in maize. The first approach involves the use of controlling elements as transposable units to identify individual maize plants in which a controlling element has been inserted at a site within standard or newly arisen cms-S nuclear restorer genes. Of 438 male-sterile plants identified, 224 have so far been testcrossed and 35 of these segregated for both male-sterile and male-fertile offspring. These prospective cases of insertion are subject to further testcrossing to determine whether they are legitimate cases of controlling element insertions into cms-S restorer genes.

We have shown that RfI, RfIII, RfIV, and RfVI are themselves transposable, and we hope to characterize them at the molecular level through an alternate route by identifying their insertions into maize genes such as Sh1 and Adh1, for which gene clones are available. We have involved the wild-type version of Sh1 in crosses with the above restorer genes. Since these restorer strains are also wild-type for the target gene locus, the F_1 individuals are homozygous for these genes and carry a restorer gene. The pollinator we used was homozygous-recessive for the target gene locus, and harvested ears were screened for kernels with the recessive phenotype. Of 934,000 kernels screened, 28 possible Sh1 kernels were found. These prospective mutants are being tested to determine whether they are insertion mutations.

In the case of Adh1-Rf combinations, the allyl alcohol pollen treatment procedure of Freeling and Cheng [Genet. Res. (1978) 31:107-129] was used to enrich for alcohol dehydrogenase-1 mutants. Ears with three or fewer kernels were selected, and plants from these are being further analyzed.

FURTHER CHARACTERIZATION OF VARIATION WITHIN WILD-TYPE ALLELES OF THE Sh1 LOCUS

J.W. Baier and L.C. Hannah

University of Florida
Gainesville, Florida 32611

Wild-type alleles of the Sh1 locus show numerous polymorphisms that can be explained as visitation by transposable elements. Sequence comparison between p17.6 [Zack et al. (1986) Maydica 31:5] and a genomic clone [Werr et al. (1985) EMBO J. 4:1373] showed a total of eight changes; seven of these could be explained by transposable elements.

Further studies of wild-type alleles of Sh1 were conducted. Sequencing of three additional alleles (p21.4, p21.6, and pBD10.5) revealed a number of polymorphisms which may be due to insertions and/or deletions. A comparison of pBD10.5, p17.6, and the Werr et al. clone revealed that pBD10.5 was not identical to either over its entirety but contained areas of similarity to each at different locations within the gene. The pBD10.5 allele can be viewed as a hybrid between p17.6 and the Werr et al. clone requiring a minimum of four crossover events along the length of the gene. Neither p17.6 nor the Werr et al. clone can be considered as a hybrid between the other two without invoking additional mutations. Furthermore, a number of base pair changes were discovered in exon 16 that suggest a possible "hot spot" for transposable element insertion. Sequence comparison among the alleles also revealed a 92-bp insertion 5' to the first exon in the clone p21.6.

THE CORRELATION BETWEEN METHYLATION AND REACTIVATION OF INACTIVE Spm ELEMENTS IN MAIZE

J. Banks, J. Kingsbury, P. Masson, and N. Fedoroff

Department of Embryology
Carnegie Institution of Washington
Baltimore, Maryland 21210

The a-m2-7991A1 mutable allele is caused by the insertion of an autonomous Spm transposable element into the maize a locus. Derivatives of the a-m2-7991A1 allele have been isolated in which the inserted element has become genetically inactive. The a-m2-8167B allele is a genetically defective derivative of a-m2-7991A1. However, it is structurally indistinguishable from, and has identical subterminal sequences to, the a-m2-7991A1 allele [Masson et al., Genetics (1987) 179:117-137].

Southern blot analysis of genomic DNA isolated from these plants was used to correlate activity of an element with methylation of sequences corresponding to a (G-C)-rich region of Spm. This region has been shown to be an important cis-determinant of excision frequency of Spm. In the fully active a-m2-7991A1 element, no methylation of these sequences was detected. In the defective a-m2-8167B element, all sequences analyzed in this region were methylated. In the inactive a-m2-7991A1 element, a population of methylated and nonmethylated sequences was observed. The inverse correlation between the extent of methylation of an inactive element and its ability to become reactivated suggests one possible role of methylation in the maintenance of an inactive state of Spm expression.

DNA MODIFICATION OF SEQUENCES WITHIN AND FLANKING MAIZE TRANSPOSABLE ELEMENTS

Jeffrey L. Bennetzen, Willis Brown, Patricia Springer, and Kathrin Schrick

Purdue University
Department of Biological Sciences
West Lafayette, Indiana 47907

Numerous recent studies have shown that modification of particular sequences within a plant transposable element is associated with element inactivation. In animal systems, methylation of specific nucleotides in a gene are correlated with decreased transcriptional activity of the modified locus. To date, the status of tissue-specific or induction-related modulations of DNA methylation in plants is not clear.

By restriction enzyme analysis, the maize genome is modified at over 75% of all 5'-CG-3' and 5'-CNG-3' cytosines. Sequences repeated at more than 1,000 copies per genome (except ribosomal DNA) are generally over 90% modified at these sites. Conversely, single copy DNAs within and near identified genes are essentially unmodified at 5'-CG-3' and 5'-CNG-3'. We have not detected any tissue-specific changes in modification correlated with alterations in gene activity.

Mutator (Mu) and other plant transposable elements preferentially insert into "undermodified" (i.e., presumably low copy number) DNA. When a Mu element becomes inactivated in a methylation-associated event, extensive 5'-CG-3' and 5'-CNG-3' modification does not extend into the flanking target DNA. In addition, heavy modification of a Mu element inserted in the first intron of Adh1 does not lead to significant additional loss of Adh1 expression.

We have also seen variations in the timing, degree, and effect of Mu element modification. Some terminal sites are modified very early during Mu element inactivation. Modification of these sites, but not several internal sites, may be lost during later stages of inactivation or along with reactivation of a Mu element. Mu elements can show tissue-specific events that appear to be due to a temporal progression of modification.

ASSAYING METHYLATION OF Mu SEQUENCES IN ZEA MAYS BY DENSITY GRADIENT CENTRIFUGATION

Anne Bagg Britt, Christine Warren, and Virginia Walbot

Department of Biological Sciences
Stanford University
Stanford, California 94305

A correlation has been reported between the activity of the maize transposons Mu and Ac and the ability to digest the sequences corresponding to these transposons with certain restriction enzymes. The pattern of restriction digestion sensitivity is consistent with the hypothesis that these uncleavable sites carry methylcytosine at the sequences CG and CNG. However, we cannot directly assay methylation of these sequences, and it is possible that some other form of DNA modification is masking these restriction sites. The assay of modification of genomic sequences by restriction digestion is also necessarily limited to a subset of restriction sites.

In order to further identify the nature, distribution, and degree of modification of transposon and other genomic sequences, we have employed CsCl density centrifugation as an alternative assay for DNA modification. Total maize DNA, digested with TaqI, was spun to equilibrium on a CsCl gradient, fractionated, and the density of the fractions assayed by refractive index. Samples were then loaded onto slot blots and the blots sequentially washed and hybridized with probes from known chromosomal, transposon, and mitochondrial sequences. The density of a DNA fragment depends on both its base composition and on the degree of modification. Knowing both the density and the sequence of a specific fragment, we can use this relationship to calculate the degree of DNA modification. Our results with active and inactive Mutator lines suggests that 70% of Mu CG and CNG sites are methylated in inactive lines, while only 47% of such sites are methylated in active lines.

Research support comes from the National Institutes of Health (GM32422) and the United States Department of Agriculture (86-00172). A.B. is a postdoctoral trainee of the U.S. Public Health Service Training Grant CA09302 in Cancer Biology at Stanford University.

PROGRESS TOWARD THE CLONING OF THE Y1 GENE FROM MAIZE

Brent Buckner, Todd L. Kelson, Donald S. Robertson, and David W. Morris

Department of Genetics
Iowa State University
Ames, Iowa 50011

The primary regulatory element in the carotene biosynthetic pathway in maize is believed to be the gene product of the Y1 locus. We have produced 28 families segregating for plants heterozygous for different putative Mu-induced y1 mutations (Y1/y1-Mum). DNA of 40 such heterozygous plants selected from these families was isolated. The approximate number of Mu-homologous elements in each plant was established by Southern blot analyses. Copy number of Mu elements ranged from six to 20 per diploid genome. The plants which had low Mu copy numbers were outcrossed to a homozygous white endosperm stock with the standard y1 allele (y1/y1). Seeds from ears of this cross were expected to be of two types: yellow (Y1/y1), and white (y1/y1-Mum). DNA was isolated from plants derived from yellow and white seeds. These DNAs were digested with restriction endonucleases that cleave outside the Mu1 element, electrophoresed, Southern blotted, and hybridized with Mu1 DNA. The EcoRI hybridization profile of sibling plants derived from the outcross of one plant, 2045-10, contained a 5.9-kbp band that appeared in all plants derived from white seeds, but was absent in all plants derived from yellow seeds. It is expected that the y1 gene, carrying a Mu element insertion, would segregate exclusively with mutant white-seeded individuals. An attempt is being made to clone the DNA segment that segregates exclusively with plants derived from white seeds.

THE MECHANISM OF Ds-INDUCED CHROMOSOME BREAKAGE

Frances A. Burr and Benjamin Burr

Brookhaven National Laboratory
Upton, New York 11973

Certain states of Ds break the chromosome where they reside. McClintock investigated two aberrant chromosomes that contained either tandem or inverted repeats between the original position of Ds and its new position. McClintock believed that the aberrant chromosomes were the result of exchanges between two sister chromatids. If this assumption is correct, it means that the commonly observed breaks are the reciprocal of duplication events. We were able to test this hypothesis by setting up the cross C1-S sh bz wx, Ac x C1-I Sh Bz Wx Ds. In two doses C1-S is able to partially overcome the inhibitory effect of C1-I, and this cross results in pale Bz Wx kernels. If Ds transposes beyond C1 and duplicates this region, then C1-I occurs in two doses and is completely inhibitory, giving rise to colorless sectors. A model involving chromosome looping and pairing between new and old sites of Ds prior to chromosome replication explains how duplication in an inverted or tandem fashion can occur. The reciprocal event of this duplication is breakage of the sister chromatid that was not duplicated. The model predicts that when C1-I is duplicated, it will be found in twin sectors paired with C sh bz wx tissue. Such twin sectors are observed in this cross.

IDENTIFICATION AND CHARACTERIZATION OF ACTIVE TRANSPOSABLE ELEMENT SYSTEMS IN SOYBEAN

Joel M. Chandlee and Lila O. Vodkin

U.S. Department of Agriculture-PGGI-PMGL
Beltsville, Maryland 20705 and
University of Illinois
Department of Agronomy
Urbana, Illinois 61801

Our laboratory has been interested in characterizing active transposable element systems in soybean since the identification of an insert element (Tgm1) in the lectin gene (Le1) of lectin-negative soybean cultivars. Our goals are to: (i) identify and characterize unstable alleles producing mottled phenotypes which suggest the involvement of transposable elements; and (ii) try to associate the instability with genomic rearrangements of Tgm1-related sequences.

We have accumulated genetic data which implicate transposable element activity in some unstable phenotypes:

(i) A mutable allele (r^m) has been identified at the R locus which is responsible for conditioning seed coat color. The allele exhibits both somatic and germinal instability in its expression such that black-seeded or brown-seeded lines that breed true can be recovered from our original stock at a relatively high frequency.

(ii) Variegated leaf mutants that arose spontaneously or after exposure of seed to gamma radiation have been recovered.

(iii) Variegated phenotypes have been recovered from F_2, F_3, and F_4 populations derived from a variety of crosses of different soybean lines.

Southern blot analysis of genomic DNAs using a Tgm1-specific probe is being conducted on lines harboring the various "mutable" alleles in order to determine whether there is an association between the instability in the expression of the alleles and genomic rearrangements of Tgm1-related sequences. Results indicate that in some cases, this indeed appears to be the case.

FOUR MUTABLE ALLELES FOR ANTHOCYANIN IN ALFALFA

W.M. Clement, Jr.,[1] L.E. Talbert,[2] E.T. Bingham,[2] and R.W. Groose[2]

[1]Vanderbilt University
Department of General Biology
Nashville, Tennessee 37235
[2]University of Wisconsin
Department of Agronomy
Madison, Wisconsin 53706

Alfalfa with mutable anthocyanin pigmentation was discovered in 1958, 1968, 1980, and 1982. Each mutable was studied independently by the four respective authors, and in a combined test of allelism. The four mutables are each due to mutation from recessive to dominant at the basic anthocyanin locus C2 in alfalfa. They are designated c2-m1, c2-m2, c2-m3, and c2-m4. Mutable phenotypes of m1, m2, and m3 are similar, and they have streaks and sectors of pigment in flower petals and seeds that are otherwise white. Reversion occurs at various times in development, and may result in periclinal chimeras.

The c2-m4 allele is unique in that it arose during tissue culture of a stock carrying stable dominant allele C2, whereas the other three mutables were discovered in plant populations. Furthermore, m4 is very stable in planta (revertant sectors in <1/1,000 flowers; revertant gametes <1/1,000) but very unstable in vitro as measured by about 25% revertant plants regenerated from tissue cultures. Most m4 reversion occurs relatively early in development and results in completely pigmented in vitro revertants, and in large sectors and whole pigmented shoots on in planta revertants.

Nonrevertants fall into three classes: (i) those that continue to revert at about 25% upon reculture; (ii) those that revert at a lower frequency; and (iii) those where m4 has stabilized as a recessive. The mutation rate at many other loci is higher in mutable stocks than in wild-type populations, implicating the existence of an active transposable element system. Molecular genetic research is in progress to clone mutable alleles, and to learn the molecular basis of the mutable phenotypes.

CHARACTERIZATION OF AN UNSTABLE ALLELE IN *ZINNIA ELEGANS*

R.K.D. Cowen

University of Illinois
Horticulture Department
Urbana, Illinois 61801

An unstable allele conditioning pigment flecking in *Zinnia elegans* corollas is described. Inflorescence flecking arose in a *Z. elegans* type following interspecific hybridization between *Z. elegans* and *Z. haageana*. Selfing of plants homozygous for corolla flecking resulted in segregations of 9 flecked to 1 solid-colored revertants. Further selfing and crossing of flecked types resulted in classes of flecked types which differ in both timing and frequency of excision. Image analysis is used as a research tool in the study of gene expression.

CHARACTERIZATION OF PALE PURPLE AND FULL PURPLE REVERTANTS OF bz-m4 Derivative 6856 AT THE MOLECULAR LEVEL

Michael F. Dowe, Jr., Cindy M. Voisine, and
Anita S. Klein

Department of Biochemistry
University of New Hampshire
Durham, New Hampshire 03824

There are a series of Ds-controlled bz-m4 alleles: bz-m4 D6856 resulted from four independent Ds transpositions into or near the bronze locus. The bz-m4 D6856 mutation was caused by a 6.7-kb insertion in a region which corresponds to the 5' untranslated leader of Bz mRNA. Ds elements are at both ends of the insertion; the elements are approximately 2 kb and are in the same direct orientation. Between the two Ds elements in a 2.7-kb segment of DNA that is actually a duplication of the 3' end of the Bz gene. The leftmost Ds element in the bz-m4 D6856 insertion is not intact; 121 bp are deleted including the TIR (Klein et al., ms. in prep.). Therefore the transposon-like structure has only three inverted repeats which could be targets for the Ac-encoded transposase.

Several revertants of the Ds-controlled allele bz-m4 D6856 have been analyzed by genomic mapping techniques. The entire complex transposon has been excised in the full purple revertant Bz'(m4 D6856):1. In contrast, the pale purple revertant Bz'(m4 D6856):7 retains the complex transposon.

Research support comes from U.S. Department of Agriculture Grant #85-02624; Hatch Project NH308.

TRANSCRIPTION OF THE MAIZE TRANSPOSABLE ELEMENT Ac IN MAIZE SEEDLINGS AND IN TRANSGENIC TOBACCO

E.J. Finnegan, B.H. Taylor, E.S. Dennis, and W.J. Peacock

CSIRO Division of Plant Industry
Canberra, A.C.T. 2601, Australia

A combination of cDNA cloning and S1 analyses of RNA isolated from maize seedlings that carry an active Ac element has been used to define the Ac transcript. The primary transcript contains four introns that are excised to give a processed message which we predict to be approximately 3.4 kb. There are a number of transcription initiation sites clustered within a 90-base region about 300 bases from one end of the element. The first ATG is 600-690 bases from the transcription start and precedes an open reading frame of 807 amino acids--the putative "transposase." The transcript extends to within 261 bases from the other end of the element. S1 analysis of RNA from a transgenic tobacco plant carrying an intact copy of the Ac element demonstrated a transcript identical to that in maize although the preferred initiation sites differ.

NITRATE REDUCTASE AS A POTENTIAL TARGET FOR TOBACCO TRANSPOSABLE ELEMENTS

M.A. Grandbastien, C. Missonier, J. Goujaud, J.P. Bourgin, A. Deshayes, H. Vaucheret, and M. Caboche

Laboratoire de Biologie Cellulaire
INRA, 78000 Versailles, France

A chlorophyll-deficient mutant of Nicotiana tabacum, named t1, expresses at high frequencies (10^{-3} to 10^{-2}) green, clear, and twinned green and clear spontaneous somatic variations. The frequency of the cellular events leading to the somatic variations is closely dependent on the stage of differentiation of the cell and the physiology of the plant [Deshayes (1979) Theor. Appl. Genet. 55:145-152]. We are attempting to demonstrate the existence of active transposable elements in the tl genotype by their molecular isolation from an appropriate locus, after selection of in vitro mutant cellular types induced by integration of these elements into this locus. We have used the cellular marker of nitrate reductase deficiency (NR^-), due to mutation of the gene coding for the apoenzyme. In the haploid tobacco genome, two homeologous loci (nia_1 and nia_2) have to be mutated in order to annul NR activity. Using a stable $nia_1^-:nia_2^-$ tobacco line [Muller and Grafe (1978) Molec. Gen. Genet. 161:67-76], we have constructed haploid monogenic lines ($nia_1^+;nia_2^-$ and $nia_1^-;nia_2^+$) carrying the tl somatic instability. By selecting for chlorate resistance, we have isolated NR^- lines ($nia_1^-:nia_2^-$) from haploid tobacco mesophyll protoplast-derived cell populations. Spontaneous chlorate-resistant clones were obtained at frequencies up to 10^{-4} with several monogenic tl lines. Plants were regenerated from these clones and confirmed to be unable to grow on nitrate as the sole nitrogen source. Molecular analyses of these plants have been performed by Southern hybridizations of genomic DNA, using as a probe a fragment of the cDNA coding for the NR apoenzyme [Calza et al. (1987) Molec. Gen. Genet. (in press)]. Preliminary results show modifications suggesting insertion of foreign sequences into the NG gene structure in some of the spontaneous NR^- lines.

A MUTABLE ALLELE OF THE w4 LOCUS FOR ANTHOCYANIN PIGMENTATION IN SOYBEAN

R.W. Groose, S.M. Blomgren, P. Keim, R.C. Shoemaker, and R.G. Palmer

Departments of Agronomy and Genetics
Iowa State University, and
U.S. Department of Agriculture-ARS
Ames, Iowa 50011

The behavior of a mutable allele in soybean [Glycine max (L.) Merr.] suggests activity of a transposable element system. Plants of the "Asgrow Mutable" line of soybean are chimeric for anthocyanin pigmentation. Individual plants may produce both white and purple flowers, as well as flowers of mutable phenotype with purple sectors on white petals. Genetic analysis has established that mutable plants carry an unstable recessive ("mutable") allele of the w4 locus that conditions anthocyanin biosynthesis. The mutable allele reverts at high frequency, both somatically and germinally, to a stable functional (dominant) form. Approximately 10% of self-progeny of mutable plants carry at least one revertant allele and exhibit wild-type pigmentation. Although most revertant alleles condition pigmentation of wild-type intensity, occasionally "pale" alleles are derived from the mutable allele. Variation in reversion frequency among sublines of the Asgrow Mutable line suggests "changes in state" of the mutable allele.

Reversion of the mutable allele in the independent cell layers of the meristem results in periclinal chimerism. Histological studies and genetic analysis of periclinal chimeral shoots indicate that the subepidermal cell layer (LII) of the growing point gives rise to the germline in soybean.

An array of new mutations for chlorophyll deficiency, partial sterility, and necrotic roots has been identified in these materials. Some of the new mutations are unstable and provide circumstantial evidence of transposition of a mobile genetic element. We are presently using cloned anthocyanin genes from other species to perform a molecular genetic analysis of the mutable allele.

PHENOTYPIC INSTABILITY IN TOMATO:

TOWARDS A MOLECULAR CHARACTERIZATION

Jacques Hille,[1] Jos Wubben,[2] Dion Florack,[3] and M.S. Ramanna[4]

[1]Free University
Department of Genetics
1081 HV Amsterdam, The Netherlands

[2]Agricultural University
Department of Molecular Biology
De Dreyen 11
6703 BC Wageningen, The Netherlands

[3]Agricultural University
Institute for Horticultural Plant Breeding
6700 AA Wageningen, The Netherlands

[4]Agricultural University
Department of Plant Breeding
6700 AJ Wageningen, The Netherlands

The cultivated tomato (Lycopersicon esculentum, 2n=2x=24) is one of the few crop species in which an important amount of classical genetic information is available. Among many other favorable aspects, the availability of a large number of good marker genes, the possibility of identifying pachytene chromosomes, and the properties of tissue culture make tomato especially suitable for both genetic studies and manipulation.

Tomato lines have been described which display a high level of phenotypic instability [M.S. Ramanna, J. Hille, and P. Zabel (1985) Theor. Appl. Genet. 71:145-152; R. Hagemann and B. Snoad (1971) Heredity 27:409-418; K. Verkerk (1959) Euphytica 8:216-222]. In one of these lines the chromosome-breakage-fusion-bridge-cycle phenomenon was observed, and it was suggested that this occurrence might induce "genomic shock," resulting in an unstable plant phenotype [M.S. Ramanna, J. Hille, and P. Zabel (1985) Theor. Appl. Genet. 71:145-152]. The other described tomato lines [R. Hagemann and B. Snoad (1971) Heredity 27:409-418; K. Verkerk (1959) Euphytica 8:216-222] were obtained following a mutagenic treatment which might also cause "genomic shock."

In this collaborative research it is attempted to characterize the phenotypic instability at the molecular level. To this end, selectable marker genes have been introduced into the tomato genome, using Agrobacterium tumefaciens, and plant transformants were regenerated. The introduced marker genes were chosen such to have subsequently a positive selection on marker gene inactivation. Recent data on these experiments are presented.

SHORT DISTANCE TRANSPOSITION OF *Tam3* ALTERS GENE EXPRESSION IN *ANTIRRHINUM MAJUS*

Andrew Hudson, Rosemary Carpenter, and Enrico Coen

John Innes Institute
Colney Lane
Norwich NR4 7UH, United Kingdom

Alleles have been described at both the *nivea* (*niv*) and *pallida* (*pal*) loci which have the transposable element *Tam3* inserted at similar positions in the gene promoters. At *pal*, *Tam3* prevents gene expression, whereas a low level of expression from the *niv* locus occurs with *Tam3* in position. We have examined a derivative of each of these alleles in which the effect of *Tam2* has changed. Expression of a novel *pal* allele, in contrast to that of its parent, occurs with *Tam3* present. However, *Tam3* has been altered at one end, probably because of an attempted excision. This suggests that the inhibitory effect of the intact *Tam3* was due to its ability to bind transposase and thus to sterically interfere with transcription. The derivative of the *niv* allele, unlike its progenitor, shows no expression with *Tam3* present. This allele contains two copies of *Tam3*. One, around 3 kb upstream of its previous site of insertion, has no effect on phenotype. The second copy is a similar distance downstream of the promoter, within the last exon of the gene, and is responsible for blocking *niv* expression. The allele has been produced by short-range transposition of copies of *Tam3* in opposite directions from the *niv* promoter.

SEQUENCE ANALYSIS OF MAIZE TRANSPOSON *Bs1* SUGGESTS EXTENSIVE USE OF MULTIPLE OVERLAPPING READING FRAMES

M.A. Johns,[1] M. Babcock,[2] M. Freeling,[3] and R. Simpson[2]

[1]Biological Sciences
Northern Illinois University
DeKalb, Illinois 60115

[2]PCRI
Dublin, California

[3]Genetics
University of California
Berkeley, California 94720

Bs1 contains 3,203 bp bounded by 302-bp perfect direct repeats, and is oriented opposite to the host *Adh1* gene. We have identified one major

(728 amino acids) and several minor (100-200 AA) open reading frames (ORFs); approximately 85% of Bs1 is apparently translated. The 5' end of the 3'-most ORF overlaps the 3' end of the major ORF, suggesting a translational frameshift mechanism. Retroviral pol genes normally contain four enzymes in the order: protease (pro), reverse transcriptase (RT), RNase H (RH), and endonuclease (endo). We have detected regions homologous to all four enzymes in Bs1. As in other retroviral homologies, the Bs1 homologies consist of short regions to similar amino acids separated by longer nonhomologous regions; the Bs1 homology regions occur in the expected order and spacing for each individual enzyme. However, the pol enzymes on Bs1 are not arranged in the standard linear order. Instead, RT is intact, but all the other enzymes are split into sections, separated by other enzyme regions. Parts of both RH and endo are apparently encoded in short ORFs that completely overlap the major RT-containing ORF. In two places, important homology regions for proteins translated in different frames share the same nucleotides. Another overlapping ORF shows strong homology to the gag (core protein) gene of ground squirrel hepatitis virus; this ORF is the only possible Bs1 gag gene. Thus, Bs1 may use all three reading frames for a distance of 252 bp, and two reading frames for an additional 834 bp, implying that Bs1 codes for up to 1,353 amino acids on an RNA molecule of 2,901 nt. The evolution of Bs1's overlapping genes might have involved convergent evolution; the expression of Bs1 is probably complex.

THE dSpm ELEMENT OF bz-m13 CONTAINS MULTIPLE ACCEPTOR SITES FOR RNA SPLICING

H.-Y. Kim, J.W. Schiefelbein, V. Raboy, and O.E. Nelson

University of Wisconsin
Laboratory of Genetics
Madison, Wisconsin 53706

The bz-m13 allele of the bronze-1 gene in maize contains a 2.2-kb defective Suppressor-mutator (dSpm) element in the second exon. Nevertheless, 5-10% of the enzymatic activity conditioned by a Bz allele is routinely detected in crude extracts prepared from plants carrying bz-m13 in the absence of an autonomous Suppressor-mutator element. Blot hybridization analyses of poly(A)$^+$ RNAs isolated from such plants reveal three bands of bronze-1-specific RNAs. One has a size (about 1.8 kb) similar to that of a Bz mRNA, and the other two are larger. One of the larger RNAs has a size expected from the addition of the dSpm to a Bz mRNA, and the other has an intermediate size. Sequence analysis of a cDNA representing a 1.8-kb RNA produced by the derivative state CS9, which resulted from an internal deletion of the dSpm in bz-m13 and which produces a higher level of the 1.8-kb RNA and five to ten times more enzymatic activity than bz-m13, shows that this RNA is produced by RNA splicing using the donor site of the single bronze-1 intron and an acceptor site within the inverted terminal repeat of the dSpm. In addition, S1 nuclease analysis of bz-m13 RNAs has identified another acceptor site 5' to that within the inverted terminal repeat. This suggests that the RNA with the intermediate size is produced by alternative splicing using this 5' acceptor site. Examination of the DNA sequence indicates that translation of the intermediate-sized mRNA would be prematurely terminated within the dSpm sequence, whereas the 1.8-kb mRNA represented by the cDNA maintains the proper reading frame.

GENOMIC INSTABILITY IN WHEAT INDUCED BY CHROMOSOME $6B^S$ OF TRITICUM SPELTOIDES

Rama S. Kota and Jan Dvorak

Department of Agronomy and Range Science
University of California
Davis, California 95616

A massive restructuring of wheat chromosomes was observed during the production of substitution of chromosome $6B^S$ from *Triticum speltoides* (Tausch) Gren. ex. Richter for chromosome 6B of Chinese Spring wheat. Chromosome rearrangements were not observed either in the amphiploid between Chinese Spring and *T. speltoides* or in Chinese Spring. No chromosome rearrangements were observed in the backcross derivatives; however, massive restructuring was observed after self-pollination of a monosomic substitution (2n=41) of chromosome $6B^S$ for wheat chromosome 6B. Chromosome rearrangements were observed in both wheat and *T. speltoides* chromosomes. The frequency of chromosome rearrangements was high among the B-genome chromosomes, moderate among the A-genome chromosomes, and low among the D-genome chromosomes. In the B genome, the rearrangements were again nonrandom, most frequently in chromosomes 1B and 5B. Chromosome arrangements were also high for the $6B^S$ chromosome of *T. speltoides*. Types of chromosome aberrations included chromosome deletions, translocations, ring chromosomes, dicentric chromosomes, and a paracentric inversion. Chromosome breaks were observed in both euchromatic and heterochromatic regions. It is speculated that the chromosome rearrangements may be caused by transposition of cryptic mobile genetic elements activated by chromosome $6B^S$ of *T. speltoides*.

MOLECULAR IDENTIFICATION OF A *Ds* INDUCED SHRUNKEN-2 ALLELE

Susan Lawrence and L. Curtis Hannah

University of Florida
2211 Fifield Hall
Gainesville, Florida 32611

The shrunken-2 gene on chromosome 3 affects the level of ADP-glucose pyrophosphorylase. A mutable allele of *sh2* has been produced from a stock containing *a1-m3* and *Ac* [Hannah and Nelson (1976) *Biochem. Genet.* 14:547]. The stock behaves genetically as if *Ds* resides at the *sh2* locus, and approximately 43 revertants have been isolated from this mutant. These revertant kernels are phenotypically wild type, but the ADP-glucose pyrophosphorylase activity in many of these revertants varies from the original stock by a number of biochemical criteria [Tuschall and Hannah (1982) *Genetics* 100:105]. Barton et al. [*Regulation of Carbon and Nitrogen Reduction and Utilization in Maize*, J.C. Shannon et. al., eds. (1986)

ASPP, p. 363] have isolated a cDNA clone that distinguishes a RFLP between closely related lines which contain either a normal or mutant allele of sh2.

As an independent test of the identity of this cDNA, the clone was used to probe genomic blots of DNA from the sh2 mutable, its progenitor, and several revertants. Since the different lines examined are produced from selfing, RFLPs are minimal and therefore identification of the cDNA clone could be substantiated. Digestion with XbaI, EcoRI or BglII yielded a single hybridizing band. With all three enzymes, the mutable line contained a restriction fragment 1.5 kbp larger than that seen in the progenitor. At the RFLP level, all revertants examined to date appear indistinguishable from the progenitor. These data suggest that the insertion is 1.5 kbp and that it has been removed from the revertant alleles. We are currently in the process of cloning the various alleles and analyzing their transcripts.

THE NUCLEOTIDE SEQUENCE OF THE bz2-mu1 ALLELE OF MAIZE

Kenneth R. Luehrsen, Margaret McLaughlin, and Virginia Walbot

Department of Biological Sciences
Stanford University
Stanford, California 94305

We have determined the nucleotide sequence of the mutable bz2-mu1 allele of maize; this sequence comprises an entire mu1.4 element with about 1 kbp of bz2 flanking sequence. The mu1.4 element is 1,370 bp long and is bounded by 9-bp direct repeats with the sequence 5' GCCAGACAC 3'. It is approximately 99% homologous to the Mu1 element from Adh1S3034 [Barker et al. (1984) Nucl. Acids Res. 12:5955], with most of the differences being base pair changes or single insertions/deletion events. Of particular note, however, is that the final 7 bp of the right terminal inverted repeat was not observed in the cloned sequences.

By removing the mu1.4 and one of the 9-bp directly repeated sequences, we have reconstructed the putative wild-type Bz2 allele. Preliminary evidence suggests that all of the Bz2-transcribed region is shown in the sequence presented, although the occurrence of large introns is not precluded. Using single-stranded probes against northern blots, we have deduced the size of the Bz2 transcript in purple tissues (approximately 840 bases) and the direction of transcription. There is no long open reading frame (ORF) in the Bz2 sequence. However, a computer search of the Bz2 allele has found regions highly homologous with the donor and acceptor splice site consensus sequences defined by Brown et al. [EMBO J. (1986) 5:2749].

Research support comes from the National Institutes of Health (GM32422) and the U.S. Department of Agriculture (86-00172). K.L. is a fellow of the American Cancer Society and M.M. was supported by a National Science Foundation postdoctoral fellowship.

DROSOPHILA P-ELEMENT TRANSCRIPT IS INCORRECTLY PROCESSED IN TOBACCO PLANTS

J. Martinez-Zapater, R. Finkelstein, and C. Somerville

Michigan State University-Department of Energy
Plant Research Laboratory
Michigan State University
East Lansing, Michigan 48824

With the goal of creating an inducible transposon tagging system for plants, we have tested expression of a modified *Drosophila* P element in transgenic plants. Tobacco plants were transformed with a binary P-element system composed of a nonmobile transposase gene and a nonautonomous mobile element. In order to make transposition an inducible event, transcription of the transposase gene is controlled by a *Drosophila* heat shock promoter. The P transposase gene used in this construction lacks the third intron, which confers tissue specificity in flies, but retains the first two introns. Expression of this chimeric gene is heat-shock inducible in transformed plants, producing two major transcripts of approximately 1.5 and 2.5 kb in length. Analysis of these transcripts shows that both are polyadenylated, but neither of the remaining introns is removed, and the most abundant transcript (1.5 kb) terminates just downstream of the second intron. In an attempt to circumvent these processing problems, we are now testing expression of a transposase gene in which all three introns have been excised.

TRANSPOSON TAGGING OF *Viviparous-1* USING ROBERTSON'S *MUTATOR*

Donald R. McCarty and Chris Carson

Vegetable Crops Department
IFAS
University of Florida
Gainesville, Florida 32611

Viviparous-1 (*vp*) is a regulatory gene required for the expression of multiple genes late in seed development. Most notably, *vp* mutations block anthocyanin synthesis in the aleurone and developmental arrest in the embryo. We have isolated and tentatively identified a *vp*-specific clone from a *vp-mum1* stock (provided by P. Stinard and D.S. Robertson, Iowa State University) using Robertson's *Mutator* as a tag. *Mu*-element number was reduced to approximately 5-10 copies/genome by outcrossing the *vp-mum1* stock to a nonmutator line. Plants from segregating populations were screened by Southern hybridization analysis using an internal 750-bp *AvaI*/*TaqI* fragment of *Mu1* as a probe. A 6.1-kbp *Bam*HI fragment was identified which correlated with the *vp-mum1* allele. The 6.1-kpb *Bam*HI fragment has been cloned from a library constructed in EMBL 3 from size-selected, *Bam*HI-cut DNA. Restriction mapping and Southern blot analysis indicate that the clone contains a 1.4-kbp *Mu1*-like element. Linkage data support assignment of this clone to the *vp* locus.

CHARACTERIZATION OF bz-m4 Derivative 6856 mRNA

L. Paje-Manalo and A.S. Klein

Department of Biochemistry
University of New Hampshire
Durham, New Hampshire 03824

The bz-m4 D6856 originated from a series of Ds transposition events involving the bronze locus. This particular mutation is the result of a 6.7-kb insertion in the region corresponding to the 5' untranslated leader of Bz mRNA. Ds elements flanking the insertion are approximately 2 kb in length and have the same orientation. There is a segment of DNA between the Ds elements that is a duplication of the 3' end of the Bz gene. The leftmost Ds element in bz-m4 D6856 has a 121-bp deletion which includes the terminal inverted repeat (Klein et al., ms. in prep.).

The bz-m4 D6856 exhibits altered developmental and tissue-specific expression of UFGT, the Bz gene product [Dooner (1981) CSHSQB 45:457-462]. The size and the position of the insertion in the mutant allele raises the possibility that transcription is directed not by the wild-type Bz promoter but by a promoter found either in Ds sequences or within the insertion. Northern analyses of BzMc(W22) and bz-m4 D6856 indicate that the sizes of the mRNAs are similar.

Research supported by U.S. Department of Agriculture Grant #85-02624; Hatch Project NH308.

CO-INDUCTION OF Uq ACTIVITY AND AN mn-TYPE MUTANT BY 5-AZA-2'-DEOXYCYTIDINE

Yong-Bao Pan and Peter A. Peterson

Department of Genetics and Department of Agronomy
Iowa State University
Ames, Iowa 50011

In this experiment, the 2'-deoxycytidine analog 5-Aza-2'-deoxycytidine was used to induce the activity of Uq, a transposable element in Zea mays. The roots of germinating seedlings from 34 sectored BF2 progeny kernels of four maize inbreds (B70, C103, C123, and 187-2) were soaked into a 30 μM 5-Aza-2'-deoxycytidine solution for 72-90 hours. Treated seedlings were then transferred into seed pots in the greenhouse and grown for another two weeks before transplanting into the field. During this period, the death of some root tips and the folding of the leaves were common. Four seedlings died; the rest recovered and looked normal. These plants, together with 15 plants derived from the same source as a control, were reciprocally crossed with an a-ruq tester. Examination of the progeny ears produced the following results: (i) one treated plant (866246U) was totally sterile; (ii) another plant (866248U) had one activated Uq element which expressed its activity only in the ear but not in the tassel. Moreover, this Uq activity co-segregated (with no exceptions) with an mn type mutant gene designated as mn866248U; and (iii) no Uq activity was detectable in the other 28 treated plants or in the 15 control plants.

A similar mn event of Uq activation happened in a single kernel at a frequency of about 6.1 x 10^{-6} in a separate experiment without any treatment. When the plant (866201X) derived from this kernel of Uq spotting was crossed by an a-ruq tester, approximately 50% of the progeny kernels were miniature and also Uq spotted, and the other 50% were normal-sized and not Uq spotted. However, tests were not available in its transmission through pollen. The mn mutant gene in this case is designated mn866201X.

THE DESTINY OF TRANSPOSED En (Spm)

Sudhansu Dash and Peter A. Peterson

Department of Agronomy
Iowa State University
Ames, Iowa 50011

Excised Ens transpose largely to nearby sites. Every En excised is not always recovered as a transposed event [Nowick and Peterson (1981) Molec. Gen. Genet. 183:440-478]. Models of transposition of plant transposable elements are generally based on observations of Mp transposition. General conclusions from the transposition of Mp indicate that most often Mp is recovered in the same chromatid from which it has been excised and that Mp transposition occurs from a replicated region of the chromosome.

With wx-844 allele [Pereira et al. (1985) EMBO J. 4:17-23] excision of En (wx --> Wx) can be accurately monitored, and, using several reporter alleles such as a-m(r), a-m1, and a2-m1, the presence or absence of En activity in the wx --> Wx sectors can be detected in the overlying aleurone tissue. With these alleles, the Suppressor (Sp) and M functions can be monitored.

Using this assay system, it is frequently observed that many clonal sectors of endosperm in which En has been excised from the wx locus lack En activity (i.e., wx --> Wx, -En sectors). Critical examination of the adjacent endosperm tissue infrequently suggests the presence or absence of twin sectors with loss of En in one, and gain of an extra En in the other sector. Several models can be adapted to account for these observations. They include: loss of En after excision; transposition of En from an unreplicated region; transposition to a replicated region; transposition from a replicated region to another replicated region; etc.

The analysis of the clonal sectors of endosperm also suggests the nature of the colored sectors seen in the a2-m1 allele and pale sectors in the case of the a-m1 allele in the presence of En. They are due to loss of En in those sectors rather than inactivation of the Sp component of En.

A CHANGE IN STATE OF THE Ac-INDUCED P-VV ALLELE IS ASSOCIATED WITH INVERSION OF Ac

T.A. Peterson, E.S. Dennis, and W.J. Peacock

CSIRO Division of Plant Industry
Canberra, A.C.T. 2601 Australia

The maize P locus controls pigmentation of the pericarp and glumes of the cob. The P-VV allele, which specifies variegated pericarp and cob, comprises the transposable element Ac situated at the P locus. We have previously isolated a DNA clone from P-VV, and we are now using P DNA sequences to study the structure and expression of P.

We have isolated an allele termed P-OVOV (orange variegated) derived from P-VV. P-VV specifies colorless pericarp with red sectors, whereas P-OVOV specifies orange pericarp with many dark red sectors, and some colorless sectors. Thus, P-OVOV represents a change in state of the P-VV allele.

Genetic tests showed that (i) P-OVOV is an allele of P; (ii) Ac activity is tightly linked to P-OVOV; (iii) P-OVOV is unstable, and produces revertant and other variant alleles; and (iv) the trans-acting functions of Ac associated with P-OVOV are not different from the progenitor P-VV allele.

Southern analysis of DNA from the progenitor P-VV allele and the P-OVOV allele shows that the Ac element associated with P-OVOV has remained within an 850-bp genomic DNA fragment at the P locus. However, the Ac element is in an inverted orientation in P-OVOV compared to P-VV.

These results suggest that the change in state of P-VV to P-OVOV may be due to inversion of Ac. An explanation of how this inversion alters P expression awaits further information on the structure of the P gene.

DNA MODIFICATION AND THE TIMING OF DEFECTIVE Spm EXCISION FROM THE bz-m13 ALLELE IN MAIZE

V. Raboy, J.W. Schiefelbein, and O.E. Nelson

Laboratory of Genetics
University of Wisconsin
Madison, Wisconsin 53706

The bz-m13 allele in maize contains a 2.2-kb defective Spm (dSpm) element inserted in the second exon of the bronze-1 locus. In the presence of a standard Spm (Spm-S), the dSpm in the initial state of bz-m13 is characteristically excised early during aleurone development, resulting in large revertant sectors. In the presence of Spm-S, bz-m13 also produces high numbers of germinal mutations. The dSpm inserts in six changes of state derived from bz-m13, in the presence of Spm-S, are characteristically excised late during aleurone development, resulting in small revertant sectors. They produce few germinal mutations. Five of these six changes of state were found to contain dSpm elements which have deletions either in one of their terminal repeats or internal deletions extending into a subterminal repetitive region. One of the six, Change of State 1 (CS1), apparently has no internal deletion. Instead, CS1 is modified, as assayed by lack of cleavage by methylation-sensitive restriction enzymes, in a GC-rich sequence that is apparently important to Spm function and dSpm response. This modification has no apparent effect on bz gene expression in the absence of Spm-S. A second class of derivatives of bz-m13 have reduced germinal mutation rates in the presence of Spm-S, but aleurone sectoring patterns indistinguishable from the initial state of bz-m13, and are thus cryptic changes of state. Restriction mapping of genomic DNA isolated from cryptic states shows that they contain dSpm inserts similar in site and position to the dSpm insertion in the initial state of bz-m13. Cryptic states might represent modification intermediates, since "CS1-like" derivatives appear more frequently in lines containing cryptic states as compared with lines containing the initial state of bz-m13.

A SEARCH FOR TRANSPOSABLE ELEMENTS IN THE TOMATO

M.S. Ramanna[1] and Jacques Hille[2]

[1]Department of Plant Breeding (IvP)
Agricultural University
6700 AJ Wageningen, The Netherlands

[2]Department of Genetics
Free University
1007 MC Amsterdam, The Netherlands

In spite of extensive genetic analysis in the past, there appears to be no published evidence for the occurrence of mutable loci in the cultivated tomato, Lycopersicon esculentum. Because of this, there are doubts as to whether there are transposable elements at all in this plant. On the assumption that tomato may not be an exception among other organisms, we are searching for transposable elements in this model system. This is a good model system because it has most of the attributes of maize and, in addition, tomato is most amenable for tissue culture and genetic transformation.

With the aim of activating transposable elements, if any, in tomato, we are applying "genomic shock" method through breakage-fusion-bridge-cycle (BFBC). Using recessive gene markers that affect chlorophyll color and anthocyanin pigmentation, which can be easily detected at seedling stage, large populations of seedlings were screened in order to detect reversions in the progeny of plants with BFBC. The marker genes include: (i) yellow virescent (yv), (ii) without anthocyanin (aw), (iii) anthocyaninless (a), and (iv) anthocyanin loser (al). Reversions from recessive to wild type have been recovered for yv, a, and al. The rates of reversions range approximately one in 3,000 to 5,000 for these loci. Apart from the reversion events, mutations affecting plant height and other morphological characteristics have also been recovered in these lines.

A notable feature of plants with BFBC is that they give rise to progeny with chromosomal structural abnormalities. These abnormalities are not confined to chromosomes involved in BFBC alone. Such chromosomal aberrants include reciprocal translocations, abnormal centromere associations, and pairing abnormalities. These abnormalities appear to be parallel to those reported in maize.

A CHROMOSOME INVERSION FLANKED BY A 5-bp DUPLICATION SUGGESTS A MODEL FOR TRANSPOSITION IN *ANTIRRHINUM MAJUS*

Tim Robbins, Rosemary Carpenter, and Enrico Coen

John Innes Institute
Colney Lane
Norwich NR4 7UH, United Kingdom

The unstable *pal*^{rec} allele of *Antirrhinum majus* contains the transposable element, *Tam3*, inserted in the promoter of the *pallida* gene. An unstable allele of altered phenotype was derived from *pal-rec* and found to have a new DNA sequence flanking the upstream end of *Tam3*. Recombination analysis indicates that the new sequence was originally present on the same chromosome as *pal* but has been brought adjacent to *Tam3* by an inversion of at least six map units of DNA (about 6,000 kb). The first 5 bp of the new sequence flanking *Tam3* is also present at the other end of the inversion. This suggests that the inversion may have resulted from an attempted *Tam3* integration event. We propose that during the normal process of transposition, excision and integration sites are brought together in a transposition complex. It is possible to explain the inversion as the result of incorrect ligation of the free ends created by *Tam3* excision and the staggered cut at the integration site. This model could also explain the observed preference of many transposable elements to integrate at closely linked loci, since these loci are more likely to come into physical contact to form the proposed transposition complex. Excision of *Tam3* subsequent to the inversion has given rise to a stable allele with a novel spatial pattern of expression. The overall result of the inversion followed by *Tam3* excision is to replace most of the *pal* gene promoter with a new sequence. This new sequence does not act as a promoter for transcription in its original position, and the novel spatial pattern of expression probably results from the altered relative position of the *pal* gene on the chromosome.

THE ORIGIN OF A COMPLEX TRANSPOSABLE ELEMENT IN *bz-m4 Derivative 6856*

Gregg Roman and Anita S. Klein

Genetics Program
University of New Hampshire
Durham, New Hampshire 03824

McClintock isolated a series of *Ds*-controlled alleles, all designated *bz-m4*. One of these, *bz-m4 Derivative 6856*, alters the developmental and tissue-specific expression of the *Bz* gene product [Dooner (1981) *CSHSQB* 45:457-462]. The genetic history of *bz-m4 D6856* accounts for the complexity of the transposable element which is inserted at the allele. The bz-m4 D6856 mutation was produced by a 6.7-kb insertion into a region corresponding to the 5' untranslated leader region of the Bz mRNA. The insertion structurally resembles a bacterial transposon: Ds elements are located at both ends of the insertion; the elements are approximately 3 kb in length and in the same (direct) orientation. Between the two Ds elements is a 2.7-kb segment of DNA that is actually a duplication of the 3' end of the Bronze gene. Based on the structure of the unique transposable element in bz-m4 D6856, it is expected that the original bz-m4 allele contained a single Ds element, and this Ds element is inserted into a 0.85-kb segment at the 3' end of the Bz locus.

Research supported by U.S. Department of Agriculture Grant #85-02624; Hatch Project NH 308.

Cin4, AN INSERT ALTERING THE STRUCTURE OF THE A1 GENE IN ZEA MAYS, REVEALS PROPERTIES OF NONVIRAL RETROTRANSPOSONS

Zsuzsanna Schwarz-Sommer, Lise Leclercq, and Heinz Saedler

Max-Planck-Institut für Züchtungsforschung
5000 Köln 30, Federal Republic of Germany

A wild-type allele of the A1 gene of Zea mays contains a 1.1-kb long insert, termed Cin4-1, which alters the structure of the transcription unit compared to other A1 alleles. The Cin4-1 element is a 5' truncated member of a family of elements occurring in 50-100 copies in the maize genome. Genomic cloning and sequence and their flanking regions allowed us to classify Cin4 as the first plant nonviral retrotransposon. Individual Cin4 elements terminate in an oligoA track of variable sizes (6-11 residues) at their 3' end. The 5' end of family members is truncated. Cin4 elements are flanked by small direct duplications the size of which varies between 6 and 16 bp. Based on the relation between the target sequence and the sequence of Cin4, we suggest and discuss a model on the mechanism of Cin4 integration via in situ cDNA synthesis on an RNA template. The longest Cin4 element analyzed so far has a long open reading frame (ORF) comprising 3,198 nucleotides. The putative 1,066 amino acids long Cin4 protein has the capacity to encode a reverse transcriptase-like protein and a DNA binding domain. The conservation pattern of these two domains and the overall organization of Cin4 is similar to that detected in nonviral retrotransposons in animals. The origin and function of Cin4 are discussed.

Mu ELEMENTS MAY CONTRIBUTE TO THE BACKGROUND MUTATION RATE IN NORMAL MAIZE STOCKS

L.E. Talbert, V.L. Chandler, and C. Faber

Institute of Molecular Biology
University of Oregon
Eugene, Oregon 97403

Sequences homologous to Mutator transposable elements are found in normal maize stocks which do not exhibit mutation rates above the spontaneous level. One class of Mu-homologous sequence contains Mu termini that are not associated with Mu1 internal sequence. Multiple copies of this class of sequence are found in all lines of maize and related Zea species we have examined, but are not found in Tripsacum dactyloides, wheat, or tobacco. Two examples were cloned from the maize inbred line B37. DNA sequencing shows that these have the structure expected of transposable elements, including Mu terminal inverted repeats flanked by short direct repeats. The internal regions of these elements are not related to Mu1 or to each other. Thus, they represent previously uncharacterized transposable elements in the Robertson's Mutator family and are present in high copy number in the genus Zea. To determine if these elements may contribute to the background spontaneous mutation rates in maize, we have exploited a series of mutant lines derived from the maize inbred P39. We find that restriction fragments homologous to probes for these Mu elements are significantly more polymorphic than restriction fragments homologous to non-Mu related DNA. Thus, this class of Mu elements appears to be associated with genomic rearrangements in the P39 derivatives. Two mechanisms whereby these polymorphisms may have been generated are transposition and recombination between Mu elements dispersed in the genome.

TRANSPOSITION OF THE MAIZE ACTIVATOR ELEMENT IN ARABIDOPSIS THALIANA AND DAUCUS CAROTA

Marie Anne Van Sluys,[1] Nina V. Fedoroff,[2] and Jacques Tempe[1]

[1]Institut de Microbiologie
Universite de Paris
91405 Orsay, France

[2]Department of Embryology
Carnegie Institution of Washington
Baltimore, Maryland 21210

Gene tagging with transposable elements has been used for identifying genes in maize [Fedoroff et al. (1984) Proc. Natl. Acad. Sci., USA 81:3825-3829; O'Reilly et al. (1985) EMBO J. 4:877-882; Wienand et al. (1986) Molec. Gen. Genet. 203:202-207]. We are interested in using the maize transposable element Activator (Ac) as a tool for isolating genes in heterologous systems. We have introduced Ac in Daucus carota and Arabidopsis thaliana using a co-transformation system where a high proportion of recipient cells received a nontumorigenic Agrobacterium tumefaciens T-DNA carrying Ac and a nopaline synthase gene, and the hairy root T-DNA from A. rhizogenes. Transformants were selected as nopaline-containing hairy roots.

We present evidence that the Ac element transposes in both transformed carrot roots and A. thaliana root-derived callus cultures.

Fertile plants were regenerated from A. thaliana-transformed callus, demonstrating the utility of the A. rhizogenes-based co-transformation system.

We are now looking for the segregation of the Ac element from the Ri plasmid T-DNA in the progeny of hairy root Ac A. thaliana transformants in order to obtain normal Ac-containing plants.

UNIQUE Ds and Mu ELEMENTS INSERTED IN EXONS OF THE Waxy GENE

Marguerite Varagona, Don Fleenor, and Susan R. Wessler

Botany Department
University of Georgia
Athens, Georgia 30602

The waxy (Wx) locus is responsible for the synthesis of amylose in immature pollen, endosperm, and embryo sac. The focus of this work is the characterization of two Wx alleles containing transposable elements. One allele, Ds wxB4, contains a 1.5-kb Ds insertion in exon 13 of the Wx gene. The wxMum5 allele, isolated from a line with mutator activity, contains a 1.5-kb insertion in exon 10. Sequence analysis has shown that each of these insertions is unique among its family of transposable elements.

The ends of the wxB4 Ds are homologous to Ac for 317 bp on one end and 314 bp on the other. A region from base pairs 530 to 780 contains stretches of Ac homology. However, a region from base pairs 780 to 1,180 is totally unique to this Ds. Experiments are underway to determine the copy number and organization of this unique region in the maize genome.

The insertion in wxMum5 has inverted repeats that are at least 145 bp in length. These repeats are 85% to 95% homologous to the terminal repeats found in Mu1, Mu1.7, and Mu3. The central region of 1.2 kb appears to be unique to this element.

Examination of the sequences conserved among families of transposable elements may help to define the regions that are important for transposition and help us to understand the evolution of each gene family.

A COPIA-LIKE ELEMENT FAMILY IN ARABIDOPSIS THALIANA

Daniel F. Voytas and Frederick M. Ausubel

Department of Genetics
Harvard Medical School, and
Department of Molecular Biology
Massachusetts General Hospital
Boston, Massachusetts 02114

A transposable element family was identified in Arabidopsis thaliana by analyzing restriction fragment length polymorphisms in ∿100 kb of DNA among 16 geographical races. A 2.2-kb insertion found in one race adjacent to the chalcone synthase gene is a member of a low-copy, interspersed family of sequences common to 11 of the 16 A. thaliana races analyzed. The insertion bears structural similarities to retroviral-like transposable elements: 1) the insertion is bounded by two ∿500-bp direct repeats, the ends of which contain short inverted repeats (5'-TGTT... ACAA-3'); 2) the insertion contains an open reading frame of 595 amino acids that shares 32% amino acid sequence homology with the Drosophila melanogaster copia element protein and includes significant homology to retrotransposon and retroviral reverse transcriptases; and 3) the insertion is flanked by 5-bp direct repeats similar to transposable element target site duplications. We propose the name Ta1 for this A. thaliana element family.

This study was supported by a grant from Hoechst A.G. to Massachusetts General Hospital.

MOLECULAR ANALYSIS OF Mut, A NEW MAIZE TRANSPOSABLE ELEMENT SYSTEM

Amanda R. Walker,[1] E.S. Dennis,[2] and W.J. Peacock[2]

[1]Division of Plant Industry
CSIRO
Canberra, Australia

[2]Department of Biochemistry
Faculty of Science
Canberra, Australia

Mut, a controlling element, was first observed by its action in trans on an unstable mutation (bz1-mut) at the bronze-1 locus [Rhoades and Dempsey (1982) MGNL No.56]. The two-element Mut system differs at both the genetic and molecular levels from other characterized transposable element systems.

The phenotype of the mutant bz1-mut in the presence of Mut is bz1 (pale) kernels finely spotted with Bz1 (dark) sectors. Unlike other transposable element systems, germinal revertants also exhibit an instability in the presence of Mut with large unstable mutant sectors arising in about 1% of kernels. This is approximately the same frequency with which large revertant sectors appear on mutant kernels. These data suggest that Mut may act as a switch rather than as a typical controlling element.

DNA sequence analysis of the mutant bz1-mut and a germinal revertant has indicated no differences between them. Compared with the functional Bronze-1 allele Bz1-McC (Furtek and Nelson, pers. comm.), both alleles contain a 176-bp insertion located in a position to disrupt normal gene function. How the germinal revertant behaves as a Bronze-1 allele is unknown.

THE BEHAVIOR OF Ac IN TRANSFORMED TOMATO PLANTS AND THEIR PROGENY

R. Khush, F. Belzile, J. Palys, and J. Yoder

University of California
Department of Vegetable Crops
Davis, California 95196

With a long-range goal of developing a transposon mutagenesis system in Lycopersicon, we have transformed the maize transposon Ac into tomato. Over 70 plants have been regenerated following Agrobacterium-mediated transformation of the cultivar VF36 with an Ac-m7:pMON200 construction. Southern hybridizations of genomic DNA prepared from axillary shoots of transformed plants were probed with isolated fragments of the donor plasmid. In a number of plants, bands of the size predicted for an empty donor site are observed. When an Ac-specific probe was used, new bands of various sizes are generated, consistent with the somatic mobilization of Ac. Progeny from selfed transformants are being examined by similar methods for evidence of mobilization in the R_2 generation. While the possibility of complicated gene rearrangements during transformation cannot be ruled out until the Ac is recloned, we feel the simplest explanation is that Ac is mobilized in Lycopersicon.

PARTICIPANTS, CHAIRMEN, AND SPEAKERS

Armitage, P.I.	University of Leicester, Leicester LE2 7RH, ENGLAND
Arney, Pamela Close	Iowa State University, Ames, IA 50010
Athma, Prasanna	University of Illinois, Urbana, IL 61801
Axtell, John D.	Purdue University, West Lafayette, IN 47907
Aycock, Harold	Dairyland Seed Company, Clinton, WI 53525
Baker, Barbara J.	U.S. Department of Agriculture, Albany, CA 94710
Banaras, Muhammad	University of Wisconsin, Madison, WI 53706
Banes, Jody	Carnegie Institution of Washington, Baltimore, MD 21210
Barton, Kenneth	Agracetus, Middleton, WI 53562
Baszczynski, Chris L.	Allelix, Inc., Mississauga L4V 101, Ontario, CANADA
Beach, Larry R.	Pioneer Hi-Bred International, Inc., Johnston, IA 50131
Bennetzen, Jeffrey	Purdue University, Lafayette, IN 47907
Benz, Wendy K.	University of Wisconsin, Madison, WI 53706
Bingham, E.T.	University of Wisconsin, Madison, WI 53706
Blomgren, Susan M.	Iowa State University, Ames, IA 50011
Brar, Gurdip S.	Agracetus, Middleton, WI 53562
Briggs, Steven	Cold Spring Harbor Laboratory, Cold Spring Harbor, NY 11724
Britt, Anne Bagg	Stanford University, Stanford, CA 94305
Buckner, Brent	Iowa State University, Ames, IA 50011
Burr, Ben	Brookhaven National Laboratory, Upton, NY 11973
Burr, Frances A.	Brookhaven National Laboratory, Upton, NY 11973
Carpenter, Jodi A.	University of Wisconsin, Madison, WI 53706
Carpenter, Rosemary	John Innes Institute, Norwich NR4 7UH, ENGLAND
Carson, Chris	University of Florida, Gainesville, FL 32611
Chandlee, Joel	USDA-PGGI-PMGL, Beltsville, MD 20705
Chandler, Vicki	University of Oregon, Eugene, OR 97403-1229
Charlesworth, Brian	University of Chicago, Chicago, IL 60637
Chen, Jychian	Yale University, New Haven, CT 06511
Chomet, Paul	Cold Spring Harbor Laboratory, Cold Spring Harbor, NY 11724
Chourey, Prem	U.S. Department of Agriculture, Gainesville, FL 32611
Clancy, Maureen	University of Florida, Gainesville, FL 32611

Clark, Janice K.	University of North Dakota, Grand Forks, ND 58202
Coen, Enrico	John Innes Institute, Norwich NR4 7UH, ENGLAND
Cone, Karen	Brookhaven National Laboratory, Long Island, NY 11973
Coors, James G.	University of Wisconsin, Madison, WI 53706
Cowen, Robin	University of Illinois, Urbana, IL 61801
Crow, James F.	University of Wisconsin, Madison, WI 53706
Dash, Sudhansu	Iowa State University, Ames, IA 50011
Dawe, R. Kelly	University of California, Berkeley, CA 94720
Day, Peter R.	Cook College, Rutgers University, New Brunswick, NJ 08903
Dellaporta, Stephen	Yale University, New Haven, CT 06511
Demopulos, James T.	University of Alabama, Birmingham, AL 35294
Dennis, Liz	CSIRO, Canberra ACT 2601, AUSTRALIA
Dietrich, Gabriele E.	Stanford University, Stanford, CA 94305
Dooner, Hugo	Advanced Genetic Sciences, Inc., Oakland, CA 94608
Dougall, Donald	University of Tennessee, Knoxville, TN 37996-1100
Dowe, Michael F.	University of New Hampshire, Durham, NH 03824
Drong, Roger	The Upjohn Company, Kalamazoo, MI 49007
Dvorak, Jan	University of California, Davis, CA 95616
Eggleston, William	University of Wisconsin, Madison, WI 53706
Eichholtz, David A.	Monsanto Company, St. Louis, MO 63198
Ellingboe, Albert H.	University of Wisconsin, Madison, WI 53706
Engels, William R.	University of Wisconsin, Madison, WI 53706
Evola, Stephen	CIBA-GEIGY Corporation, Research Triangle Park, NC 27709
Fedoroff, Nina	Carnegie Institution of Washington/ Embryology, Baltimore, MD 21210
Finkelstein, Ruth	Michigan State University, East Lansing, MI 48824
Finnegan, E. Jean	CSIRO, Canberra Act 2601, AUSTRALIA
Foss, Henriette M.	Iowa State University, Ames, IA 50011
Freeling, Michael	University of California, Berkeley, CA 94720
Fromm, Michael	U.S. Department of Agriculture, Albany, CA 94710
Gerdes, James	University of Wisconsin, Madison, WI 53706
Grandbastien, Marie Angele	Laboratoire De Biologie Cellulaire, C.N.R.A., Versailles 78000, FRANCE
Groose, Robin W.	USDA-ARS & Iowa State University, Ames, IA 50011
Hannah, L. Curtis	University of Florida, Gainesville, FL 32611
Harberd, Nicholas	University of California, Berkeley, CA 94720
Hardeman, Kristine	University of Oregon, Eugene, OR 97403
Hehl, Reinhard	U.S. Department of Agriculture, Albany, CA 94710
Helgeson, John P.	University of Wisconsin, Madison, WI 53706
Hille, J.	Free University, De Boelelaan 1027, THE NETHERLANDS

Howes, Neil	Agriculture Canada, Winnipeg R3T 2MP, Manitoba, CANADA
James, Martha	Iowa State University, Ames, IA 50011
Jen, George C.	CIBA-GEIGY Corporation, Research Triangle Park, NC 27709
John, M.E.	Agracetus, Middleton, WI 53562
Johns, Rick	Northern Illionis University, DeKalb, IL 60115
Kearney, Brian	University of California, Berkeley, CA 94720
Keim, Paul	Iowa State University, Ames, IA 50011
Kermicle, Jerry L.	University of Wisconsin, Madison, WI 53706
Kidwell, Kimberlee Kae	University of Wisconsin, Madison, WI 53706
Kim, Hwa Yeong	University of Wisconsin, Madison, WI 53706
Kim, Won K.	Research Station Agriculture/Canada, Winnipeg R3I 2M9, Manitoba, CANADA
Kimpel, Janice	Agrigenetics Advanced Science Company, Madison, WI 53716
Klein, Anita S.	University of New Hampshire, Durham, NH 03824
Kota, Rama S.	University of California, Davis, CA 95616
Kris, Morena Seitz	University of Wisconsin, Madison, WI 53706
Lassner, Michael	University of California, Davis, CA 95616
Lawrence, Susan	University of Florida, Gainesville, FL 32611
Lim, Johng K.	University of Wisconsin, Eau Claire, WI 54701
Luehrsen, Kenneth	Stanford University, Stanford, CA 94306
Lynch, Tim	Michigan State University, East Lansing, MI 48824
Martinez-Zapater, Jose M.	Michigan State University, East Lansing, MI 48824
Mathews, Dennis E.	University of Wisconsin, Madison, WI 53706
McCabe, Dennis	Agracetus, Middleton, WI 53562
McCarty, Donald R.	University of Florida, Gainesville, FL 32611
Mohammad, Kamaruzaman	Virginia Polytechnic Institute, Blacksburg, VA 24061
Moreno, Maria A.	Iowa State University, Ames, IA 50011
Muhawish, Salah	Iowa State University, Ames, IA 50011
Murray, Elizabeth	Agrigenetics Advanced Science Company, Madison, WI 53716
Muszynski, Michael G.	Iowa State University, Ames, IA 50011
Natarajan, Guru Murti	Iowa State University, Ames, IA 50011
Nelson, Oliver E.	University of Wisconsin, Madison, WI 53706
Neuffer, M. G.	University of Missouri, Columbia, MO 65203
O'Reilly, Catherine	Sunderland Polytech, Sunderland SR1 3RG, ENGLAND
Osborn, Thomas C.	University of Wisconsin, Madison, WI 53706
Ott, Russ	The Rockefeller University, New York, NY 10021-6399
Paje-Manalo, Leila	University of New Hampshire, Durham, NH 03824
Pan, David	University of Wisconsin, Madison, WI 53706
Pan, Yong-Bad	Iowa State University, Ames, IA 50010
Patterson, Garth	University of Oregon, Eugene, OR 97403
Peschke, Virginia	University of Minnesota, St. Paul, MN 55108
Peters, N. Kent	Ohio State University, Columbus, OH 43210
Peterson, Peter	Iowa State University, Ames, IA 50011

Peterson, Thomas A.	CSIRO, Canberra ACT 2601, AUSTRALIA
Phillips, Ronald	University of Minnesota, St. Paul, MN 55108
Phillis, Randy	University of Wisconsin, Madison, WI 53706
Phinney, Bernard O.	University of California, Los Angeles, CA 90024
Presting, Gernot G.	University of Wisconsin, Madison, WI 53706
Quatrano, Ralph S.	E.I. DuPont de Nemours & Company, Wilmington, DE 19898
Raboy, Victor	University of Wisconsin, Madison, WI 53706
Rabson, Robert	U.S. Department of Energy, Washington, DC 20545
Ramanna, M.S.	Department of Plant Breeding, Wagenningen 6700AJ, THE NETHERLANDS
Ray, Ian M.	University of Wisconsin, Madison, WI 53706
Reiter, Robert	University of Wisconsin, Madison, WI 53706
Riemenschneider, Don	U.S. Forest Service, Rhinelander, WI 54501
Rifaat, Mahmoud M.	Virginia Polytechnic Institute, Blacksburg, VA 24061
Robbins, Tim P.	John Innes Institute, Norwich NR2 7UH, ENGLAND
Robertson, Donald	Iowa State University, Ames, IA 50011
Robertson, Hugh	University of Wisconsin, Madison, WI 53706
Roman, Gregg	University of New Hampshire, Durham, NH 03824
Romao, Jose	Coimbra University, PORTUGAL
Rosenbaum, Max J.	University of Wisconsin, Madison, WI 53705
Roth, Brad	U.S. Department of Agriculture, Albany, CA 94710
Rudert, M.R.	CIBA-GEIGY Corporation, Research Triangle Park, NC 27709
Russell, Dave	Agracetus, Middleton, WI 53562
Russell, Julie A.	University of Wisconsin, Madison, WI 53706
Saedler, Heinz	Max-Planck-Institut fur Zuchtungforschung, Egelsplad, D-5000 Koln 30, FEDERAL REPUBLIC OF GERMANY
Schiefelbein, John	University of Wisconsin, Madison, WI 53706
Schlappi, Michael	F. Miescher Institute, Basel 4002, SWITZERLAND
Schmidt, Bob	Brookhaven National Laboratory, Upton, NY 11973
Schmidt, Daria	University of Wisconsin, Madison, WI 53706
Schwarz-Sommer, Zsuzsanna	Max-Planck-Institut fur Zuchtungforschung, Egelsplad, D-5000 Koln 30, FEDERAL REPUBLIC OF GERMANY
Schwartz, Drew	Indiana University, Bloomington, IN 47405
Scofield, Steven R.	Plant Breeding Institute, Cambridge CB2 2LQ, ENGLAND
Sellmer, James C.	University of Wisconsin, Madison, WI 53706
Shattuck-Eidens, Donna	NPI, Salt Lake City, UT 84108
Shepherd, Nancy	Experimental Station Dupont, Wilmington, DE 19801
Sheridan, William F.	University of North Dakota, Grand Forks, ND 58202
Shoemaker, Randy C.	Iowa State University, Ames, IA 50011
Simcox, Kevin	Purdue University, Lafayette, IN 47904
Sisco, Paul	North Carolina State University, Raleigh, NC 27695

Smith, Harriet J.	U.S. Department of Agriculture, Washington, DC 20251
Sommer, Hans	Max-Planck-Institut fur Zuchtungforschung, Egelsplad, D-5000 Koln 30, FEDERAL REPUBLIC OF GERMANY
St. Clair, Dina	University of Florida, Gainesville, FL 32611
Stadler, Joan	Iowa State University, Ames, IA 50011
Starlinger, Peter	Universitat zu Koln, Institut fur Genetik, D-5000 Koln 41, FEDERAL REPUBLIC OF GERMANY
Sullivan, Elizabeth J.	University of Wisconsin, Madison, WI 53706
Sullivan, Thomas D.	University of Wisconsin, Madison, WI 53706
Sundaresen, V.	Cold Spring Harbor Laboratory, Cold Spring Harbor, NY 11724
Swain, William	Agracetus, Middleton, WI 53562
Tekkala, Reddy	University of Iowa, Iowa City, IA 52242
Temin, Rayla G.	University of Wisconsin, Madison, WI 53706
Tracy, William F.	University of Wisconsin, Madison, WI 53706
Treat, Calvin	University of Wisconsin, Madison, WI 53706
Tuttle, Annmarie B.	CIBA-GEIGY, Garner, NC 27529
Umbeck, Paul	Agracetus, Middleton, WI 53562
Van Sluys, Marie-Ann	INRA, University Paris XI, Institut Microbiologie, Orsay 91405, FRANCE
Varagona, Marguerite	University of Georgia, Athens, GA 30602
Velguth, Peter H.	University of Minnesota, St. Paul, MN 55108
Vodkin, Lila	University of Illinois, Urbana, IL 61801
Voytas, Daniel	Harvard University, Boston, MA 02114
Walden, D.B.	University of Western Ontario, London N6A 5O7, Ontario, CANADA
Walbot, Virginia	Stanford University, Stanford, CA 94305
Walker, Amanda (Mandy)	CSIRO, Canberra ACT 2601, AUSTRALIA
Weck, Edward A.	Northrup King Company, Stanton, MN 55081
Wessler, Susan	University of Georgia, Athens, GA 30602
White, Frank	Kansas State University, Manhattan, KS 66506
Williams-Karoly, Christie	University of Wisconsin, Madison, WI 53706
Willis, D. Kyle	University of Wisconsin, Madison, WI 53706
Wilson, Claire M.	Council for Research Planning in Biological Sciences, Washington, DC 20036
Yang, Ning-Sun	Agracetus, Middleton, WI 53562
Yerk, Georgia L.	University of Wisconsin, Madison, WI 53706
Yoder, John	University of California, Davis, CA 95616
Yoneyama, K.	Max-Planck-Institut fur Zuchtungforschung, Egelsplad, D-5000 Koln 30, FEDERAL REUPLIC OF GERMANY
Zhang, Hong	Michigan State University, East Lansing, MI 48824

INDEX